The
EVOLUTION
CONTROVERSY

THE EVOLUTION CONTROVERSY

A SURVEY OF COMPETING THEORIES

THOMAS B. FOWLER
AND DANIEL KUEBLER

B

Baker Academic
Grand Rapids, Michigan

Published by Baker Academic
a division of Baker Publishing Group
P.O. Box 6287, Grand Rapids, MI 49516-6287
www.bakeracademic.com

Printed in the United States of America

Library of Congress Cataloging-in-Publication Data
Fowler, Thomas B., 1947–
    The evolution controversy : a survey of competing theories / Thomas B. Fowler
and Daniel Kuebler.
        p.   cm.
    Includes bibliographical references and index.
    ISBN 10: 0-8010-3174-5 (pbk.)
    ISBN 978-0-8010-3174-8 (pbk.)
    1. Evolution (Biology) 2. Natural selection. 3. Evolution (Biology)—Religious
aspects—Christianity. I. Kuebler, Daniel, 1971– II. Title.
QH367.F75 2007
576.8′2—dc22   2007011459

To Nellie, for your constant support and encouragement,
and to Joseph, Patrick, and Carolyn, for all the future holds.

D. R. K.

To Maika, for your love and devotion,
and to Isabella, may your talents and efforts
make the world a better place.

T. B. F.

CONTENTS

Illustrations

Figures

Tables

11

PREFACE

Most people are aware that there is some type of controversy surrounding evolution. However, given the complex and expansive nature of the subject matter it can be difficult to pinpoint what the main issues are or what exactly is at stake. Compounding this problem is the fact that most of the literature on the subject is written by partisans who have a vested interest in the outcome. Not surprisingly, these authors often do not present the scientific data in an impartial fashion. They tend to bolster their own case by ignoring certain pieces of data, while simultaneously building a straw man out of their opponents' position.

In the course of our lectures and discussions on the topic, many people have asked if there is a book that could help them make better sense of the controversy, a book that provides an unbiased scientific overview of the leading theories about evolution. Unfortunately there was no single book we could recommend. This book is an attempt to address this deficiency.

Because the evolution controversy is so multifaceted, and the literature on it is so voluminous, no single book, including this one, can do justice to every important question. With this in mind, we have attempted to provide an objective look at (1) the relevant scientific facts regarding evolution and (2) the scientific merits of the major schools of thought regarding evolution. Other issues such as the theological and cultural aspects of the controversy are discussed only insofar as they impact the scientific arguments. Our book is intended for those who are aware of the scientific controversies swirling about the theory of evolution and who seek an objective reference that will help them critically evaluate the myriad articles and books on the subject that appear every year.

Toward this end, we address many of the questions that students, colleagues, and friends routinely ask about the subject:

- Is evolution a fact or a theory?
- Why is the Neo-Darwinian theory of evolution so widely accepted by the scientific community?
- Is the Neo-Darwinian theory as well verified as other scientific theories, such as the heliocentric theory of the solar system?
- Does the evidence for Neo-Darwinism justify labeling its critics as "yahoos"?
- Do Creationist and other attacks on Neo-Darwinism pose a danger to the independence and integrity of science?
- Do Creationists want to discard all scientific evidence and replace science textbooks with the Bible?
- Is the idea of Intelligent Design just another cover for Creationism, a Trojan horse clumsily trying to break into the scientific community?
- Is science restricted to naturalistic (as opposed to supernatural) explanations by virtue of its nature as a process, or is it a matter of how we choose to define science?
- What would it mean for science if all naturalistic explanations of evolution broke down?
- To what extent is the accepted theory of evolution necessary for the teaching of science, for the day-to-day conduct of it, and for research?
- What, if anything, should be taught in the classroom about Creationism and other competing theories?

We will try to answer these questions impartially, but as befits an objective examination, we will not take a position on which is the "best" or "true" theory; readers will have to make that judgment for themselves. It is our hope that this book will equip readers with the background necessary to read and analyze books, articles, and other materials on evolution and to draw their own informed conclusions. In order to accomplish this, the book is structured around the following topics:

- The historical development of the theory of evolution, up to the present time
- The observable facts that need to be explained
- The principal points in dispute

- The kind of reasoning employed in discussing the subject, and common errors
- The major schools of thought and their positions
- The evidence and principal arguments for and against each school
- The public policy implications of the evolution controversy

In addition, there is an extensive bibliography listing books, articles, videos, web sites, and organizations of all persuasions.

What Do We Mean by "Objectivity"?

Most authors on the subject of evolution claim—or at least believe—that their works are objective, whereas those of their opponents are hopelessly biased. We too are claiming objectivity, and indeed are going much further and asserting that this book is more objective than any other that we know of on the subject. To buttress our claim, we offer our definition of objectivity in the context of the evolution controversy:

- None of the schools of thought regarding evolution will be assumed, a priori, to have a "corner" on the truth.
- All arguments will be considered on an equal basis, regardless of religious or philosophical persuasion of the authors.
- Arguments will be evaluated strictly on their scientific merits and must stand on their own.
- There is no presumption that one of the schools of thought must emerge as the winner. All may be found defective. In such a case, intelligent men and women should suspend judgment on the evolution question until further evidence (and possibly a new theory) becomes available.
- There is no presumption that science can explain *all* observed phenomena associated with evolution and the history of flora and fauna. This must be the conclusion, rather than the premise, of any argument or discussion. However, it is permitted as a working hypothesis, one that may or may not be verified.
- We have received no funding in any form to write this book. Furthermore, we are not answerable to any of the schools of thought, nor are we vulnerable to any pressure from them as regards employment, grants, tenure, or any other venue.

15

While it is impossible to come to the evolution debate without some preconceived notions or philosophical bias, we have made every effort not to let such notions or bias color the way we present or discuss the data in the book. Rather than try to lead the reader to some preordained conclusion, we seek to provide the reader with the resources to find the truth of the matter.

A Few Final Notes and Then . . . Bon Voyage!

To carry out our program, which is to fully explore the subject and put all important issues on the table, we ask some questions that certain schools would prefer to leave unasked. We explore issues from which others shy away, and we prod the reader to think about the subject in ways that some schools may find uncomfortable. We believe readers will be grateful for this frankness and honesty.

Much of the discussion, including much of the criticism and many of the examples, concerns the dominant school of thought, Neo-Darwinism. This is inevitable given that school's position; it does not mean that Neo-Darwinism has more problems than other schools, or that its problems are more serious—or serious at all, for that matter. It is the quality of arguments, not their quantity, that will decide the issue. Indeed, as we will stress throughout the book, the fact that four schools of thought are covered does not mean that they are somehow equally valid as science, or that the reader can simply choose any one he or she likes because no objective assessment can be made of them. One purpose of the book is to help the reader discern which schools have made a viable scientific case for their position.

This book has been structured so that it can be read from cover to cover, but it is not necessary to use it this way; most chapters can stand alone. Readers interested in only a particular aspect of evolution, such as the theory of Intelligent Design, can read just those chapters relevant to their interests and refer to the others as necessary for background. Please note, however, that any book such as this, which examines much interrelated material, cannot follow a linear path; some repetition is inevitable, and in some cases the discussion may assume some background the reader does not as yet possess. In these cases, we recommend consulting the index, the glossary, or, if necessary, some of the references given.

Given the voluminous material on evolution, a concerted effort over many years has been necessary to distill the essentials of the controversy and present them in a fair and balanced way. This process involved many choices; inevitably some will object that we omitted an important issue, or that we covered issues that are not so critical, or that we have presented one issue or another with some bias or prejudice. We will certainly be mindful of such criticism and make such revisions as seem appropriate in future editions. What we are certain of now is that we would not have been able to do this job nearly as well had it not been for the many individuals who helped us fine-tune the book both from a stylistic and from a scientific standpoint, and ferret out instances of bias. We would like to thank Mark Tanouye, Dean Kenyon, John Coleman, Jonathon Sanford, Eric Smith, Denise Masi, Carolyn Rathburn, Howard Bernett, Robert Young, Mark McShurley, and Al Costreba for their efforts in this regard. Not all of them agree with every point we make in this book, but their comments and critical reading have been invaluable. It is through their efforts that the book has reached its present state; any errors that remain are solely the responsibility of the authors. We would also like to thank Emily Buck and Stephanie Langham for their help in providing a number of illustrations for the book.

We close with one final cautionary note: the roiling and highly charged atmosphere surrounding evolution means that nearly every argument or claim has a counterargument or counterclaim, which in turn have their own counterarguments and counterclaims, and so on. We have attempted to sort things out here, but none of the parties to the dispute is likely to be entirely pleased with the result. Bear in mind that the subject of the book is the evolution *controversy*, not any particular school of thought about it. The book is not written for people who have already made up their minds about the facts and have no desire to review the debate in an unbiased fashion. We hope, however, that the broad range of information presented will be of value even to them.

At the end of the book, you may well come to agree with one of the schools of thought, or you may decide that not enough is known to justify a definitive answer, or perhaps you may find that you need to do more research. Provided that you believe you have been given the information necessary to make this decision, or have been shown the path to such information, we will be content that we have done our job. Information, not indoctrination, is what is needed, and that is what this book hopes to provides.

Thomas B. Fowler
Daniel Kuebler

PART 1

BACKGROUND
ON THE
CONTROVERSY

1

INTRODUCTION

Drawing the Battle Lines

Charles Darwin's theory of organic evolution, which emerged from his long meditation on living organisms and the history of life, undoubtedly has the strangest and most amazing history of any major scientific theory. From his remarkable insights into the subject, published in his 1859 classic, *The Origin of Species*, evolution has developed into the dangerously combustible subject that it is today. Hailed as one of science's greatest and most far-reaching achievements by some, condemned as a fraud by others, it is more controversial today than in Darwin's own time. Never before has a legitimate scientific theory become surrounded by so many extra-scientific trappings, triggered such visceral reactions, caused so much ink to be spilled, or led so many otherwise rational people—including scientists—to abandon their reasoning powers wholesale and lose sight of the ultimate objective, which is the pursuit of truth. Why is this happening? What is at stake? What makes the theory of evolution so unique? The answer is not far to seek.

The majestic sweep of evolution, together with its enormous explanatory and integrative potential, makes it one of the pivotal intellectual issues of our time. Darwin's theory—assuming it to be correct—illuminates such an enormous range of phenomena that its impact extends far beyond a narrow specialty of biology. It has the potential to influence many, if not most, of the ways in which we organize our experience and ground our beliefs. In the sciences, it affects all of biology because it is one of the keys to our understanding of the relationships among organisms; it also draws upon and to some extent influences chemistry, geology, physics, and even engineering, all of which are used to explicate aspects of evolutionary history. In the humanities, it is often used as a paradigm for understanding historical, social, and political change. Perhaps more important, however, it affects the way we view ourselves in the universe, our belief (or nonbelief) in the supernatural, and our relationships with the rest of the natural world.

Evolution, or more precisely as we shall see, the Neo-Darwinian theory of evolution, makes certain claims in these areas that appear to be in direct conflict with the views held by some of the world's major religions. The impact of Darwin's theory on religion therefore can be devastating:

> Darwin's dangerous idea cuts much deeper into the fabric of our most fundamental beliefs than many of its sophisticated apologists have yet admitted, even to themselves. . . . The kindly God who lovingly fashioned each and every one of us (all creatures great and small) and sprinkled the sky with shining stars for our delight—*that* God is, like Santa Claus, a myth of childhood, not anything a sane, undeluded adult could literally believe in. *That* God must be either turned into a symbol for something less concrete, or abandoned altogether.[1]

The potential for conflict is therefore very real, and in the eyes of some, there must be a fight to the death. As usually posed by the media, the question is one of evolution *versus* religion, with biblical fundamentalists arrayed on one side and scientists on the other—a reprise of the infamous Scopes "Monkey Trial" of 1925.[2] Hence the widespread interest in evolution and the heated controversies surrounding it.

Scientific controversies often become venomous and spill over into the general culture when they concern fundamental beliefs of society, or at least beliefs deemed essential at the time. This happened with Galileo and the controversy over the geocentric universe in the early seventeenth century. More recently, it happened when cherished views of the deterministic nature of scientific laws (and nature itself) were challenged by quantum mechanics in the first decades of the twentieth century. But evolution is undoubtedly the longest running and most bitterly fought of these controversies, and the one with the highest stakes. The reader may even have observed the "emblem war": fish

1. Daniel Dennett, *Darwin's Dangerous Idea* (New York: Simon & Schuster, 1995), 18.

2. Engagingly but inaccurately portrayed in the 1955 play *Inherit the Wind*, by Jerome Lawrence and Robert E. Lee (New York: Ballantine Books, 2003), and the 1960 motion picture of the same name, by which most people know the story.

and Darwin auto emblems, by which drivers exhibit their predilection—something unheard of in any other scientific controversy (see fig. 1.1).

Figure 1.1. Dueling auto emblems, symbols of the culture war swirling around the subject of evolution.

In the midst of this debate, we would like to look past the noisy and nasty arguments that dominate popular literature on the subject and penetrate to the core of the matter. Let us begin our search with a brief historical overview.

Capsule History of the Evolution Controversy

The notion of fixed and immutable species, coupled with the belief in a recent creation of the earth and the heavens, dominated Western thought from Roman times until the middle of the eighteenth century. By that time, evidence had accumulated that suggested an old earth. Fossils began to be recognized as species of animals no longer living, and gradually the notion emerged that these extinct animals were related in some way to modern, living animals. In about 1830 natural selection, as a *conservative* principle, was recognized by Edward Blyth, Charles Lyell, and others. This principle—in essence, the survival of the fittest—was used as an example of divine providence ensuring the conservation (survival) of species. Charles Darwin took that idea and by conjoining it with ever-present random variations exhibited by organisms, endowed it with a new function, the promotion of innovation. In doing so he created a theory intended to explain the growing body of facts about the history of life on earth. His famous book, *The Origin of Species*, was published in 1859 and triggered a raging controversy. Under Darwin's theory, small changes (mutations) are constantly arising in living things, and natural selection chooses the better ones, those that improve the species' ability to prosper in certain environments. Gradually these small improvements accumulate to yield a whole new species, and eventually new genera, orders, and phyla. Darwin made several predictions based on his theory, indicating what sort of discoveries should be made in the future about the history of life on earth, especially regarding the fossil record. Some of these predictions were not fully borne out, causing doubts to arise in scientific circles about the theory in the years after Darwin's death.

The principles of genetics discovered by Gregor Mendel in the last years of the nineteenth century, and rediscovered by others in the early twentieth century, became critical to the prominence attained by Darwin's notion of evolution. In the 1930s and 1940s, this new information was used to refurbish and solidify Darwin's original theory, leading to the "new synthesis," or "Neo-Darwinism," as it is usually termed. This new form of Darwin's theory quickly gained the allegiance of most scientists and intellectuals, and it has maintained that allegiance into the twenty-first century.

However, by the 1960s dissatisfaction with the theory was growing among a number of groups. Some religious groups were alarmed at the theological and moral implications that had become associated with the theory and its contradictions with the accounts of natural history as recorded in the Bible. This led to the modern Creationist movement, which immediately began to attack the scientific case for evolution. Gradually it developed alternative explanatory paradigms, nearly all of which are built upon a "young earth" hypothesis. Drawing upon its religious base, the Creationist movement currently claims many adherents.

In addition to Creationist challenges, many mainstream biologists became concerned with what they perceived as serious discrepancies between Neo-Darwinian theory and the fossil record. Others were disturbed by what they understood to be extraordinarily implausible events required by the theory. Still others became convinced that the proposed mechanisms of the theory could not bear the explanatory burden placed upon them. Another group became disgruntled with what they regarded as violations of established scientific methodology and norms for proof, claiming that the theory was riddled with tautologies, ad hoc arguments, and just-so stories.[3] These groups, here collectively termed the "Meta-Darwinian" school, though small in number, came into prominence from the 1970s onward. They seek to go beyond the Neo-Darwinian theory by supplementing it with other natural explanations, while rejecting Creationism and all supernatural explanations.

In the 1990s, building on much of the same evidence that led to the Meta-Darwinian school, a group of scientists proposed that the extraor-

3. Just-so stories is a reference to a book of that name published by Rudyard Kipling in 1902, consisting of a collection of fables purporting to explain how the leopard got his spots, how the camel got his hump, and other biological facts. It has since become a derogatory term used to refer to the type of narrative stories often appearing in books on evolution, stories that describe hypothetically how evolutionary changes, such as the origin of feathers or of flight, may have occurred.

dinary complexity of organisms and their component physiological systems cannot be explained by naturalistic means at all. Thus arose the Intelligent Design school, which, unlike the Creationist school, did not insist upon a young earth.

At the present time, then, there are four prominent schools of thought: the Neo-Darwinian school, which clearly dominates in terms of scientific and academic adherents; the Creationist school, which enjoys much popular though little academic support; the Meta-Darwinian school, which is well represented in the academic and scientific community and rapidly growing; and the Intelligent Design school, which is also growing in numbers although its influence in academia is still quite small. The schools and their relationships are depicted in figure 1.2.[4]

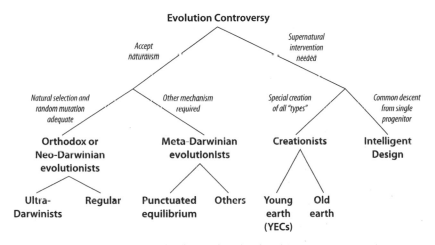

Figure 1.2. Relationship of schools of thought in the evolution controversy.

4. Where do theistic evolutionists fit within this scheme? Theistic evolutionists are found in all camps except the Creationist camp. Theistic evolution is an attempt to harmonize or reconcile theological doctrines and the Bible with some particular scientific understanding of evolution. It comes in many varieties, including continuous creative activity, "front-loaded" activity, and completely indirect or hands-off approaches. Theistic evolution is thus not a *new scientific theory* to explain observed facts, but rather an *interpretation* of some *existing* scientific theory. It is a fascinating subject, of great interest, but cannot be properly understood or evaluated until the groundwork has been laid—the job of the present book. It makes sense to discuss theistic evolution only after one knows the scientific theories that have been proposed to explain observable facts, and the strengths and weaknesses of each. For example, if Neo-Darwinism is found to be a weak theory, then why spend time devising a theological interpretation of it that is in accordance with the Bible? Or, if Creationism is found to be nonviable because it cannot solve the problem of the age of the universe, then searching for a theistic evolution approach might be found worthwhile.

What Is Evolution? What Is Natural Selection?

If we are going to discuss the evolution controversy, we should at the very least know what we are talking about. So let us begin with a definition of evolution. As we shall soon discover, this is not just a simple lexicographical exercise. Rather, it is an investigation that goes to the very heart of the controversy, because "evolution" has more than one meaning. To start, we turn to the undisputed source for historical lexicographical information, the *Oxford English Dictionary* (2nd ed.). The first meaning, attested as early as 1647, is that of unrolling, the "spreading out before the mental vision (of a series of objects); the appearance in orderly succession of a long train of events." This is what one could call the basic or "wide" meaning of evolution; it corresponds to what is least controversial about evolution, the simple notion of change over time.

Harking back to Darwin (1859), a second definition of interest to us focuses on what causes this change. It states that evolution is the origination of species of animals and plants by a naturalistic process of development from earlier forms, and not by a process of "special creation." This is what one could call the "narrow" sense of evolution—in essence, the Neo-Darwinian theory, as presently understood. That theory, which is the dominant theory of evolution today, is founded upon two principles: the *common descent* of all organisms from a single ancestor, and *natural selection* coupled with random mutations as the mechanism of innovation that drove the common descent. The thrust of evolution, as understood by the Neo-Darwinian theory, stems from the presumed ability of natural selection to harness and direct random events, so that increasing levels of order can emerge. What is perhaps most significant about this mechanism (and the whole notion of evolution behind this definition) is that it generates order *without need of human or divine guidance*. Moreover, as envisioned by the Neo-Darwinian school, evolution is also considered to be a *fundamental organizing principle* of science. This means that it has a dual role: it is both a theory to explain the history of life on earth and a conceptual framework for integrating the results of other disciplines.

As we shall see, "evolution" is sometimes used with *different* meanings not only in the same work but also in the *same* discussion or argument. Later in the chapter we shall distinguish three meanings, or tiers, of evolution, which unfortunately give the word "evolution" a slippery character. This can easily be exploited by unscrupulous proponents of a school to persuade people to accept a conclusion based on logically invalid arguments. We shall review instances of such arguments and discuss related problems later in this chapter. For now, note that all schools

26

accept that some evolution in the first or wide sense has occurred; they differ in regard to the narrow sense.

While it is important to define what is meant by the term "evolution," it is equally important to explicitly state what is meant by the expression "natural selection." In the Darwinian definition of evolution described above, natural selection is synonymous with innovation, the building of new organisms. This has not always been the case and indeed is misleading. As early as the 1830s, the concept of natural selection was used by Edward Blyth to explain how a population maintained itself under changing environmental conditions. Blyth, a Creationist writing for the *British Magazine of Natural History*, observed that "The [animal] best organized must always obtain the greatest quantity [of food]; and must, therefore, become physically the strongest and thus be enabled, by routing its opponents, to transmit its superior qualities to a greater number of offspring."[5] While this may seem remarkably similar to Darwin's view, for Blyth natural selection was an inherently conservative force, indicative of the wisdom of the Creator in conserving and sustaining species, whereas Darwin and his successors concentrated on its innovative potential, in conjunction with what we now term "random mutations." Rather than seeing natural selection as a mechanism merely to maintain the fitness of a species, Darwin and later researchers concentrated on how natural selection could in theory promote innovation and lead to new life-forms. It was natural selection's presumed ability to deliver long-term improvement, rather than short-term stability, that captivated their interest.

It is important to keep this distinction in mind: no one disputes that natural selection occurs; it is what it can do that is in dispute. Everyone agrees that it has the ability to protect a population against adversity and maintain its stability in the face of changing environmental conditions, as Blyth described. This takes place in two dimensions: to weed out the deleterious mutations that constantly occur and to adjust the distribution of phenotypes so that organisms match as well as possible the population's current environment.[6] Both of these are necessary to preserve the fitness level of a population; without that, all populations of organisms would rather quickly die out. Even if there were never a single beneficial mutation—not one—natural selection would still be essential for life. What is in dispute is whether natural selection (together with random mutation) is capable of generating improvements in fitness, or

5. Quoted in Luther D. Sunderland, *Darwin's Enigma* (Green Forest, AR: Master Books, 1988), 18.

6. "Phenotype" refers to an organism's physical structure, as opposed to its genotype, which is the sum of the genetic information in its cells. Not all genetic information is expressed in the phenotype, but it does remain for transmission to offspring.

even whole new species or higher taxa. This is what Darwin argued, and it is still the main issue at the heart of the evolution debate. To reiterate: the existence of natural selection as a preservation process is not in dispute; only the question of its ability (together with random mutation) to innovate and create new species and higher taxa is disputed.

The Three Tiers of Evolution

One of the essential keys to identifying the fallacious reasoning used by partisans of the evolution debate (on all sides) is a relatively simple but profound distinction among what we have termed the three *tiers*, or levels, of evolution. This distinction also helps us to better understand the schools of thought and differentiate them, and to unravel some of the knottier problems of the evolution controversy. The three tiers that we wish to distinguish are *historical evolution, common descent evolution,* and *strong Darwinian evolution.* Failure to recognize the tiers and their roles has led to completely fallacious claims about the theory of evolution, and is probably the single most important factor contributing to confusion about the subject. Many authors who are otherwise well versed in the subject stumble over this point, with disastrous consequences for their arguments. Let us examine these tiers in more detail:

1. **Historical Evolution:** Belief that the timescale of earth's history—that is, the chronology of events worked out by geologists, physicists, paleontologists, and astronomers—is approximately correct, and that organisms appeared at the times usually assigned to them, with some living on and others such as the dinosaurs becoming extinct. No implication is made that one species gave rise to any other, only that there was a historical sequence. There is a considerable body of independent evidence from various disciplines supporting the notion of historical evolution.
2. **Common Descent:** Partial explanation of historical evolution (tier 1) by the hypothesis that shared characteristics among organisms indicates that they have descended from a single, original ancestor (or possibly multiple ancestors). Belief in this level of evolution is not a commitment to natural selection (or any other process) as the exclusive causal mechanism in the development of all life-forms, or the belief that *any* purely naturalistic process can account for such development. The evidence for this level of belief is completely different from the evidence for tier 1. It consists primarily of identifying similarities in structure, biochemical functions, and genetic coding of organisms.

28

3. **Strong Darwinian Evolution:** Complete explanation of common descent by the hypothesis that natural forces *alone* are responsible for the emergence of all organisms. For the Neo-Darwinian school, it is natural selection acting upon random mutations that supplies the force; for Meta-Darwinian evolutionists, additional natural mechanisms are presumed to be involved.

The levels are shown schematically in figure 1.3. Now, armed with the important distinction between the three tiers of evolution, we are prepared to examine the four major schools of thought regarding evolution.

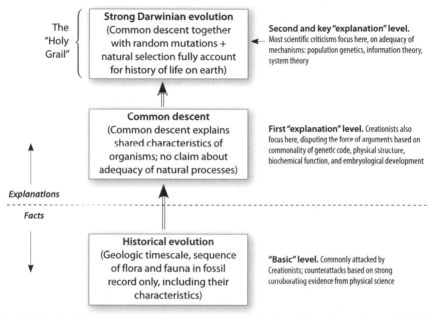

Figure 1.3. Schematic indicating the three tiers of evolution. Note that each level builds upon the previous one. For example, historical evolution must be demonstrated before one can even talk about common descent. Likewise, the verification of common descent is a precursor to a discussion of strong Darwinian evolution.

The Major Schools of Thought Regarding Evolution

Categorization of the vast literature on and the varied thought about evolution is perforce a difficult task. No set of boundaries will satisfy everyone involved in the evolution controversy. Nonetheless the division into four schools, presented in this book, is based on real differences and supplies a convenient jumping-off point for subsequent discussions. The boundaries of these four schools are not hard and fast, and may overlap in certain respects. In fact, individual scientists may fit into more than one school. What follows is a brief synopsis of the four schools.

1. **The Neo-Darwinian School.** At the dawn of the twenty-first century, the dominant school of thought is unquestionably Neo-Darwinism. Virtually all major scientific institutions and government funding agencies, and nearly all college and university faculty members, and most practicing scientists have allegiance to this school. This is a formidable array of scientific talent, by any standard. The school defends what is termed the "Neo-Darwinian" theory of evolution, with its central tenets of common descent and natural selection, as discussed above. It believes that these tenets, proposed by Darwin but modified and codified in the "new synthesis," are adequate to explain, in a naturalistic fashion, all the flora and fauna we see in the world, as well as how they arose from predecessors.

The school explicitly rejects supernatural explanations and relegates other natural mechanisms to minor importance, at best. Because of the dominance in the academic world of the school's adherents, their fairly uniform set of beliefs, and their attitude toward other schools, they could be called "orthodox evolutionists," but we shall use the name that is in common use, "Neo-Darwinians." In actuality, they divide into two camps: what we may term the ordinary Darwinists, and the so-called ultra-Darwinists. The latter are distinguished by their somewhat more radical belief that genes rather than organisms are the fundamental units of selection. According to this belief, it is the genes that "act" to preserve and disseminate themselves, and organisms are merely reduced to the sum of their genes.[7] For most of the discussion of this book, however, that distinction is not important.

2. **The Creationist School.** The scientific Creationists,[8] usually simply referred to as Creationists, naturally believe that some sort of "special creation" is necessary to account for the history of life on our planet. According to their understanding, available evidence points to such special creation and refutes the Neo-Darwinian theory. Creationists believe that the observed similarities and commonalities in organisms are absolutely necessary for creatures to *function* and thus are a reflection of divine providence rather than the product of evolutionary processes. They make no secret of the fact that their religious beliefs bound the range of permissible scientific hypotheses, and indeed inform them to a considerable extent, contrary to the usual

7. As in Richard Dawkins, *The Selfish Gene* (Oxford: Oxford University Press, 1976).

8. Members of this school use the adjective "scientific" in their name to emphasize the fact that they base (or claim to base) all their arguments and theories on scientific evidence, rather than on the Bible. As far as the evolutionists are concerned, "scientific Creationist" is an oxymoron.

procedures of science. They justify this by arguing that both the Bible and science give truths, and these truths cannot contradict. Creationists may be conveniently divided into two camps. First and largest is that of the young earth Creationists (YECs), those who believe that the earth is of rather recent vintage and who advocate a number of theories to account for its features, such as a "flood geology." They reject the conclusions of geologists, paleontologists, and others about the age of the earth and its implications, and maintain that all observed features can be better explained by means of their hypotheses, which postulate high-energy processes acting over much shorter intervals. They do not dispute that some evolution in the broad sense has occurred over a relatively short time interval (about 10,000 years); indeed, they require it to account for repopulation of the earth after Noah's flood. They even hypothesize that new forms of life have arisen from the original created "types" or "kinds."[9]

What the young earth theory excludes is evolution in the third or Neo-Darwinian sense, which requires vast time spans to operate and generates new biological information. This leads to the so-called Creationist lawn model as opposed to the Neo-Darwinian tree model of evolution (see fig. 1.4). But since anyone with a backyard telescope can observe galaxies that are tens of millions of light years away, the "young earth" theory must confront a formidable chronological challenge at the outset.

A second Creationist camp, known as "ordinary" Creationists, accepts the common dating scheme for the earth and the universe but rejects other doctrines essential to standard evolution theory, specifically, common descent from a single ancestor and the adequacy of natural selection to account for all biological innovations. This camp is not proposing an entirely new scientific theory to explain observed biological, geological, and astronomical facts, and so will not be discussed in detail in the present book.

3. **The Intelligent Design School.** The third school already has a name of which it is proud and by which it is known, so we shall use it: Intelligent Design. This school rejects (on scientific grounds) the adequacy of natural selection *alone* to account for the complexity found in nature. However, nearly all members of the school believe in some form of historical evolution and many even accept "common descent evolution." Most also accept the scientific work done to date in geol-

9. Creationists also accept common descent in a limited sense, namely that new species can arise from the original created "kinds" through degeneration (loss of information). These separately created kinds thus function as common ancestors to a variety of species. For instance, dogs and wolves all descended from a common ancestor aboard Noah's ark.

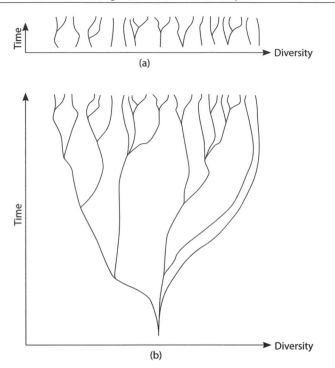

Figure 1.4. (a) Creationist "lawn" versus (b) Neo-Darwinian "tree" models of evolution.

ogy, biochemistry, paleontology, genetics, and evolutionary biology, but their interpretation is much different from the Neo-Darwinists'. What defines this school is its belief that one can identify, through scientific research, proof of Intelligent Design in the natural world. Moreover, they believe that an Intelligent Designer is required for certain steps in evolution to occur, although exactly which ones are unclear at this point. Their thought centers on the extreme complexity and highly optimized nature of biochemical mechanisms, processes, and structures, and the impossibility (in their view) of such complexity arising from any type of natural process, whether guided by natural selection or not. One of the descriptive terms they use is "irreducible complexity" (IC). Members of the Intelligent Design school believe that they can, on the basis of probabilities, distinguish objects or processes that have arisen by chance from those that have been designed, in much the same way that archaeologists detect design in centuries-old artifacts. However, as we shall see, attaching probabilities to the likelihood that a biological organism or structure arose by chance is tricky business, and this remains one of the main obstacles the Intelligent Design school faces in gaining acceptance as a legitimate scientific theory. On account of its belief in a designer, this school is sometimes confused with that of Creationism or rather lumped in with it. The two do indeed share some views; however, the two schools differ in important respects, and we therefore treat them separately.

4. **The Meta-Darwinian School.** The fourth school is by far the most heterogeneous, and unlike the other schools, it does not have a unified front. In fact, speaking of a "Meta-Darwinian school" is a bit misleading because the Meta-Darwinians consist of a variety of rival camps, all of which, however, share a common theme. The common theme is that the Neo-Darwinian explanation of evolution is inadequate, and therefore other naturalistic mechanisms must be found to supplement it. From this commonality, the various camps within this school part company, often drastically. While admitting that the Neo-Darwinian paradigm is capable of explaining *some* observed changes, the various members differ on the amount that they believe the paradigm can actually explain. Some believe that it can explain nearly everything, though other mechanisms must also be recognized as important in shaping natural history. Others believe that it can explain only small changes in organisms and that the heavy lifting of evolution must be assigned to totally different mechanisms. The members diverge even further in the particular mechanisms that they accept to supplement perceived deficiencies in the Neo-Darwinian paradigm.

Various groups that occupy this tent include proponents of the punctuated equilibrium theory, advocates of endosymbiosis theories, advocates of the self-organizing ability of complex systems, proponents of the creative potential of developmental mutations (evo-devo), and supporters of the neutral theory of selection. Despite the variety of mechanisms advocated, all members of this school believe that entirely naturalistic explanations can overcome the difficulties faced by Neo-Darwinian theory. In recent years, many Neo-Darwinists have acknowledged the utility of some of these mechanisms, while still maintaining the primacy of natural selection and random mutation. As a result, there can be considerable gray area between the two schools. Because both Meta-Darwinists and Neo-Darwinists seek entirely naturalistic mechanisms to explain the history of life, we shall use the generic term "evolutionist" when we wish to refer to a member of either school.

The position each school takes in regard to the three tiers of evolution is summarized in table 1.1. The positions the schools take on a number of contentious issues in the evolution debate are listed in table 1.2. It is important to keep in mind that the positions attributed to the four schools are based on certain generalities; individual members will differ in respect to some of these positions.

Table 1.1. Distinguishing the four schools with respect to the tiers of evolution

	Creationism (young earth)	Creationism (other)	Intelligent Design	Meta-Darwinism	Neo-Darwinism
Tier 1: Historical evolution (short-term)	✓	✓	✓	✓	✓
Tier 1: Historical evolution (long-term)	✗	✓	✓	✓	✓
Tier 2: Common descent (degenerative, from "kinds")	✓	✓	✓	✓	✓
Tier 2: Common descent (broad sense)	✗	✗	✓	✓	✓
Tier 3a: Strong Darwinian evolution (natural forces alone)	✗	✗	✗	✓	✓
Tier 3b: Strong Darwinian evolution (natural selection only)	✗	✗	✗	✗	✓

✗ = rejects   ✓ = accepts

Table 1.2. Distinguishing the four schools with respect to evolutionary hypotheses

	Creationism	Intelligent Design	Meta-Darwinism	Neo-Darwinism
Origin of life by natural physical processes*	✗	✗	✓	✓
Common descent (from single progenitor)	✗	✓	✓	✓
Existence of natural selection (as preservation mechanism)	✓	✓	✓	✓
Ability of natural selection and random mutations to give rise to new species (but not higher taxa) through degeneration	✓	✓	✓	✓
Ability of natural selection and random mutations to innovate (give rise to new species and higher taxa through creation of new information)	✗	✗/✓**	✓***	✓
Adequacy of natural selection and random mutations to explain all of evolution	✗	✗	✗	✓

✗ = rejects   ✓ = accepts

* The origin-of-life issue is not, strictly speaking, a question of evolution, but it is a separate question on which the schools differ.

** This is the least-defined aspect of the Intelligent Design school. It appears that some members of this school rule out the creation of any type of new information (William Dembski), while others admit that some novel structures may emerge if there is a functional gradual pathway that exists to form them. The line between what can and cannot be achieved by natural selection is thus poorly defined by the Intelligent Design school.

*** There is considerable difference of opinion among Meta-Darwinian evolutionists on this issue, although most tend to agree with this statement to some extent.

A Pretty Picture

On the surface, the controversy regarding evolution is surprising because, given the number and position of Neo-Darwinists, there should be no evolution controversy at all. The overwhelming majority of biologists either subscribes to the factuality of the Neo-Darwinist position or believes in some form of naturalistic evolution (they are Meta-Darwinists). The other theories are for the most part consigned to the fringes of academia, if not excluded outright. In fact, the controversy is often portrayed as the "bad guy Creationists" versus the "good guy scientists," with the former advocating wholesale rejection of everything incompatible with a literal interpretation of the Bible and adoption of a "young earth" theory and the latter scrupulously extracting all possible information from available sources and drawing irrefutable conclusions about evolution. Who could take such a dispute seriously? Surely one side would demolish the other. The proponents of evolution, vastly superior in numbers, publications, and institutional support, should win hands down.

What's Wrong with This Picture?

The first indication that something might be awry in this picture is that Creationists have won evolution debates often enough that most supporters of Neo-Darwinism will not go head-to-head with them in such public forums. Eugenie Scott, executive director of the National Center for Science Education (NCSE), even advises evolutionists not to debate Creationists at all: "If your local campus Christian fellowship asks you to 'defend evolution,' please decline . . . you probably will get beaten."[10] Walt Brown has offered for over 25 years to have a detailed, written, publishable scientific debate, with no takers.[11] Major scientific figures such as Isaac Asimov and Steven Jay Gould flatly refused to participate in debates with Creationists. Since no reputable physicist would hesitate to debate opponents of the heliocentric theory, conservation of energy, or quantum mechanics, the behavior of evolution advocates does seem peculiar. Refusal to debate crackpot theories held by a few fringe thinkers is one thing; refusal to debate a theory held by half of the population (see below) is quite another.

The second indication of problems is the fervor with which Creationists and other critics of the reigning Darwinian theory are denounced and their

10. Eugenie C. Scott, "Monkey Business," *The Sciences* (January/February 1996), 25.
11. Walt Brown, *In the Beginning*, 7th ed. (Phoenix: Center for Scientific Creation, 2001), 193; also on his web site, www.creationscience.com/onlinebook/FAQ416.html#wp1675667.

arguments summarily dismissed. For a discipline such as science, which prides itself on both skepticism concerning even well-established theories and the impartial examination of facts, this is strange indeed. Of course, it is the duty of every scientist to defend the splendid accomplishments of science against frauds, charlatans, and crackpots; the problem is where to draw the line between meritless criticism and serious challenges.

The third warning flag is the relatively small number of Americans who accept the theory of evolution in an unqualified sense: only about 10%, according to recent Gallup polls,[12] despite its ubiquity in textbooks and classroom instruction,[13] and the relatively large number who identify themselves as Creationists (45–50%). Table 1.3 gives the relevant statistics; similar numbers have emerged from other polls,[14] and the numbers have remained nearly constant for two decades. The second category in the table does not distinguish those who accept Intelligent Design from those who prefer some type of theistic evolution. But if we split the difference and allow about 20% for Intelligent Design, and 20% for theistic evolution, we have a total of about 30% in the evolution camp, 20% in the Intelligent Design camp, and 47% in the Creationist camp. If the Intelligent Design group is added to the Creationist group, the "anti" naturalism camp outnumbers the "pro" naturalism camp by about 2 to 1 (67% vs. 30%). Not surprisingly, this is reflected in the numbers of those who believe that Creationism should be taught *along with* evolution (68%), while 40% believe that Creationism should be taught *instead of* evolution. It is clear that, regardless of merit, the scientists have failed to win the battle for public opinion.

The fourth sign comes from the growing number of biologists who have become dissatisfied with the Neo-Darwinian paradigm. While still a small minority, they have become much more candid regarding the difficulties facing evolutionary theory, even if they think that all will be well in the end:

Despite our ignorance of the overwhelming majority of life-forms which exist on Earth today (indeed, most biochemical and genetic generalizations are still derived from just three organisms: the rat, the fruit fly, and the common gut bug *Escherichia coli*), *and our inability to do more than offer informed speculations about the processes that have given rise to them*

12. "Public Favorable to Creationism," Gallup News Service, February 14, 2001, http://poll.gallup.com/content/default.aspx?ci=2014&pg=1; "Reading the Polls on Evolution and Creationism," Pew Reseach Center Pollwatch, http://people-press.org/commentary/display.php3?AnalysisID=118.

13. The percentage in each category depends on nuances in how the questions are phrased, but all polls are in general agreement.

14. CBS News/New York Times poll, November 18–21, 2004, reported by American Scientific Affiliation, www.asa3.org/ASA/topics/Evolution/index.html.

over the past 4 billion years, we biologists are *beginning to lay claims to universal knowledge* of what life is, how it emerged, and how it works. . . . Biology also makes claims as to who we are, about the forces that shape the deepest aspects of our personalities, and even about our purposes here on Earth.[15]

If our knowledge at present is so fragmentary, and our ability to verify theories so limited, certainty would seem to be out of the question, regardless of the assertions one might wish to make; and a modicum of caution should be injected into evolutionary claims.

Table 1.3. Public acceptance of theories about evolution (numbers given represent the percentage of the population that agrees with the survey statement)

Survey statement	1982	1993	1997	1999	2004
God created human beings pretty much in their present form at one time within the last 10,000 years or so [essentially Creationism]	44%	47%	44%	47%	45%
Human beings developed over millions of years from less advanced forms of life, but God guided this process [either Intelligent Design or theistic evolution]	38%	38%	39%	40%	38%
Human beings have developed over millions of years from less advanced forms of life. God had no part in this process [essentially Neo-Darwinism]	9%	11%	10%	9%	13%
No opinion	9%	7%	7%	4%	4%

Of course, none of these anecdotes *proves* anything about the Neo-Darwinian theory of evolution. Indeed, some of its proponents have made the point that the controversies within evolutionary theory are an indication of its vitality and the ongoing efforts to resolve difficulties. But it does suggest that all may not be right in the Neo-Darwinian camp, with possible problems lurking in the construction of the theory, the evolutionary mechanisms it proposes, or the evidence put forward to support it. It may also suggest that some hold the theory for reasons that go beyond the strictly scientific.

Is Darwin's Theory Really Vulnerable? Are Other Theories Really Viable?

Modern science is the product of enormous effort over many centuries. It is a marvelous system for unlocking nature's secrets, and from it have emerged many grand and beautiful theories with breathtaking

15. Steven Rose, *Lifelines, Biology beyond Determinism* (Oxford: Oxford University Press, 1998), 4–5; italics added.

explanatory and predictive powers. To achieve these results, scientists have developed a superb (though not perfect) process of peer review, largely self-correcting, that subjects theories, hypotheses, and experimental results to relentless scrutiny. In the words of the National Academy of Sciences:

> Science is a particular way of knowing about the world. In science, explanations are limited to those based on observations and experiments that can be substantiated by other scientists. Explanations that cannot be based on empirical evidence are not a part of science. . . . In the quest for understanding, science involves a great deal of careful observation that eventually produces an elaborate written description of the natural world. Scientists communicate their findings and conclusions to other scientists through publications, talks at conferences, hallway conversations, and many other means. Other scientists then test those ideas and build on preexisting work. In this way, the accuracy and sophistication of descriptions of the natural world tend to increase with time, as subsequent generations of scientists correct and extend the work done by their predecessors.[16]

In addition to providing us with magnificent theories of the natural world, science is now also an essential part of our society's infrastructure, as we depend on science and its applications to solve problems for us and even to feed our engines of economic growth. To accomplish their mission, scientists must work hard, and they can scarcely be expected to spend their time or our society's resources debating the merits of theories and ideas that do not pass through the normal scientific channels. Such ideas and theories abound, and it would be ridiculous—and a complete dereliction of duty—to give equal time in the classroom to all ideas maintained by fringe groups in any given area of science, such as the Flat Earth Society. There is no question that most dissent from established theories must be ignored, or the process of science teaching and research would become so bogged down as to be impossible. So if scientists say that evolution via natural selection is the scientific explanation of the diversity of life, its characteristics, and its history, what is the problem?

That is a fair question. Why should Creationism, Intelligent Design, or any other competing theory be accorded any consideration, time, or respect? Two such reasons are alleged: (1) Neo-Darwinism contains significant difficulties despite its widespread acceptance by the scientific community, and (2) science is going outside its realm of competence.

---

16. National Academy of Sciences, *Science and Creationism*, 2nd ed. (Washington, DC: National Academy Press, 1999), 1.

We shall consider whether the theory does have significant problems in chapters 5 through 8. For now, let us observe that there are historical precedents of theories that had to be abandoned despite widespread acceptance: the geocentric theory of the solar system, the caloric theory of heat, the corpuscular theory of light, the phlogiston theory of combustion, the aether, spontaneous generation, and the theory of humors. Historical precedents, however, are not proof of or even evidence for problems with the Neo-Darwinian theory.

With regard to the accusation that science is stepping outside its realm of competence, we must begin, as we shall often do, by making some important distinctions. Consider the following categories, which are often confused:

- Scientific facts, laws, and theories
- Extra-scientific inferences or long-range extrapolations drawn from those facts, laws, and theories
- Philosophical systems erected upon the extra-scientific inferences and extrapolations

Strictly speaking, the first category consists of facts that can be observed by anyone with suitable equipment, the laws that explain those facts, and the theories that incorporate the laws and give us a picture of some aspect of nature. Insofar as a theory explains some set of observations and explains everything it is supposed to explain, makes no predictions not borne out by observation, and is not otherwise contradicted by observation, then it belongs in this category and can justifiably be called a scientific theory.

The second category includes inferences and long-range extrapolations drawn from established scientific facts, laws, and theories, sometimes by scientists themselves, sometimes by others. They are not scientific statements but affirmations that may have some plausibility based on what is truly scientific. For example, Newton's laws of motion are completely deterministic. Many, including scientists, inferred that such laws therefore governed not only the motion of objects but also everything in the universe, including man, and that therefore man was completely determined in his actions. Consequently, he had no free will, and in effect, morality was a meaningless notion. Of course this was never proven scientifically (it never reached the status of scientific fact, law, or theory) but was merely an extrapolation. Several links in this argument can be shown to be defective, and Newton's deterministic laws were replaced in the last century. But the conclusion, which fitted many agendas and had been drawn by some prior to Newton's time, was seized upon as justification for or reinforcement of the belief that man has no free will.

The third category consists of what we might term worldviews or philosophies based upon inferences from the second category. Scientists themselves, as well as others, often are involved in the creation of such philosophical systems. To return to the example of Newton, whole philosophies were built on his laws and functioned (as such worldviews often do) as surrogate religions. Notable in this category is "Laplace's demon" and the antireligious worldview built around it.[17] Philosophical systems tend to be capable of explaining any and all facts within themselves—that is, they cannot be refuted and are therefore not scientific. The danger in this, of course, is that such philosophical systems will become so enmeshed with the scientific theories upon which they are erected that they foreclose even legitimate criticism of these theories, and thus impede the progress of science itself. That this can happen with respect to Creationism is rather obvious. However, it can also happen in the case of the Neo-Darwinian school. Consider the following statement by George Gaylord Simpson:

> Although many details remain to be worked out, it is already evident that all the objective phenomena of the history of life can be explained by purely naturalistic or, in a proper sense of the sometimes abused word, materialistic factors. They are readily explicable on the basis of differential reproduction in populations [natural selection], and the mainly random interplay of the known processes of heredity [random mutations]. Therefore, man is the result of a purposeless and natural process that did not have him in mind.[18]

The history of science provides us with a warning about this type of argument. Consider the following remark of Albert Michelson (1852–1931): "The more important fundamental laws and facts of physical science have all been discovered and these are now so firmly established that the possibility of their ever being supplanted in consequence of new discoveries is remote."[19] This is of course the very same Michelson whose name will forever be enshrined in physics because of the famous Michelson-Morley experiment (1881). What is especially ironic is the fact that this very experiment (which concerned a detail of Newtonian physics) was in large measure responsible for the overthrow of the science that Michelson thought so well established. So details *are* important because

17. "Laplace's demon" is a hypothetical creature who was supposed to know the position and momentum of all the particles in the universe, and thus be able to predict all of the future and retrodict all of the past.

18. George Gaylord Simpson, *The Meaning of Evolution*, rev. ed. (New Haven: Yale University Press, 1967), 345.

19. Quoted in Robert Youngson, *Scientific Blunders* (New York: Carroll & Graf, 1998), 20.

they can destroy theories, and it is rather surprising that a scientist such as Simpson would trivialize them. But in the case of evolution that is exactly what he is prepared to do in order to advocate his philosophical belief in the purposelessness of nature.

If there is any doubt that philosophical bias has crept into the picture on the side of evolutionists, consider the remarks of Michael Ruse, a key witness in the 1981–82 Arkansas "balanced treatment" case:

> Evolution is promoted by its practitioners as more than mere science. Evolution is promulgated as an ideology, a secular religion—a full-fledged alternative to Christianity, with meaning and morality. I am an ardent evolutionist and an ex-Christian, but I must admit that in this one complaint—and Mr Gish [a prominent Creationist] is but one of many to make it—the literalists are absolutely right. Evolution is a religion. This was true of evolution in the beginning, and it is true of evolution still today. . . . Evolution therefore came into being as a kind of secular ideology, an explicit substitute for Christianity.[20]

This is a startling admission by any standard, one that should remove any doubt in the reader's mind about whether there can be strong ideological components in all positions in the evolution controversy.

Those who criticize the scientific establishment's position on evolution often do so because they claim—with some justification, as the foregoing quotations illustrate—that evolution is a case in which genuine scientific status is claimed for affirmations that belong in the category of metaphysics. The result is that a philosophy or even a surrogate religion is being illegitimately foisted upon the public through legitimate scientific and educational channels, and moreover is being subsidized by public funds. This may be true—Neo-Darwinists may have philosophical commitments that influence their science—but it does not imply that Neo-Darwinists' theory is inadequate as a scientific explanation of the history of life. Determining the adequacy of their theory (or any other school's) can be done only via a thorough scientific evaluation of the evidence.

The remaining chapters in part 1 of the book are designed to prepare the reader to do just that, namely, to evaluate the adequacy of the theories put forth by the major players in the debate. We begin by reviewing the history of the evolution controversy in chapter 2. This should help to orient the reader with respect to the different schools of thought. In chapter 3 we examine the major pieces of evidence that must be explained by any theory of evolution, indicating along the way how the various schools incorporate this evidence into their explanatory framework. Chapter 4

20. Michael Ruse, "How Evolution Became a Religion: Creationists Correct?" *National Post*, May 13, 2000, B1, B3, B7.

is a discussion of the major points in dispute in the evolution debate. This chapter should clarify many important matters that are central to the debate. The heart of the book, though, is found in part 2, where we examine the four schools and their respective positions in turn (chaps. 5–8), indicating their strengths and weaknesses with respect to the evidence and as scientific theories. While the chapters in part 2 build on the introductory material in part 1, they can be read in isolation to get a feel for the arguments put forth by a particular school. Finally, in part 3 we consider public policy issues and summarize the major points made in the course of the book.

2

A BRIEF HISTORY
OF EVOLUTIONARY THOUGHT

Thought about organic evolution has a long history, stretching back to ancient Greece. In this chapter we trace that history and look at the major currents of thought on evolution and how they have come together to create the situation we have today. The history of evolution falls into six major phases that we will discuss in turn.

Up to 1650	*Ancient speculation*: guesses and hunches regarding possible evolutionary scenarios; domination of special creation and young earth ideas
1650–1800	*Emergence of modern science*: recognition and formulation of first scientific theories about geology and the history of life to explain fossils; continued domination of special creation idea
1800–1859	*Laying the foundations*: development and refinement of ideas about changing life-forms; concomitant development of geology and other sciences; recognition of natural selection as a process
1859–1910	*Darwin's triumphal entry*: publication of *The Origin of Species*, spread of Darwin's ideas, early battles over evolution, and growing doubts
1910–1960	*"New synthesis" period*: incorporation of genetics into evolution and restructuring of evolution around mathematical population genetics; domination of Neo-Darwinism
1960–present	*Modern battles over evolution*: rise of modern Creationist movement; critiques of key evolution ideas by Intelligent Design and Meta-Darwinian schools; disagreements within Neo-Darwinian school

Ancient Speculation (to 1650)

Two schools of thought about nature developed in ancient Greece: believers in the relative immutability of species and advocates of a more flexible view. In the first camp was Aristotle (384–322 BC), who attempted to understand nature systematically through firsthand observation. Though he was not an experimenter, his meticulous observations included a multitude of dissections and led to a forward-looking scheme for classifying organisms. His anatomical studies gave him an appreciation of the interdependence of the various organ systems in a living organism and how this interconnectedness was critical for the proper functioning of the organism.[1] This observation strongly influenced his belief in the relative immutability of species. This is depicted in his famous *scala naturae* (the ladder of life, or great chain of being), a linear ordering of all inanimate and animate objects based upon increasing levels of complexity (fig. 2.1). Humans occupy the top rung, and organisms on the ladder are fixed in position. While Aristotle believed that organisms were capable of change as they struggled to obtain perfection within their respective classes, he concluded that they were incapable of evolving, that is, moving up or down the ladder.

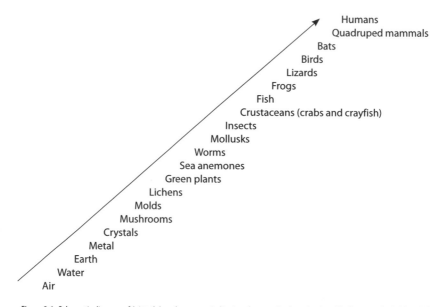

Figure 2.1. Schematic diagram of Aristotle's *scala naturae*, indicating the organization of various life-forms on the ladder of life.

The opinions of two other ancient philosophers, Anaximander (ca. 610–546 BC) and Lucretius (94–55 BC), stand in contrast to Aristotle's view. Both believed that species were quite mutable and, for the most part, were randomly assembled. Anaximander believed that life began in the sea, with simpler forms giving way to more complex forms. He

1. Aristotle, *Parts of Animals* (645a: 31–37), Loeb Classical Library, trans. A. L. Peck (Cambridge, MA: Harvard University Press, 1968), 101.

proposed that humans arose inside of fishes (which he believed to be the precursors of mammals), were reared, and once capable of taking care of themselves, moved onto dry land. Although his ideas were rather fanciful and no mechanism for evolution was ever discussed, many give him credit for being the first full-fledged evolutionist.

Lucretius, the first-century BC Roman philosopher and poet, can be credited with the first well-developed evolutionary story. In his famous poetic work *De rerum natura* (*On the Nature of Things*), Lucretius struck upon the idea of survival of the fittest or natural selection as the driving force behind evolution. Though he had no idea how this would actually work, the crux of his reasoning is remarkably similar to Darwin's theory.[2] Lucretius's argument contained all the Darwinian essentials: a random set of variations that produced various life-forms coupled with the differential selection of these forms, such that those most fit to survive in their respective environments flourished. He says:

> Many were the monsters also that the earth then tried to make, springing up with wondrous appearance and frame. . . . Whatever you see feeding on the breath of life, either cunning or courage or at least quickness must have guarded and kept it from its earliest existence. . . . But those to which nature gives no such qualities, so that they could neither live by themselves at their own will, nor give some usefulness for which we might suffer them to feed under our protection and be safe, these certainly lay at the mercy of others for prey and profit, being hampered by their own fateful chains, until nature brought that race to destruction.[3]

Lucretius further believed that the initial generation of life was spontaneous, springing up from the earth once heat from the sun and moisture were supplied. Because both Lucretius and Anaximander saw nature as having no purpose or design, there was no need to invoke a deity as the organizer or prime mover of life.

The Emergence of Modern Science (1650–1800)

Early Fossil Discoverers: Extinct Species and the Implied Change in Nature

Despite the antiquity of ideas surrounding evolution, the hard evidence required to give credence to any evolutionary theory took much longer

2. Benjamin Wiker, *Moral Darwinism: How We Became Hedonists* (Downers Grove, IL: InterVarsity Press, 2002), 63.

3. Lucretius, *De rerum natura*, Loeb Classical Library, trans. W. H. D. Rouse (Cambridge, MA: Harvard University Press, 1966), 5:837–77.

to surface, particularly since science and popular culture in the Western world were dominated by a biblically informed worldview, centered on special creation. It was not until the seventeenth and eighteenth centuries, when scientific efforts were redoubled across Europe, that apparent problems with special creation emerged. The two disciplines with the greatest impact on the advancement of evolutionary ideas were paleontology and geology.

Although fossil finds were abundant prior to and during this time, many believed that the fossils actually had grown within the rocks; otherwise marine fossils would not be found in dry mountainous regions. Leonardo da Vinci (1452–1519) was one of the most vocal opponents of this view, believing it ludicrous to assume that fossils could grow trapped inside rocks without food or air. However, it was not until the mid-1600s that the brilliant Danish biologist/geologist Nicolaus Steno (1638–86) demonstrated conclusively that specific fossils were the remnants of once-living organisms.[4] Despite his success, unrecognizable fossil forms with no living correlates were still a source of confusion and debate.

In his work, Steno also noticed that different fossils were contained within different rock layers, which he concluded corresponded to different periods of geological history. This led Steno to propose his theory of superposition, one of the central principles of geology: layers of rock are arranged in a time sequence with the oldest layers of rock at the base and newer layers at the top (see fig. 2.2).[5] Using this principle, researchers could "read" the natural history of the earth and examine changes in the fossil record over time. Steno was careful to read the history of the earth as being no more than 4,000 years old, to maintain the biblical worldview, but his advocacy of geological change over time was revolutionary.

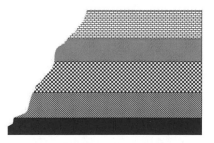

Figure 2.2. The supposition of different rock layers can be seen in this diagram. The oldest layers are toward the bottom.

4. Nicolaus Steno, *The Prodromus*, trans. J. G. Winter (Norwood, MA: Norwood Press, 1930), 219.
5. Ibid., 229–30.

Despite Steno's excellent work, debate over fossils continued for centuries because of two fundamental questions fraught with theological implications: (1) How did the fossils come to be buried within the rock layers? (2) What exactly did the unrecognized fossil forms represent? Many researchers, Steno included, believed that various fossil layers were formed in successive floods, the most notable of which was the biblical flood of Noah, as recounted in Genesis 7—a theory still advocated by Creationists today. This belief was reinforced by the presence of fossilized fish and mussels on arid mountaintops. Unfortunately, the question of the age of the rock and fossil layers was a problem that science at the time was ill equipped to address.

The presence of unrecognized forms was approached in two ways: dismiss them as aberrations, spurious chance mutants, or templates used by God in the creation of real species, or take them at face value, but postulate that they represent species still present in remote unexplored regions of the earth. The latter was the approach of English naturalist John Ray (1627–1705), who is considered the father of modern systematic zoology. Ray was also a prime advocate of the tradition that became known as "natural theology," which sees all of nature as a demonstration of God's power and majesty. To Ray and those who shared his beliefs, species created by an all-powerful and perfect God should not be subject to extinction or extensive change. As a result, he believed the unusual fossil forms must still be in existence somewhere on the earth.

Although Ray could not accept the position that species had gone extinct, his contemporary Robert Hooke (1635–1703) did so and went one step further. Hooke's microscopic examination of fossil shells and living mollusk shells added to the mounting evidence that fossils were indeed once living organisms. The continual discovery of additional unrecognizable fossil forms, coupled with the presence and then absence of these fossils in certain rock layers, convinced Hooke that species had not only gone extinct but that new species had been created.[6]

Hooke's ideas found validation in the work of the father of vertebrate paleontology, the renowned French comparative anatomist Georges Cuvier (1769–1832). Cuvier's ability to reconstruct an organism based on fragmentary fossils was legendary. Drawing upon his extensive studies, Cuvier was able to demonstrate that species had gone extinct. Using a detailed anatomical comparison, Cuvier was able to prove conclusively that many large vertebrate fossils, such as the mastodon, were separate species from other living vertebrates. As for the possibility that these

6. Margaret Espinase, *Robert Hooke* (Berkeley: University of California Press, 1962), 76–77.

novel species would be found in remote regions, Cuvier dismissed this as wishful thinking, exclaiming that if such large vertebrates existed, they would surely have been discovered already even in places as remote as the American west. To Cuvier and many of his contemporaries, the notion that species had gone extinct was an empirical fact. However, Cuvier was not an evolutionist. He believed that organisms were interdependent wholes resistant to any amount of substantial change, as expressed in his principle of the correlation of parts:

> Every organized individual forms an entire system of its own, all the parts of which mutually correspond, and concur to produce a certain definite purpose, by reciprocal reaction, or by combining towards the same end. Hence none of these separate parts can change their forms without a corresponding change in the other parts of the same animal.[7]

In such delicately balanced systems, alterations would compromise the functionality of the entire system, compelling species to "keep within certain limits fixed by nature"; therefore evolution could never occur—an argument still employed by critics of evolution today. Cuvier also argued against evolution, particularly gradual evolution, based on evidence from the fossil record, namely the lack of transitional forms—another argument still used. He believed that the discontinuous fossil record better corresponded with the idea of massive catastrophic extinctions followed by God's repopulation of the world with new, distinct species, rather than the idea of gradual evolution.

Functionalism versus Formalism

Cuvier's one-time friend Étienne Geoffroy St. Hilaire (1772–1844) took a different approach in his study of organisms, focusing upon the overall form of organs, such as the arm, instead of the particular interactions between the parts. Rather than seeing the human arm as merely a complex system of interacting parts, he saw it as similar in form to arms of other primates, legs of horses, wings of bats, and fins of fishes (fig. 2.3). What Geoffroy saw was a single archetypal form, which could have been modified to flourish in various environments.[8] Cuvier disagreed, maintaining that structures were similar in form only to the degree that they had a similar function. Thus to Cuvier the forelimbs of various vertebrates, an example of what is known as a homologous

7. Georges Cuvier, *Essay on the Theory of the Earth*, trans. Robert Jameson (New York: Kirk & Mercein, 1818), 99.

8. Étienne Geoffroy St. Hilaire, *Philosophie anatomique*, vol. 1 (Paris: Bailliere, 1818), 9.

structure (see chap. 3), implied only a similarity of function, not any evolutionary relationship between organisms.

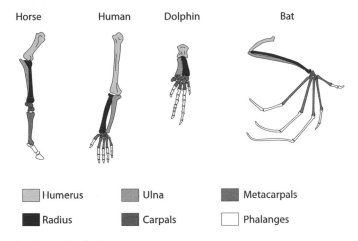

Figure 2.3. A classic example of homologous structures. The front leg of a horse, the arm of a human, the front fin of a dolphin, and the wing of a bat share a similar sequence and arrangement of bones as indicated by the shading. The proximal end of the appendage begins with a single bone, the humerus. This is followed by two bones, the radius and ulna, which become fused in the horse. The carpals and wrist bones are next, followed by the metacarpals and then the phalanges at the distal end.

Species Change

The father of taxonomy, Swedish naturalist Carl Linnaeus (1707–78), first published his famous classification of living things, the *Systema Naturae*, in 1735. It is a system that, despite many modifications, is still used today. Linnaeus was responsible for the convention of naming an organism by its genus and species, and for establishing a comprehensive hierarchical classification scheme that grouped organisms into species, species into genera, genera into orders, orders into classes, and classes into kingdoms. However, the purpose of his classification scheme was not to uncover evolutionary relationships (Linnaeus subscribed to the creation story of Genesis). But using his results, others interpreted the changing fossil record, the success of hybridization experiments, and the lack of clear morphological delineations between certain species and subspecies as evidence that evolution had occurred. For example, Georges-Louis Leclerc, Comte de Buffon (1707–88), maintained that species had been altered over long expanses of time by changes in the environment. According to Buffon, the result of this process was that species had experienced significant transformations.[9] Though unable

9. Georges-Louis Leclerc, Comte de Buffon, *Natural History*, trans. J. S. Barr (London: Barr, 1792).

to find a viable mechanism, he nonetheless devised a line of reasoning that has become a standard argument for common descent, namely the presence in some creatures of vestigial parts that appear to serve no purpose to the organism. Degeneration played a major role in Buffon's view of evolution, though he believed that there could be change for the better, especially with human intervention.[10] Buffon stopped short of advancing a radical new evolutionary thesis, but he did tread on dangerous ground for his time by pointing out the anatomical similarities between man and apes, and questioning the proposed biblical age of the earth.

Lamarckian Inheritable Change

Another early evolutionist was the French botanist Jean-Baptiste Lamarck (1744–1829), who formulated the first plausible and well-developed (albeit mechanistically incorrect) theory of evolution.[11] Lamarck's theory, based upon the inheritance of acquired characteristics, has many similarities to Darwin's theory. However, the mechanism of change is quite different. Lamarck believed that as organisms struggle to adapt to changing environments, they use certain structures or parts preferentially. Lamarck believed these parts would be altered for the better during an organism's life span, based upon the innate efforts of the organism. He further believed that these alterations could be passed on to the organism's offspring. This is termed "the inheritance of acquired characteristics." Over several generations, it would lead to marked improvements.[12] Disuse, however, would lead to atrophy and disappearance.[13] The classic example of Lamarck's theory is its proposed origin of the giraffe's neck, which would have increased gradually over a giraffe's life span by stretching. The longer neck then could be passed on to its offspring. Over many generations this could lead to significant evolutionary change.

Lamarck's theory was based upon an idea that was to become a central tenet in Darwinian theory, namely that organisms could evolve as a result of the accumulation of *gradual beneficial* changes. But he was able to find only circumstantial evidence for his theory,[14] and it was not well received during his lifetime, in large part due to the opposition of

10. A. R. Hall, *The Scientific Revolution 1500–1800: The Formation of the Modern Scientific Attitude*, 2nd ed. (Boston: Beacon Press, 1962), 300.

11. University of California Museum of Paleontology, History of Evolutionary Thought, www.ucmp.berkeley.edu/history/lamarck.html.

12. J. B. Lamarck, *Philosophie zoologique*, trans. H. Elliot (New York: Hafner Publishing, 1963), 119.

13. Ibid., 115.

14. Ibid., 116.

Cuvier. It did, however, enjoy a resurgence during the latter half of the nineteenth century, as a foil to Darwin's ideas, before it was ultimately discredited when the mechanism of inheritance was rediscovered at the beginning of the twentieth century.

Paley and Design

As described previously, the school of natural theology held that an understanding of nature inevitably led to belief in and understanding of the Creator. While this school has its roots in the writings of seventeenth-century naturalist John Ray, it was most powerfully articulated in the works of William Paley (1743–1805). Paley believed that species were distinct entities designed and specially created by God, and that God had taken care to give organisms the adaptations needed to prosper in their environment. Paley believed that if an all-powerful God had taken care to establish the proper arrangement of nature, surely a blind evolutionary process could result only in disaster.

Paley championed his ideas in his famous 1802 work, *Natural Theology*, which was well known to Darwin, who studied it during his university days. In it, Paley argued that organisms were designed because they were complex objects comprising many parts that interacted for a purpose, and they appeared to be designed specifically to fit within their environment. Paley's most famous argument for design was his description of the complexities of a watch, as compared to a stone:

> [W]hen we come to inspect the watch, we perceive—what we could not discover in the stone—that its several parts are framed and put together for a purpose, e.g., that they are so formed and adjusted as to produce motion, and that motion so regulated as to point out the hour of the day; that if the different parts had been differently shaped from what they are, of a different size from what they are, or placed after any other manner or in any other order than that in which they are placed, either no motion at all would have been carried on in the machine, or none which would have answered the use that is now served by it.[15]

Noticing that organisms are endowed with a similar complexity, Paley concluded by analogy that they too must have been designed. Paley also cataloged many examples of how organisms are uniquely adapted to their specific environment, and he inferred that only God could have been responsible for this seemingly providential arrangement. Paley's ideas made a great impression on the young Darwin.

15. William Paley, *Natural Theology* (New York: American Tract Society, 1802), 9–10.

Laying the Foundations (1800–1859)

Typical of the much more integrative approach to knowledge in the nineteenth century (as compared to today) is the fact that two of the ideas that had particular impact on Darwin's theory came from outside the realm of biology. The first was from the British geologist Charles Lyell (1797–1875), best known for his book *Principles of Geology* (1830–33), in which he advocated the principle of uniformitarianism. This principle was first proposed by James Hutton in 1785 and states that geological features are predominantly the result of slow geological forces of the same kind and intensity as those found acting in the present day, rather than catastrophic floods. Because uniformitarianism demanded long time spans for creation of today's impressive geologic features, Lyell (and subsequently Darwin) advocated a much older earth—tens to hundreds of millions of years older than the proposed biblical age of the earth. Though Lyell advocated an old earth, he did not embrace Darwin's theory. As a geologist, Lyell realized that species had gone extinct and that new species had taken their place, but he regarded the absence of intermediary forms as problematic to both Lamarck's and ultimately to Darwin's ideas. Echoing earlier thought, he believed that gradual changes could improve existing biological structures, but that slow accumulation of slight variations could not account for novel structures. Nonetheless, the belief inherent in uniformitarianism—that vast amounts of time coupled with gradual, nearly imperceptible processes, might just make the "impossible" perfectly plausible—was something Darwin incorporated into his theory.

The second major nonbiological influence upon Darwin's thinking came from the work of the British political economist Thomas Malthus (1766–1834). Malthus sought to explain and remedy the decline of living conditions in nineteenth-century England, a decline he attributed to overpopulation. He observed that many animals and plants produce more offspring than can possibly survive,[16] which leads to fierce competition for survival among the offspring because the dwindling food supply cannot support them all. When this happened in the human population, Malthus believed it could lead to the poverty and poor nutrition seen in nineteenth-century England. Malthus's solution for England was forthrightly eugenic, as he advocated reducing childbirth among the lower classes. However, it was Malthus's observations, rather than his "solutions," that garnered Darwin's attention and admiration:

16. Thomas Malthus, "A Summary View of the Principle of Population 1830," in *On Population: Three Essays* (New York: Mentor Books, 1960), 13.

[B]eing well prepared to appreciate the struggle for existence which everywhere goes on from long-continued observation of the habits of animals and plants, it at once struck me that under these circumstances favourable variations would tend to be preserved, and unfavourable ones to be destroyed. The results of this would be the formation of a new species. Here, then I had at last got a theory by which to work.[17]

It was this principle, competition for limited resources, that Darwin eventually conjoined with favorable variations to create his theory.

In addition to these influences from outside biology, it is probable that Darwin was influenced by British naturalist Edward Blyth (1810–73), who gave the notion of natural selection a very clear statement in the 1835 edition of the *Magazine of Natural History*, a periodical known to have been read by Darwin:

The original form of a species is *unquestionably* better adapted to its natural habits than any modification of that form; and, as the sexual passions excite to rivalry and conflict, and the stronger must always prevail over the weaker, the latter, *in a state of nature, is allowed but few opportunities of continuing its race.* In a large herd of cattle, the strongest bull drives from him all the younger and weaker individuals of his own sex, and remains sole master of the herd; so that all the young which are produced must have had their origin from one which possessed the maximum of power and physical strength; and which, consequently, in the struggle for existence, was the best able to maintain his ground, and defend himself from every enemy . . . [and] transmit its superior qualities to a greater number of offspring.[18]

Despite his formulation of how natural selection might work, Blyth believed natural selection to be a profoundly conservative principle, one that seeks to optimize the "form" of a species with respect to its "natural habits," that is, its environment. In modern terms, if the environment changes, natural selection will select those individuals from the gene pool that are best fitted to survive under the new conditions. As the gene pool typically has enough information stored to allow some variation in nearly all physical characteristics, natural selection working in conjunction with this information ensures survival of the species under a range of environmental conditions. So what about evolution in the stronger sense? It was considered but rejected by Blyth, Lyell, and others, partly

17. Charles Darwin, "Autobiography," in *The Life and Letters of Charles Darwin*, ed. F. Darwin, 1887, www.victorianweb.org/science/darwin/darwin_autobiography.html.
18. Edward Blyth, "Varieties of Animals," *Magazine of Natural History* 8 (1835): 40–53. Text available in Loren Eiseley, *Darwin and the Mysterious Mr. X* (New York: Harcourt, Brace, Jovanovich, 1979), 97–111.

because of the fossil evidence, but also because of perceived limitations of species to adapt to their environment near its boundaries. If species can adapt well in a limited geographic area, why can they not do so toward the edges of that area? Surely that would be less difficult than the great changes needed to explain the origin of higher taxa such as families, orders, classes, and phyla.[19] For Blyth, this fact about species demonstrated that their capacity to adapt was far too limited to account for the panorama of natural history. This argument, and others based on the fossil record, held Darwin back for twenty years.

Darwin's Triumphal Entry and Early Battles over Evolution (1859–1910)

The scientific landscape in the mid-nineteenth century was thus primed for Darwin's simple yet controversial theory. Key ideas—including competition for survival, natural selection, gradual imperceptible change accounting for the complexities of the natural world, and the adaptive fit organisms have for their environment—are found in the works of his immediate predecessors. One aspect of Darwin's genius was his ability to synthesize these various lines of thought into a coherent, well-developed theory, although Darwin's work was more than a mere synthesis of the ideas of others. His theory relied equally on the large amounts of time he spent collecting and examining fossils and specimens, a task that allowed him to marshal volumes of circumstantial evidence in favor of his theory, much of which he gathered on his famous voyage as naturalist aboard the HMS *Beagle* (1831–35). Two things in particular struck Darwin during his travels: (1) Living creatures in a specific geographical area are often closely related to fossil specimens from the same region. This he took as strong evidence that the structurally similar living forms had descended from the extinct forms. (2) Radiation of similar types is observable in the different regions of South America. Nowhere was this more striking than on the Galapagos Islands. Here Darwin found that each island had species peculiar to it, but these species were quite similar to ones from neighboring islands and the mainland. Drawing from his experiences, Darwin began to formulate his theory of evolution in his notebooks during the late 1830s. However, he did not formally introduce his theory to the public until 1858.[20] During the intervening years he labored painstakingly over draft manuscripts, cautiously sharing them

19. Fred Hoyle, *Mathematics of Evolution* (Memphis: Acorn Enterprises, 1999), 105.
20. Charles Darwin, "Notebooks on Transmutation," in *Darwin: A Norton Critical Edition*, ed. Philip Appleman (New York: W. W. Norton, 1970), 73–74.

with his trusted friends, one of whom was Lyell. He was also wrestling with the objections he knew his theory would meet. And, given the scope of his idea, he was reluctant to publish prematurely, because such a move might have disastrous consequences on the acceptance of his thesis.

By 1855 Alfred Russell Wallace (1823–1913) had independently struck upon a similar theory of organic evolution (although not as well developed) and was ready to publish his work. In 1858, at the urging of Lyell, a sketch of Darwin's work and Wallace's paper were presented jointly to the Linnaean Society of London. Despite the joint presentation, it was Darwin who would become inextricably linked to the theory of evolution due to the publication one year later of *On the Origin of Species by Means of Natural Selection, or the Preservation of Favoured Races in the Struggle for Life*, a manuscript that laid out his theory in much more detail and forever changed science.

Since the conservative function of natural selection was well known and had been recognized since the mid-1830s, what was Darwin's contribution? It was twofold. First, Darwin thrust a new function onto natural selection. He made it the promoter of innovation, so that working in conjunction with random changes in hereditary material, natural selection would choose those changes that were beneficial to the organism.[21] Darwin of course knew nothing of genes and molecular genetics, but he surmised that new, slightly better organisms constantly arose, and that these would have an advantage over their counterparts. Second, Darwin drew upon his enormous storehouse of firsthand knowledge about organisms and their environments for insight and examples. This allowed him to speak with an authority that others could not muster.

For all the impact that it has had on science, it is remarkable that *The Origin of Species* contained no experiments. Rather, it comprised a simple and elegant thesis, supported by a mountain of circumstantial evidence and observations. The thesis Darwin proposed in *The Origin of Species* was as follows: (1) organisms within a species vary and this variation is heritable; (2) organisms struggle within their environment for survival; and (3) those with advantageous variations will survive and reproduce and will be disproportionately represented in the next generation. This thesis, he believed, could account for all the diversity of the natural world. He summed up this principle, popularly known as "survival of the fittest," in the introduction to *The Origin of Species*:

21. Luther Sunderland, *Darwin's Enigma* (Green Forest, AR: Master Books, 1998), 18.

As many more individuals of each species are born than can possibly survive, and as consequently there is a frequently recurring struggle for existence, it follows that any being, if it vary however slightly in any manner profitable to itself, under the complex and sometimes varying conditions of life, will have a better chance of survival and thus be *naturally selected*. From the strong principle of inheritance, any selected variety will tend to propagate its new and modified form.[22]

Darwin extrapolated that those propagated variations eventually accumulate and result in new species and higher taxa.

Since direct evidence for species change was lacking in nature, Darwin turned to observations from animal husbandry, in which breeders breed domestic animals for certain traits such as wool quality or temperament. Based on the fact that breeders, through cumulative selection, were able to induce a limited amount of change in domestic animals, Darwin believed the same could be true in the wild if nature had a "selective breeder." Darwin believed that natural selection could act as the "selective breeder" by continually weeding out the less fit and rewarding the innovators. As a result, species would continually adapt in order to better survive in their dynamic environments. In Darwin's mind, this principle, not divine providence, was the reason that organisms appeared to be uniquely suited for their environments.

Although Darwin admitted that only limited change had been observed in animal husbandry, he posited that this process could be extrapolated over vast amounts of time, so that monumental change could occur.[23] Of course, the mere fact that breeders observe limited amounts of change in species during a few generations in no way *proves* that large amounts of change can occur over long periods of time (macroevolution). Not surprisingly, this was an early objection to the theory and remains a mainstay of the theory's critics.

While proof of the efficacy of natural selection and random mutation[24] to explain macroevolution (strong Darwinian evolution) is lacking in *The Origin of Species*, Darwin did posit a wealth of circumstantial evidence that evolution (i.e., historical evolution and common descent) had indeed occurred. Such evidence is consistent with Darwin's theory, though it can be explained in other ways.[25] Still, the fact that Darwin proposed

22. Charles Darwin, *The Origin of Species by Means of Natural Selection, or the Preservation of Favoured Races in the Struggle for Life* (New York: Mentor Books, 1958), 29.

23. Ibid., 56–57.

24. Darwin of course was unaware of modern genetics, but he did assume that there was some vehicle for transmitting information from one generation to the next. It is variations in this vehicle that he had in mind as the raw material needed for natural selection.

25. Evidence cited can also be explained by special creation if one posits that the common architecture is similar in these cases because the designer struck upon an exceedingly

a comprehensive naturalistic explanation for an enormous amount of observational detail gave the theory much credibility.

Despite the impressive amount of circumstantial evidence Darwin compiled in support of the theory, it still had its problems. With characteristic honesty, Darwin acknowledged these problems, such as the lack of transitional forms in the fossil record, the prolonged time needed for such a process to take place, the difficulties in assembling novel structures through a gradual process, and the sudden appearance of fossil forms. He even admitted that the absence of fossil forms before the sudden explosion of the Cambrian period was a major problem, stating, "the case at present must remain inexplicable; and may be truly urged as a valid argument against the views here entertained."[26]

On the positive side, Darwin, being a good scientist, was bold enough to make predictions regarding his theory of natural selection that he felt future research would confirm. These predictions, if verified, would support his proposition that his proposed mechanisms could indeed create new forms of life, rather than merely alter already existing species. The principal predictions and their status are found in table 2.1.

Table 2.1. Darwin's predictions and assumptions and their current status

Predictions/assumptions	Status
There is a constant "struggle for existence" among all organisms, and only the fittest survive.	*True*. This is a feature of most plant and animal life, although defining the "fittest" is difficult without hindsight.
Further paleontological research will disclose massive numbers of transitional forms.	*Uncertain*. While 150 years of research has found many transitional forms, researchers have certainly not found the number Darwin expected. The transitional forms problem remains very controversial.
Environmental pressures can induce heritable changes in an organism.	*Uncertain*. There is some evidence that this may occur in rare instances, but the primary mechanism remains random mutation.
All change must be the result of accumulation of incremental changes.	*Uncertain*. The fossil record suggests many discontinuities; it is unclear whether small changes alone can accumulate to produce large-scale changes required for evolution.
Structural complexity (as revealed through phyla with more complex organization) is the result of accumulated changes over long periods.	*Uncertain*. The structural complexity of nearly all animal phyla appears to emerge rapidly in the Cambrian period.
Nature will preserve even the slightest variation that proves beneficial.	*False*. Population genetics calculations show that single mutations, even if positive, usually have only a small chance of survival.

robust and functional plan. However, if a naturalistic explanation of the similarities is available, there would be no reason to discard it and invoke special creation. It is also true that most of the evidence in *The Origin of Species* could have been used (at the time) to support Lamarck's theory, but Lamarck never developed his theory with the same volume and type of evidence mustered by Darwin, and so it never received the same credibility.

26. Darwin, *The Origin of Species*, 317.

We see in table 2.1 that some of the predictions made by Darwin have yet to be borne out and one has been refuted by new research. New versions of the theory, incorporating genetics and other discoveries, have been developed to deal with some of these problems. But in the decades after Darwin published *The Origin of Species*, these problems were regarded as quite serious (and some still consider them to be so today).[27] In fact, in the late 1800s Darwin's theory was nearly supplanted by a theory based not on gradual changes but on incremental leaps or saltations.

Darwin's Critics

Tempestuous events followed the publication of *The Origin of Species*. Darwin was attacked by many, both inside and outside science. Many of Darwin's scientific contemporaries, although sympathetic to his naturalistic bent, had serious concerns regarding the validity of his theory on empirical grounds. The most significant problem raised was the lack of gradual transitions within the fossil record and within living species. Darwin argued that the discontinuities in the fossil record were an artifact of its spotty nature. Because very few organisms had been preserved as fossils and because researchers had looked for fossils only in limited locations, the fossil record should by its very nature be discontinuous. He explained the lack of living transitional forms in nature a bit more cleverly. He argued that transitional forms, those that were intermediary between two more robust forms, would eventually be selected against in competition with their more robust cousins. Over time they would disappear altogether, leaving large gaps between various living forms.

Although these explanations were reasonable, they were highly speculative and many critics were unpersuaded. Thomas Huxley (1825–95), known as "Darwin's Bulldog" for defending the theory of evolution, was nonetheless troubled by the dearth of evidence for any real transition forms between natural groups. He was of the opinion that evolution was driven by leaps or saltations, an opinion he famously made known

27. Early additional support for Darwin's theory was proposed by German embryologist Ernst Haeckel (1834–1919). Haeckel proposed the law of recapitulation, which stated that during development organisms pass through the various forms of their evolutionary ancestors, the infamous "ontogeny recapitulates phylogeny" principle. Haeckel even produced a famous set of pictures purportedly demonstrating how embryos pass through similar phases, pictures that continue to be used as support for evolution to this day. Unfortunately, it was all faked. Despite the fact that the fraud was unmasked almost immediately by L. Rutimeyer, professor of zoology and comparative anatomy at the University of Basil, and Haeckel was censured by his own university colleagues, his faked diagrams inexplicably became a mainstay in evolution books for nearly a century.

to Darwin in a letter that stated, "You have loaded yourself with an unnecessary difficulty in adopting *natura non facit saltum* [nature does not make leaps] so unreservedly."[28]

Another significant problem raised was Darwin's generalization from artificial selection to natural selection. The traits that breeders varied were things like leg length, body size, and coat color—hardly the characteristics that define new families and genera. Adaptive traits such as hoofs turning to claws or a heart losing or gaining a chamber were never observed. Thus if artificial selection could alter only *superficial* traits, there was no reason to assume that natural selection should be able to alter *adaptive* ones. On this point, Huxley was particularly adamant.

In addition to Huxley's concerns, Darwin's theory immediately ran into problems regarding the supposed age of the earth. Because Darwin relied on a gradual mechanism, his theory required immense lengths of time to operate. Based on the assumption that the earth was originally molten, and has been gradually cooling, Lord Kelvin calculated its age to be 20 million to 40 million years, far too short for evolutionary mechanisms to work.[29] This calculation led to a direct clash between the advocates of natural selection and the physicists of the late nineteenth century.[30] Fortunately for Darwin, later estimates of the age of the earth based upon radiometric measurements—approximately 4.5 billion years—are much more amenable to his theory.

Darwin and Heredity

Darwin's theory tenuously weathered all criticism during his lifetime, but as the nineteenth century drew to a close, new challenges began to mount, particularly those dealing with the mechanism of inheritance. Darwin's theory hinges upon the inheritance and accumulation of advantageous traits, but exactly how this occurred was unclear because little was known about the mechanisms of inheritance. Ignorance on the matter spawned speculation, and most biologists speculated that inheritance was the result of the blending of the traits of the parents. For example, if one parent was tall and the other was short, these traits would blend in children who would exhibit an intermediate height. But

28. Thomas Huxley, *Life and Letters of Thomas Huxley*, vol. 1, ed. Leonard Huxley (New York: D. Appleton, 1902), 189.

29. Of course, Kelvin neglected the earth's internal heat source, namely radioactive decay, because radioactivity had not yet been discovered, and his hypothesis about the earth as originally molten is incorrect.

30. Loren Eiseley, *Darwin's Century: Evolution and the Men Who Discovered It* (Garden City, NY: Doubleday, 1958), 233–41.

for Darwin's theory, blending inheritance was problematic because if blending occurred to a significant extent, small advantageous changes in one individual would gradually be lost. In fact, in each successive generation these advantageous changes would be diluted and ultimately destroyed by blending with individuals that lack such a change. Instead of the gradual spread of minute beneficial mutations as Darwin argued, there would be a rapid suppression and elimination of all slight changes, making Darwin's theory highly improbable. This was, in fact, pointed out to Darwin in the 1860s.[31]

To address this problem, Darwin postulated a theory of inheritance based on inheritance molecules, which he termed "gemmules," that were manufactured by all the various body parts during the organism's life cycle. These molecules were capable of reproducing themselves and were passed on in large quantities to future generations. In these subsequent generations, the gemmules directed the development of the organ from which they were derived, a theory he termed "pangenesis." In addition, these gemmules could be modified based on the conditions of the body, thereby leaving open the possibility that acquired characteristics could be inherited. While Darwin's theory of pangenesis did correctly propose the existence of discrete units of heredity, the theory was in many respects an ad hoc explanation that allowed for anything from blending inheritance to discrete inheritance to Lamarckian inheritance. No rationale was given to explain exactly why one type of inheritance would occur in any particular instance. Nor did Darwin provide a mechanism by which the gemmules were transported in the correct quantities from generation to generation.

Because of these difficulties, Darwin's theory needed a unit of inheritance that was permanent, independent from the unit inherited from the other parent, and inherited in a regular pattern. The rediscovery at the dawn of the twentieth century of work on genetics by Gregor Mendel (1822–84) provided Darwin's theory with just such a unit. In doing so, however, it raised a whole new set of issues.

Continuity or Discontinuity?

Gregor Mendel, an Augustinian monk, was interested in determining how variation was transmitted within a species from one generation to the next. Using the common edible pea plant for his experiments, Mendel crossed and tested approximately 30,000 plants in his garden in order to examine this question. The data Mendel presented on his work in an 1866 paper indicated that physically discrete units, which

31. Eiseley, *Darwin's Century*, 210.

were later termed "genes," governed inheritance. These genes did not blend into oblivion, but in fact propagated unaltered from generation to generation (except for rare cases of mutation), independently of the corresponding gene from the other parent. While Mendel's work provided Darwin's theory with some much-needed empirical support with regard to heredity, it also demonstrated the discontinuous nature of some heritable traits. It was this revelation that proved to be a stumbling block for the theory.

Unlike traits such as human height and hair color, the traits that Mendel chose to study (short vs. long stems, brownish vs. white seed-skin, wrinkled vs. smooth peapods) did not show a continuous degree of variation but rather varied discontinuously, that is, one got either a wrinkled or a smooth peapod. Because of this, Mendel's data seemed to argue *against* Darwin's idea of gradual continuous variation and to support the notion that evolution had occurred through the leaps and bounds associated with discontinuous variation or saltations.

Though overlooked during his lifetime, Mendel's work was rediscovered at the beginning of the twentieth century by the Dutchman Hugo de Vries (1848–1935), among others, and it was he who most clearly realized the importance of Mendel's work for the study of evolution. De Vries spent much of his career investigating the mode of inheritance in plants, and his observations led him to believe in a discontinuous form of evolution in which natural selection played only a minor role. In breeding experiments with evening primroses, de Vries was able to generate sudden abrupt mutations, such as twisted stems, in a single generation. De Vries drew two conclusions from this work: (1) the genetic elements that determine stem shape must be discrete and independent units, which he called "pangenes"; and (2) the abrupt appearance of novel features in his primroses implied that new species are formed as the result of chance large-scale mutations in a pangene, or the sudden addition of a new pangene, rather than a gradual accumulation of minor changes.[32] Given this, he believed that gradual change was no basis for developing a theory of evolution. Rather, speciation in nature must arise by a different process, saltational change, which he believed to be random, plentiful, and able to proceed in any direction. Natural selection was relegated to the secondary role of merely sorting through the saltations that had occurred and discarding the disadvantageous ones. While Darwin envisioned that all organisms were connected in an insensibly graded continuum, de Vries's mutational or saltational theory left large gaps between species.

32. Hugo de Vries, *The Mutation Theory*, vol. 1, trans. J. B. Farmer and A. D. Darbishire (Chicago: Open Court, 1909), 118–19.

The Englishman William Bateson (1861–1926) came to favor a similar saltational theory as a result of his studies as a naturalist. Darwin had stressed that organisms would change gradually in the face of changing environments, eventually giving way to new species. Bateson, seeing the logic in this argument, set out to verify it but was unable to do so. He spent years doing field research in Asia and in Egypt, tracking animals and plants in varying environments. What he found was organisms randomly fluctuating around a relatively stable form. To Bateson, the minor variations that Darwin had discussed at great length in *The Origin of Species* seemed to be taking species nowhere, even during drastic environmental changes such as prolonged floods and droughts. As a result, he became convinced that the variation within populations was only of minor significance, and that discontinuous variation was the mechanism of evolutionary change.[33] Bateson still held Darwin in high regard because Darwin had performed the invaluable service of making the study of evolution a viable discipline. Nonetheless, he became increasingly frustrated with the unscientific speculations advanced in support of Darwin's theory:

> In these discussions we are continually stopped by such phrases as, "if such and such a variation took place and was favorable," or, "we may easily suppose circumstances in which such and such a variation if it occurred might be beneficial," and the like. The whole argument is based on such assumptions as these—assumptions which, were they found in the arguments of Paley or of Butler, we could not too scornfully ridicule. "If," say we with much circumlocution "the course of nature followed the lines we have suggested, then, in short, it did." That is the sum of our argument.[34]

Bateson's criticisms are echoed by twentieth-century biologists such as Stephen J. Gould, who see such just-so stories as evidence of an overwrought explanatory paradigm that is deficient in hard empirical evidence.

But despite such criticisms, by the turn of the twentieth century naturalistic evolution was triumphant, and all theories of special creation had been soundly routed, at least among the academic community. The only real question about evolution then remaining for the academics was whether Darwinian gradualism or some new saltational theory would emerge as the comprehensive explanation for the history of life.

33. William Bateson, *Materials for the Study of Variation* (London: Macmillan, 1894), 16.
34. Bateson, *Materials for the Study of Variation*, v.

Genetics and the "New Synthesis" Period (1910–60)

The early 1900s probably marked the low point in the history of Darwin's theory. For a brief period it appeared that Darwin's vision of gradual evolution would be discarded and replaced by the type of saltational theory favored by de Vries and Bateson. By the 1950s, though, Darwin's ideas had been skillfully integrated with a modern understanding of Mendelian genetics, which enabled credible responses to be given to many objections, and opened new areas for exploration. This monumental achievement was the work of a number of dedicated and gifted men in the first half of the twentieth century, and their efforts became known as the "new synthesis," or what is commonly referred to today as Neo-Darwinism. To understand how the synthesis occurred, one must start with the great American geneticist Thomas Hunt Morgan (1866–1945), whose work in genetics turned that field from a stumbling block for Darwin's theory into the discipline most responsible for establishing Neo-Darwinism as the reigning orthodoxy.

Morgan began his scientific career skeptical of the Darwinian position. Like de Vries and Bateson, he felt that evolution was discontinuous, and he invoked Mendel's work and many of the same arguments de Vries and Bateson had used to justify their beliefs. He believed that selection of the normal variants found within a population was of minor importance in evolution.[35] According to Morgan, saltational mutations created novelty; selection did not.

Morgan realized that to properly understand evolution one would have to gain an understanding of inheritance. Beginning in 1910, he began to investigate the inheritance of traits within fruit flies, and his opinions regarding both Darwinian gradualism and Mendelian genetics shifted. In performing mutation experiments on fruit flies, Morgan was impressed with the number of minor mutations that could occur within a species. He found mutants with variable wing sizes, reduced eyes, or stubby bristles. Some of these mutations, inherited in a Mendelian fashion, could, through selective breeding, be gradually modified. For example, through cumulative selection he could gradually modify the wing size such that it became progressively smaller in each subsequent generation (fig. 2.4). In addition to uncovering many important genetic principles, such as the location of genes on the fly chromosomes, Morgan's work was enough to convince him that evolution proceeded in a gradual manner by working on minor mutations. He came to believe that the gradual accumulation of minor mutations could improve or modify organisms

35. T. H. Morgan, *Evolution and Genetics* (Princeton, NJ: Princeton University Press, 1925), 1279.

to the extent needed to drive evolution, and that this process could be described by Mendelian genetics.

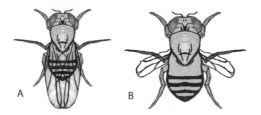

Figure 2.4. Fly mutants studied in Morgan's lab had alterations in wing size similar to that seen in a normal (A) and short-winged fly (B).

Morgan never actually witnessed the transformation of a fruit fly into a new species via the accumulation of minor mutations, only the emergence of new variants (some of which emerged quite abruptly) of a particular species of fruit fly. Even so, based upon these observations, he extrapolated that if minor mutations were allowed to accumulate indefinitely, new species would arise. But just as Darwin never demonstrated speciation in *The Origin of Species*, Morgan never demonstrated it in his work. What he did do, however, was develop a practical language for describing exactly how evolutionary change could occur. The value of this cannot be overstated, as evidenced by the influence that genetics has had in the development of evolutionary theory. Due to Morgan's influence, the accepted wisdom today is that most evolutionary change is the result of gradual genetic change.

In the ensuing decades every other discipline that studied evolution—paleontology, anatomy, and embryology included—had to come to terms with the burgeoning field of genetics. What really mattered in the study of evolution were the genetic changes that occurred at the chromosome level, and inferring this via paleontology or gross anatomy was impossible.[36] This marked a critical turning point in evolutionary theory and biology in general. The study of organisms via embryology, zoology, and botany was forced to take a backseat as genetics began to dominate the evolutionary debate.

Population Genetics

While Morgan established the heritability and spread of relatively minor genetic mutations in the lab, there was no evidence that the accumulation of such mutations in the wild could lead to the origin of new

36. Julian Huxley, *Evolution: The Modern Synthesis* (New York: Harper & Brothers, 1942), 38.

species, much less higher taxa such as families, classes, or phyla. Since this is the fundamental question of evolutionary biology, a vigorous examination of the ability of accumulated minor mutations to bring it about was needed. Rather than focusing on this question—obviously a difficult problem—researchers instead focused their efforts on an easier problem, namely how minor mutations, such as those spread by Morgan throughout a laboratory population, could spread through a population in the wild without the aid of human intervention. To answer this question, an entirely new discipline at the border of mathematics and biology emerged, known as population genetics. This discipline, which examines the theoretical flow of genes through natural populations, soon became the dominant player in studies of evolution.

The earliest of the population geneticists was Sir Ronald A. Fisher (1890–1962), who in 1930 published his influential work *The Genetical Theory of Natural Selection*. Fisher advocated a brute force view of evolutionary change in which natural selection transformed large populations in gradual incremental steps. Using mathematical models to support his ideas, Fisher argued that all evolutionary changes were either adaptive and were maintained in the population or they were maladaptive and eliminated. In addition, Fisher argued that nearly all evolutionary change was of a small magnitude. Fisher believed this because his mathematical models demonstrated that large mutations would have a very low probability of improving the organism, mainly because the large change would be too drastic for the organism to handle. Small changes, however, were much more likely to impart improvement as they would be easier to assimilate into the organism without a massive disruption of its form and function. As Fisher wrote, "the chance of improvement, for very small displacements tends to the limiting value ½ [50%], while it falls off rapidly [close to 0%] for increasing displacements"[37] (fig. 2.5). For this reason, Fisher was opposed to all views of saltationism. Fisher's model predicted that if a small adaptive mutation appeared, it would eventually spread through the entire population because it endowed its bearers with a selective advantage.

Fisher did not address the question of whether there are limits to what incremental change can accomplish, as many breeding experiments suggested. Fisher simply assumed that there was always room for improvement by means of small changes (as opposed to saltations).

Although Fisher's model explained how the genetic makeup of a specific species could change over time, it did not explain how the *number* of species would increase, such as the simplest case of a single species

37. R. A. Fisher, *Genetical Theory of Natural Selection*, 2nd rev. ed. (New York: Dover, 1958), 43.

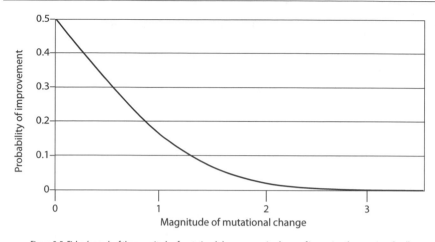

Figure 2.5. Fisher's graph of the magnitude of mutational change versus its chance of improving the organism. Small changes are much more likely to improve the fitness of an organism than are large changes.

diverging into two separate species. Fisher argued insightfully that for this to occur gene flow must be disrupted between two populations. Then, over time, their genetic makeup could diverge enough so that they could no longer produce fecund offspring. A physical barrier that splits the population is one way to disrupt gene flow. The end result is two reproductively isolated populations, each of which can pursue separate evolutionary paths, possibly becoming separate species. Exactly how much change is needed to allow speciation to occur is a question Fisher did not address.

J. B. S. Haldane (1892–1964), Fisher's contemporary, also believed that the spread of minor adaptive mutations through a population could lead to new species, but he did not reckon this sufficient to explain every aspect of natural history. In populations many features arose that appeared to be nonadaptive; that is, they gave the bearer no clear advantage. Such features, which often were used to distinguish between two species, could not have arisen as an adaptive mutation that spread through the population, as Fisher proposed.[38] But rather than abandon Darwinism, Haldane allowed room for other factors to influence evolution, such as internal developmental constraints, the random variation found in nature, and possible rapid speciation via hybridization. In this manner, Haldane was not compelled to find an adaptive function for every characteristic of an organism.

Haldane also differed from Fisher in his view on the *tempo* of evolution, favoring a rapid discontinuous pace over a slow and continuous

38. J. B. S. Haldane, *The Causes of Evolution* (London: Longmans Green, 1932), 113–14.

one.[39] In modeling this mathematically, Haldane started with much smaller population sizes than did Fisher, reasoning that adaptive traits would be able to spread much more rapidly in this case. In fact, Haldane believed that the large stable populations Fisher described would be the most resistant to evolutionary change because of the difficulty of a gene successfully conquering such a large population. Small isolated subpopulations, however, would have more rapid gene flow, allowing novel mutations to spread more readily. Although both Haldane's and Fisher's models are mathematically viable, determining which (if either) accurately describes what happens in nature has proved difficult. Despite more widespread acceptance of Fisher's views, the exact tempo of evolutionary change is still a matter of considerable debate.

The third of the founding fathers of population genetics, and by far the most controversial, was Sewall Wright (1889–1988). Wright concurred with Haldane that large populations would be resistant to evolutionary change, but he realized that many genes interact with one another in surprisingly complex ways. As a result, one gene could affect a variety of traits. Given the many interactions between the different gene combinations that are found in a large population, such populations must reach a stable equilibrium (genetic homeostasis) that would be largely resistant to change. Wright argued that smaller populations would be less genetically stable and therefore more liable to respond positively to the injection of new mutations. Although a smaller population size helped, Wright realized that size could not be decreased indefinitely because extremely small populations were subject to the damaging effects of inbreeding. Because of this, he reasoned that these small populations would have to reintroduce the genetic novelty that had arisen within them back into the larger population for evolution to proceed.

These small subpopulations lead into the most controversial aspect of Wright's work, his shifting balance theory. According to Wright, each small subpopulation within a species consists of a specific combination of genes, albeit spread throughout many organisms. This subpopulation would have a particular fitness that would correspond to a specific location it would occupy on the varied fitness landscape seen in figure 2.6. These subpopulations compete against one another, with some groups prevailing over others because they have more advantageous gene combinations. The formation of these groups is random, so that certain combinations of genes come together, not because any selection has occurred, but rather because a certain population has become isolated from the parental population. This random drift allows novel combina-

---

39. This is very similar to the tempo proposed by Gould and Eldredge's theory of punctuated equilibrium. See chap. 8 for a discussion of their theory.

tions to be tested by natural selection working on different subpopulations.[40] Some of these isolated groups might get lucky and prosper for a time because of novel advantageous gene combinations or the presence and rapid spread of novel mutations. However, as the environment inevitably changes, the fitness level of the group might begin to decline. Wright believed it was imperative for such a small group to periodically reestablish gene flow with the larger population in order to survive in a changing environment.[41]

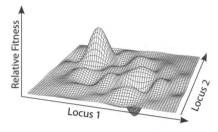

Figure 2.6. A fitness landscape for two loci demonstrating fitness peaks as well as fitness valleys. Because the environment is changing continually, the fitness landscape is constantly changing as well. Good gene combinations today may not be good combinations tomorrow.

The notion that groups, rather than genes or organisms, were the subjects of selective pressure was antithetical to Fisher's position, and it turned Wright and Fisher into lifelong rivals. Fisher's view was the one that eventually won the day and became part and parcel of modern Neo-Darwinian theory, so much so that any whiff of group selection was seen as heretical. Recently, though, the tide has begun to turn, as Stephen J. Gould (1941–2002), David Sloan Wilson, and others have attempted, with some success, to bring higher levels of selection, such as group and species selection, back into mainstream evolutionary thought (see chap. 8).

Fisher, Haldane, and Wright all published their ideas in the 1930s, and the mathematical precision of their genetic models set the stage for the synthesis of modern Darwinian thought during the 1940s. Despite their differences, all three men assumed that the accumulation of small changes could lead to the formation of new species, although

40. Sewall Wright, "The Roles of Mutation, Inbreeding, Crossbreeding and Selection in Evolution," *Proceedings of the 6th International Congress on Genetics*, 1932, 358–59.

41. Note, however, that this problem is far more difficult than Wright's comments, and the accompanying figure, suggest. Because the number of ways a genome can vary is quite large, the "space" that must be searched to find suitable peaks is vast—so vast that the trial-and-error method Wright proposes would be able to search only a miniscule fraction of it, even over evolutionary time spans.

their mathematics has been challenged recently by Sir Fred Hoyle and others.[42] The different models they proposed demonstrated how genes moved through populations based on their selective advantage. While this was and remains extremely important work, the "genes" they investigated were not actual genes, but rather mathematical abstractions. Hence their models do not take into account many important aspects of real organisms, such as how new genes affect development, morphology, or anatomy—all extremely critical parameters in terms of survival. Because of this inherent limitation, their work furnishes no *biological* proof that small mutational change, mediated by natural selection, can account for novel structures, much less new species or higher taxa. For that, other real-world studies are needed.

The New Synthesis

In the late 1930s, there was a call to synthesize the approaches of all the disciplines investigating evolution. In 1942 the Committee on Common Problems of Genetics, Paleontology, and Systematics was formed to allow dialogue between the various disciplines, and although the committee finished its work seven years later, the development of modern evolutionary thought, or what came to be known as the modern or new synthesis, continued rather informally through the 1960s.

The principal items incorporated into the new synthesis were the following:

- Populations (population genetics) as the foundation for species change
- Random mutation induced by radiation, chemicals, and various chance factors as the basis of new or changed genetic information
- No inheritance of acquired characteristics
- No saltations

With these givens, what emerged from the new synthesis was rather restrictive. Instead of an expansive theory integrating the ideas from many disciplines and open to new interpretations of the evidence, the new synthesis developed a narrow, hard-line view of evolution (Neo-Darwinism), dominated by the work of population genetics.[43] Many observers from other

42. See www.evolutionprimer.net.
43. This is not atypical of the way science proceeds, of course. Scientific theories are almost always the product of a long and difficult road, and as such are defended tenaciously. At the same time, they are subjected to ruthless scrutiny and relentless testing, so that any empirical deviation from theory, however small, can be found and its impact on the theory assessed.

disciplines (paleontology and systematics) complained that the mathematical theory was being used to inform the biology, rather than the other way around. These dissenters, who believed that the new synthesis was based too heavily on genetically informed mathematical population dynamics, increasingly became marginalized by advocates of the new synthesis.

The ideas and the hardening of the new synthesis can be reflected in the careers of two of its most towering figures, the geneticist Theodosius Dobzhansky (1900–1975) and the naturalist Ernst Mayr (1904–2005). Dobzhansky, a student of T. H. Morgan, initially advocated large-scale chromosomal rearrangements and random genetic drift[44] as the predominant factors in evolution. Over time he began to discount these mechanisms and gave primacy to natural selection.

As an advocate of a slow, gradual process (natural selection), Dobzhansky had to deal with the observation of discontinuity in the fossil record. His solution was much like Darwin's: he asserted that this was just a by-product of a spotty and incomplete fossil record. He thought that if all life-forms that had ever existed were lined up, side by side, one would have a virtually continuous sequence. Unfortunately, because many if not most species have not been preserved in the fossil record, this was not a testable hypothesis. Regarding the nature of gradual change, Dobzhansky had observed that small changes of genetic material within fruit flies could, over time, lead to significant change in the morphology of the fly. This, coupled with the theoretical work of Fisher, gave him the confidence to extrapolate the observations of small-scale change (microevolution) to the evolution of large-scale difference among organisms (macroevolution).

> There is no way towards an understanding of the mechanisms of macroevolutionary changes, which require time on a geological scale, other than through a full comprehension of the microevolutionary processes observable within the span of a human lifetime, often controlled by man's will and sometimes reproducible in laboratory experiments. . . . A geneticist can approach macroevolutionary phenomena only by inference from the known microevolutionary ones.[45]

Many, like Dobzhansky, began to assume that population genetics had the answers to the problem of macroevolution, despite the fact that this position rested upon an unproven long-range extrapolation.[46]

---

44. This occurs as the genetic composition of a population changes randomly from one generation to the next. Offspring are genetic samples of their parents and this sample may not be completely representative of the genetic makeup of the parental population. This is particularly true of small populations.

45. Theodosius Dobzhansky, *Genetics and the Origin of Species*, 3rd ed. (New York: Columbia University Press, 1951), 16–17.

46. Ibid., 12.

The second intellectual giant of the synthesis, Ernst Mayr, had a unique role because he was a naturalist by trade, working in a field dominated by geneticists and mathematicians. The influence of Mayr's acceptance of the Neo-Darwinian paradigm is hard to overstate, given his standing outside the field of population genetics. While Mayr held the founding fathers of population genetics in high regard, he cautioned that mathematical models could not capture completely what was actually happening in nature. For example, Fisher's models assumed a closed population of organisms. Mayr, based on his experience as a naturalist, knew that most populations were open, with individuals entering and leaving the population over time. This open gene flow is an important aspect of evolution that most of the mathematical models had failed to capture. As a result, Mayr's acceptance of the Neo-Darwinian model stems not from his allegiance to a particular mathematical model but rather from observations of actual bird populations, data that are much more appealing to naturalists and systematists.

Mayr observed that different degrees of speciation can be observed in various bird populations in the wild. For example, we can find two distinct populations of birds that as a whole have not diverged from each other, while we can find another population in which small peripheral isolates have diverged to the point that they are considered subspecies. We can also find everything in between, such as hybrid zones in which populations have acquired new morphological characteristics but have not yet obtained reproductive isolation. Mayr argued that if we can see the various stages of speciation in the wild, then speciation must progress at a gradual and continuous rate. These ideas were in accord with the models proposed by Fisher and Haldane. Having corroborated the models of population genetics with his observations of nature, Mayr applied gradual change to the whole of evolution:

> The proponents of the synthetic theory maintain that all of evolution is due to the accumulation of small genetic changes, guided by natural selection, and that transpecific evolution [the origin of new species] is nothing but an extrapolation and magnification of the events that take place within populations and species.[47]

In the end, the synthesis more or less established population genetics as *the* field of evolutionary study. Disciplines that examined whole organisms, such as paleontology, were marginalized by the reduction of evolution to the interaction between favorable and unfavorable genes. In the view of Neo-Darwinism's critics, relatively simple mathematical

47. Ernst Mayr, *Populations, Species, and Evolution: An Abridgement of Animal Species and Evolution* (Cambridge, MA: Harvard University Press, 1970), 351.

models, which could never fully explain something as multidimensional as a living organism, were permitted to dictate almost completely our understanding of evolution. The problem with this was summed up nicely by one critic: "The whole real guts of evolution—which is, how do you come to have horses and tigers, and things—is outside the mathematical theory."[48] Unfortunately, dissent from the mathematically informed synthesis was not taken lightly.

Dissidents

Despite the dominance of the new synthesis, there was, indeed, dissent. Most of the dissenters, though, were dealt with swiftly and marginalized. The best example of this can be seen in the career of the German-born geneticist Richard Goldschmidt (1878–1958). Goldschmidt, like Mayr, was wary of pushing mathematics too far in the study of biology, believing that the mathematical models were useful only insofar as they mimicked what happened in nature. Drawing from his own work, the study of variation in moths, Goldschmidt concluded that small genetic mutations (the stuff of population genetics), though able to provide for impressive variation within species, were not the engine that drove evolution. In Goldschmidt's view, examples of successful microevolutionary change were simply a case of a species adapting within the confines of its inherent genetic and morphological limitations. He saw no evidence that such change could account for the presence of novel structures and species.

Unlike the Neo-Darwinists, he was not ready to equate microevolution and macroevolution. Since he discarded the Neo-Darwinian mechanism as ineffectual, Goldschmidt had to propose something else that could create new species. Goldschmidt advocated large-scale mutations that had far-reaching effects on the organism for the more formidable task of creating new species, though these did not have to occur in one step.[49] Ironically, this was an idea taken from the early work of Dobzhansky. Goldschmidt believed that a new pattern of genomic organization arising from wholesale chromosomal rearrangement would provide the novelty needed to create species rapidly, and he was encouraged by the presence of such drastic mutations in fruit flies that lead to legs growing where antennae should (fig. 2.7). However, Goldschmidt could never verify that such mutants, which were infamously called "hopeful monsters," actually led to the formation of new species. Moreover, his views were in direct

48. C. Waddington, "Discussion [of paper by Murray Eden]," in *Mathematical Challenges to the Neo-Darwinian Interpretation of Evolution*, ed. P. S. Moorehead and M. M. Kaplan (Philadelphia: Wistar Institute Press, 1967), 14.

49. Richard Goldschmidt, *The Material Basis of Evolution* (New Haven: Yale University Press, 1982), 206.

opposition to Fisher's mathematical argument, which demonstrated the immense likelihood that large mutations would reduce an organism's fitness, most likely killing it.

Goldschmidt's rejection of both gradualism and natural selection left him the subject of much criticism, based on the extremely low probability that the random rearrangements he proposed could be successful.[50] Goldschmidt, naturally, did not believe that the mutations were as random as his critics made them out to be. He thought robust developmental pathways could help to channel seemingly chaotic large-scale mutations into producing organized adult organisms—what he called "emergent evolution"—and that certain small mutations that affected the development of the organism could have profound phenotypic effects. Although Goldschmidt was soundly routed, some of his ideas regarding both developmental mutations and developmental constraints have been resurrected by the Meta-Darwinian school (see chap. 8).

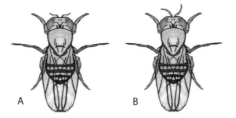

Figure 2.7. Homeotic fly mutants demonstrate large aberrations in structural organization such as legs growing from the head. One such mutant (B) is the result of a defect in a single developmental gene. A normal fly (A) is seen on the left.

While Goldschmidt took the synthesis to task based upon what he felt was an unjustified extrapolation from microevolution to macroevolution, another German, the paleontologist Otto Schindewolf (1896–1971), challenged the synthesis over the discontinuous fossil record. Schindewolf believed that the gaps in the fossil record were real; they did not represent incomplete or missing links. As a result, he too embraced the idea of rapid rather than gradual transformation of species. According to Schindewolf, any organism that was the founder of a new group would have to be equipped with a drastic structural reorganization, and he turned to Goldschmidt's thesis of large-scale mutations as the mechanism to provide this.

To help address such critics, in 1953 H. B. D. Kettlewell (1907–79), at the instigation of Oxford geneticist Edmund Ford, set out to demonstrate

50. Ernst Mayr, *Systematics and the Origin of Species* (New York: Columbia University Press, 1982), 155.

the effects of natural selection in the wild.[51] He took a particular species of moth, *Biston betularia*, released large numbers of light- and dark-colored varieties, and then observed where they alighted. By counting the number of survivors in areas with dark- and light-colored trees, Kettlewell showed how color affects survival. With his experiment, all that Kettlewell could have hoped to demonstrate was the role of natural selection in optimizing a population for particular environmental conditions, that is, changing the frequency of dark- versus light-colored moths. As such, it would have been a good demonstration of a key biological mechanism in action. Unfortunately it was marred by either fraud or shoddy technique, depending on one's point of view.[52] But regardless of that, it proved nothing about evolution in the strong Darwinian sense, because there was no innovation, no creation of new species, and no new information—just a change in relative frequencies of moth colors. Regrettably, the experiment is often presented as solid evidence for strong Darwinian evolution.

Modern Battles over Evolution (1960–Present)

During the last forty years, despite numerous successes, the Neo-Darwinian school has increasingly found itself under attack from scientists and nonscientists alike. By the early 1960s, the materialistic philosophies associated with Neo-Darwinism were setting off alarm bells in certain Christian groups,[53] which led to a resurgence of young earth Creationism. Most early Creationist efforts tended to be rather amateurish, but over the years the quality has improved steadily. With increasing numbers of converts, including among scientists, some modern Creationist research and analysis is fairly sophisticated, if limited in quantity.

In addition to continual and mounting criticism from Creationists, Neo-Darwinists have had to contend with prominent scientists who have publicly broken ranks regarding Darwin's theory. This has spawned the new school of thought that we have termed the "Meta-Darwinian school." Members of this school have proposed new mechanisms for evolution, from the self-organizing qualities of matter to endosymbiosis. Because the Meta-Darwinists still advocate natural mechanisms, their critiques of the Neo-Darwinian paradigm have held considerable sway among the academic community, given that many Meta-Darwinists hold prominent positions within academia.

---

51. The story of Kettlewell and his famous experiment has recently been thoroughly documented in Judith Hooper, *Of Moths and Men: An Evolutionary Tale* (New York: W. W. Norton, 2002).

52. Ibid., 262.

53. For the most part, these were evangelical Christian groups.

To further compound matters, others who were not willing to embrace all the premises of modern Creationism (particularly the young age of the earth), but who were nonetheless dissatisfied with the naturalistic explanations for evolution, started their own movement called the Intelligent Design school. The Intelligent Design movement, aided by a number of publications and think tanks, has quickly become a formidable participant in the public debate.

In summary, recent challenges to Neo-Darwinism have come from all sides on a variety of topics and have caused organizations such as the National Academy of Sciences (NAS) and the American Association for the Advancement of Science (AAAS) to publish position papers in response to the criticism. These modern critics have taken many old arguments, which the new synthesis, in their opinion, failed to address properly, and have reformulated and augmented them with modern data. Critics have also developed some new arguments based on more recent developments such as information theory and the stunning advances in molecular biology. The modern battles center on six key issues:

- The discontinuous nature of the fossil record given the gradual, incremental nature of evolutionary change under Neo-Darwinism
- The extent to which the complexity of biological systems exceeds that which can be accounted for by the processes envisioned by Neo-Darwinism
- The ability of organisms, which represent functionally integrated wholes, to tolerate the changes needed for macroevolution
- The validity of extrapolating from microevolution (with its observable mechanisms) to macroevolution
- The ability of random processes (i.e., random mutations) to create new genetic information
- The age of the earth

All of these issues will be addressed at length when we take an in-depth look at each of the four schools in chapters 5 through 8. For now, we turn to the considerable amount of data that any school of evolutionary thought must be able to explain.

3

A REVIEW OF THE EVIDENCE

Aristotle opens his *Metaphysics* with a phrase that has echoed down the centuries: "All men by nature desire to know."[1] This dictum is rather more difficult in practice than in theory, because "to know" imposes stringent requirements, beginning with careful observation and recording of observations. Of particular importance is the need to minimize prejudging or imposing preconceived ideas on the observations. Equally important, however, is separating observations from inferences, extrapolations, and explanations. These come later and should not be confused with the raw observation itself.

In this chapter, therefore, we shall set down the raw evidence that must be explained by any theory of evolution. By "raw evidence" here we mean observations of things and patterns present in the modern-day world, observations that can be duplicated by anyone with ample time and resources. This evidence about life and its history falls into six categories:

1. Traits and adaptations
2. The fossil record
3. Variation and diversity in populations
4. Physiological and developmental similarities
5. Genetic makeup
6. The nature of living species

1. Aristotle, *Metaphysics*, Book A, 980a21, trans. W. D. Ross, in *The Basic Works of Aristotle* (New York: Random House, 1941), 689.

As we examine the raw evidence for each category, we shall keep interpretations to a minimum. However, we shall indicate common inferences the different schools make regarding the raw evidence, when appropriate. Unfortunately, both observations and inferences in the field of evolution tend to be labeled indiscriminately as "facts." While inferences and predictions are critical for the advancement of science, these inferences must be identified as such and subjected to constant scientific scrutiny. Confusing inferences with facts is a major source of the acrimony in the debate.

An illustration of this problem is seen in a case that was extremely controversial in its day, that of the geocentric versus the heliocentric theories of the solar system. Prior to the time of Copernicus (1473–1543) and Johannes Kepler (1571–1630), scientific thought was dominated by the geocentric theory, or Ptolemaic system, which was developed in the ancient Near East and elaborated by Claudius Ptolemy (ca. 100–170). Some of the direct observations that the theory sought to explain included the apparent rotation of everything about the earth (as seen by an earthbound observer looking at the heavens), the diurnal pattern of the rising and setting sun with its variation over the course of a year, the unchangeable appearance of the stars, and the absence of any sensation of movement of the earth.

In order to explain these and other observations, the Ptolemaic theory proposed two hypotheses that were the *immediate inferences* made from the direct observations:

1. The earth is stationary.
2. The sun, moon, and stars are in constant motion.

Both of these immediate inferences were empirically testable and therefore could be (and eventually were) falsified based upon further scientific investigation. But in addition to the immediate inferences, some *long-range extrapolations* were made from the observations, extrapolations accepted not necessarily because they accorded well with the observational data, but because they fit well into other philosophical or religious belief systems. Examples of these extrapolations are as follows:

- The sun, moon, and stars are eternal.
- They are also perfect and incorruptible.
- They move in circles.
- The stars are all located on a distant crystal sphere.

These long-range extrapolations quickly acquired the position of unassailable "facts" for which any competing theory would have to account.

Note that no real scientific explanation of observed phenomena was given—that had to await Newton's laws—only an "explanation" based on *extrapolations* from the very observations to be explained, coupled with a (dubious) philosophical principle (that circular motion is perfect). The process by which this situation came about is illustrated in figure 3.1.

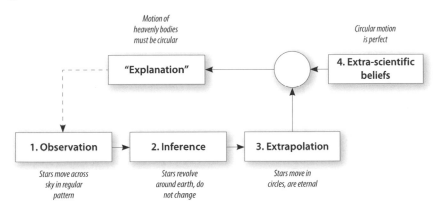

Figure 3.1. Improper use of inference and extrapolation in scientific explanation.

The philosophical notion that heavenly bodies had to move in circles held such sway that it was maintained even in direct opposition to certain other observations. Many heavenly bodies move across the sky in one direction and then reverse their motion, moving in the opposite direction, a phenomenon known as "retrograde motion." To incorporate this into the Ptolemaic system, it was necessary to introduce increasingly complex minicircular cycles, termed "epicycles," into the planets' orbits. These epicycles were ad hoc assumptions and did not flow from the theory in any natural way. As observations of retrograde motion became more refined, the epicycles that were proposed to account for them became increasingly complex, making the theory more and more unwieldy.

While it might have eventually collapsed under its own weight, the Ptolemaic system was demolished by Galileo when he trained his telescope on the planets, and specifically on Jupiter, where he was able to observe four of Jupiter's moons circling it on a regular basis. As time passed, all the long-range extrapolations associated with the Ptolemaic system were shown to be false, and certain direct observations were shown to be misleading (though not false) because earthbound observers could not sense motion nor could they observe stellar parallax with equipment then available.

So what lessons from this case are applicable to the study of evolution?

- Direct *observations* (empirical facts) must be carefully distinguished from *inferences*, *extrapolations*, and *explanations*.
- Long-range extrapolations *must* be treated with skepticism.
- Even direct observations can be misleading, and immediate inferences can be false, especially when there is a predisposition to favor a particular set of inferences.
- Observations that do not fit the accepted explanatory paradigm should not be ignored or discarded.
- *All* theories can explain *some* observations; the key is to make sure that each theory is forced to confront *all* the observations.

What Observations Must a Theory of Evolution Explain?

With this lesson in mind, it is time to investigate the observations that any theory of evolution must explain. This is a daunting task given the mounds of data that exist regarding our natural world, from the genetic and molecular level up to the ecological level. These observations span a range of disciplines and subdisciplines and fill innumerable journals and books. Picking out those observations most relevant to evolutionary theory is by no means an easy task, and we lay no claim to compiling an exhaustive list. What we focus on are the observations that we think are the most fundamental and therefore most critical for any viable theory of evolution to incorporate successfully.

In addition to presenting the raw data, from time to time we identify inferences that are made from these data by some of the schools. We have given inferences made by multiple schools to avoid any appearance that only one school makes inferences while all the others merely "stick to the facts." This does not imply that all such inferences are equally valid or scientifically justified. In this chapter, however, we present the inferences only as illustrations; we do not discuss their value. The validity of these inferences as well as the ability of any school to adequately deal with the data presented here are topics treated in later chapters.

Traits, Organisms, and Their Environment

The Fit between an Organism and Its Environment

No observation is more central to the biological investigation of evolution than the manner in which organisms seem to be adapted to their environment. Study an organism in its natural habitat for any amount of time, and one cannot help but be amazed at the many intricate details in the structure and behavior of the organism that allow it to flourish there. Even casual observers of nature notice this. It is a fact that has been used to "prove" (infer) the existence of a designer by Creationists as well as the efficacy of natural selection by Darwinists. The modified pectoral fins on skates and rays that allow them to propel themselves along the flat ocean bottom, the unique form of stick insects that allows them to avoid detection, the ultraviolet patterns on flowering plants that attract prospective pollinators, algae that can change the amounts of pigment they produce in response to various light conditions—these are but a few examples. In all of these cases, it is clear that if the characteristic were lacking, the organism would be at a major disadvantage. If a stick insect no longer looked like a stick but instead like a big juicy bug, its chances of becoming dinner would skyrocket. This *adaptive nature* of many traits is thus a key observation that must be explained by any theory of evolution.

Inefficiently Designed Traits

Organisms possess certain traits that, although adaptive in nature, do not appear to be of an efficient design. Many in the Neo-Darwinian school believe such traits are useful for demonstrating naturalistic evolution. They have *inferred* that such inefficient designs demonstrate the random nature of evolution and refute the notion of an Intelligent Designer.[2] The vertebrate eye is the favorite example. The vertebrate eye is arranged such that the cells that sense light—the photoreceptors—are arranged behind two other layers of cells (fig. 3.2). This arrangement allows light to be scattered before it hits the photoreceptors, and it leads to a blind spot in each eye where the nerve bundle leaving the eye has to pass through the photoreceptor layer. Creationists have responded with counterarguments, to the effect that the observed arrangement is necessary to protect the light-sensitive cells from damaging ultraviolet radiation, and to ensure that the retinal pigment epithelial (RPE) cells can receive the required large quantities of blood

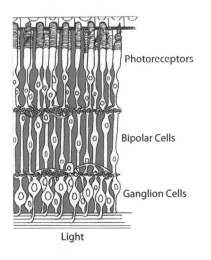

Photoreceptors

Bipolar Cells

Ganglion Cells

Light

Figure 3.2. Cross section of human retina. Light enters from the bottom of the figure, while the photoreceptors, the cells that sense the light, are at the top.

---

2. Any such inference assumes that we are able to discern how an Intelligent Designer would design.

from the underlying choroid layer.[3] Creationists have accordingly inferred that the design of the eye points to an intelligent creator.

Another example, put forth by Stephen J. Gould, is the panda's hand, which would appear to benefit from having a real opposable thumb like that of humans. Instead, as Gould describes it:

> The panda's "thumb" is not, anatomically, a finger at all. It is constructed from a bone called the radial sesamoid, normally a small component of the wrist. In pandas, [it] is greatly enlarged. . . . The panda's true thumb is committed to another role, too specialized for a different function to become an opposable, manipulating digit. So the panda must use parts on the hand and settle for an enlarged wrist bone and a somewhat clumsy, but quite workable solution.[4]

The inferred inefficiency, though, much like the appearance of the earth being stationary, may be illusory. There may be a biomechanical advantage for the panda in having the wrist bone used as an opposable thumb. As Gould hints, there might be competing design constraints, such that adding a true opposable thumb to the hand of the panda would adversely affect another beneficial function of the fingers or the hand. It is possible that this design, as awkward as it looks to Gould, is the optimal design for the panda, given its environment and behaviors. Such would be the inference of the Creationists. No detailed tests have been performed on these traits to answer this question.

Vestigial Organs

Organisms possess a number of structures that are functionless, as far as we know, although they have a definite function in closely related organisms. Such structures include nonfunctional eyes that appear on cave animals, nonfunctional legs that appear on certain snakes, and even the human appendix and coccyx vertebrae. Such structures, termed "vestigial," are inferred to have been useful somewhere in the organisms' evolutionary past and to be in the process of degenerating because they serve no current function. While it seems that the organisms in question retain no advantage by having these vestigial organs, the organs still emerge during development, which is an inefficient use of precious resources. In fact, in some cases these organs emerge during development only to be subsequently dismantled such that they never appear in the adult. Two

3. Carl Wieland, "Seeing Back to Front," *Creation Ex Nihilo* 18 (1996): 38–40; Wieland "An Eye for Creation: An Interview with Eye-Disease Researcher Dr. George Marshall, University of Glasgow, Scotland," *Creation Ex Nihilo* 18 (1996): 19–21.

4. Stephen J. Gould, *The Panda's Thumb: More Reflections in Natural History* (New York: W. W. Norton, 1980), 22–24.

examples are teeth that appear during the development of baleen whales and extra digits that appear during the development of chickens.

While many of these vestigial structures are probably functionless, particularly those that appear only during development, some supposedly vestigial structures do appear to have a function. For example, the coccyx appears to be necessary for the attachment of certain muscles in humans that aid in defecation. On the basis of such apparent exceptions, Creationists infer that *all* such structures have a function, and only our lack of knowledge prevents us from perceiving what that function is. This inference is made despite the fact that most vestigial organs offer their owner no known functional advantage.

A summary of the foregoing observations is presented in table 3.1. A similar table appears at the end of each of the next five subsections detailing the observational data to be explained by evolution theories. The table includes inferences from the observations and the position of each of the four major schools regarding the inferences. The validity of many of the inferences will be discussed in later chapters.

Table 3.1. Inferences based on traits of living organisms

		Position of evolution schools			
Observation	*Inference*	Neo-Darwinism	Meta-Darwinism	Intelligent Design	Creationism
1. Adaptive nature of many traits	Selected for in certain environments	✓	✓	✓	
	Work of an Intelligent Designer			✓	✓
2. Nonoptimal or inefficient design	Result of organism's evolutionary history	✓	✓		
	Illusory; lack of knowledge makes one assume design is nonoptimal			✓	✓
3. Vestigial traits	Result of organism's evolutionary history	✓	✓		
	Illusory; lack of knowledge makes one assume trait is vestigial			✓	✓

The Fossil Record

Progression in the Fossil Record

The fossil record is one of the most contentious points in the evolution controversy, and it has been so since the publication of *The Origin of Species* in 1859. What one school believes to be raw data concerning the fossil record, another school discounts as an interpretation of the data. This is particularly true in regard to the dating of fossils. It is best, then, to start with the least controversial aspect of the fossil record (if in fact there is a least controversial aspect), namely, the progression shown in the fossil record.

Study of the different strata or layers of rock found on the surface of the earth shows that each layer contains a representative group of fossils. Generally, more primitive and simpler forms are found in the deeper layers and more advanced forms appear in the higher layers. The usual inference from this raw data is that each layer and its representative fossils correspond to a period of time in the earth's history, with deeper layers being more ancient. The sequence of rock layers is generally referred to as the "geologic column," and the column is often presented with its associated fossils. The geologic column and the fossil sequence (see fig. 3.3), according to the commonly accepted chronology, are discussed below. For example, reptiles are not seen in rocks dating to the Silurian period (approximately 420 million years ago), but they are found in rocks dating to the Permian period (approximately 260 million years ago) and beyond.[5] Likewise, members of our genus, *Homo*, are not found until the upper layers of the Tertiary period (approximately 2 million years ago). This temporal pattern in the fossil record is helpful in dating new fossil finds, and is taken to imply an evolutionary sequence of life-forms. The reality of the column is accepted even by most Creationist geologists; what they dispute is the set of dates associated with the strata. Their inference from the column is that the life-forms are *not* an evolutionary sequence, but that they and their associated rock layers resulted from a great flood.

Extinct Creatures and Rapid Mass Extinction

Another striking observation from the fossil record is that many fossil species appear in a group of contiguous strata and then disappear abruptly from the record. The immediate inference, for all but the Creationists, is that such fossilized species have become extinct. Biologists estimate that over 99.9% of all species have become extinct over the course of the earth's history. As evolutionary biologist Ernst Mayr notes, "the history of the earth is a history of extinction."[6] Any theory of evolution must explain why species eventually become extinct.

5. Note: Dates cited in this and subsequent sections of the chapter are not direct observations but inferences based on the standard geologic model. Some would call them "facts" because they believe the standard model to be exceedingly well verified. We will not debate this point, but leave it to the reader to judge.

6. Ernst Mayr, "The Emergence of Evolutionary Novelties," in *Evolution after Darwin*, vol. 1: *The Evolution of Life*, ed. S. Tax (Chicago: University of Chicago Press, 1960), 75–76.

Geologic column				Age MYBP*	Life
Eon	Era	Period	Epoch		Life-forms/ geologic events
Phanerozoic	Cenozoic	Quaternary	Holocene		
			Pleistocene	1.8	
		Tertiary — Neocene	Pliocene	5.2	Earliest *Homo*
			Miocene	23.8	
		Tertiary — Paleocene	Oligocene	33.5	First apes
			Eocene	55.5	First whales First horses
			Paleocene	65	{ Asteroid impact? Extinction of dinosaurs
	Mesozoic	Cretaceous	Late	98.9	
			Early	144	
		Jurassic	Late	160	First birds
			Middle	180	
			Early	206	First mammals
		Triassic	Late	228	
			Middle		
			Scythian	251	Permian mass extinction
	Paleozoic	Permian		290	First mammal-like reptiles
		Carboniferous	Pennsylvanian		First reptiles
			Mississippian	354	First amphibians First insects
		Devonian		409	First land plants
		Silurian		439	First fish with jaws
		Ordovician		495	
		Cambrian		543	First shelled organisms "Cambrian explosion" First multi-celled organisms
Protero- zoic	Precambrian			2500	
Archaean					First bacteria
Hadean				3800	Origin of life? Oldest rocks
				4800	Formation of earth

*MYBP: Millions of years before present

Figure 3.3. The geological timescale is depicted under "Geologic column" and the first appearance in the fossil record of specific organisms under "Life."

Furthermore, the fossil record shows that most species that have gone extinct have not gone alone but have disappeared with much company. The fossil record documents many cases in which large numbers of species simply vanish; even entire genera and families disappeared. The most famous of these is the mass extinction event (thought to be caused by an asteroid slamming into the Yucatan Peninsula) that claimed the dinosaurs at the end of the Cretaceous period, 65 million years ago. The dinosaurs were not the only organisms to disappear during this event; based upon estimates from the fossil record, about 75% of the species present at that time went extinct as well. Many other mass extinction events are documented in the fossil record. The one occurring at the end of the Permian period (245 million years ago) was the most devastating, as it is estimated to have wiped out over 95% of the species present at that time.[7]

For Creationists, the rock layers do not correspond to the usual time sequence, so the observed pattern of fossils cannot be due to ancient extinction events. Rather, they argue that the pattern is due to the fact that the life-forms of each period were all contemporary but occupied different ecological zones; thus they were buried in separate layers by the great flood.

Discontinuous Nature of the Fossil Record

One of the most contentious aspects of the fossil record has to do with the size and nature of the gaps in the record. One of the predictions of Darwin's theory was that many intermediate fossil forms would be found, for example fossils bridging the gap between dinosaurs and birds or between fish and tetrapods. Indeed, some such intermediate forms have been found and Darwinists point to these as a vindication of Darwin's theory. Despite this, it is clear that significant gaps remain in the fossil record, a fact that many Darwinists acknowledge[8] and that Creationists use to infer that Darwinism is incorrect. These gaps, or discontinuities, were known by Darwin. Recognizing that the gaps had consequences for his theory, he proposed that they were an artifact of an incomplete fossil record and that the gaps would be closed with further research and exploration.

Today, many gaps *have* indeed been narrowed or closed. For example, a number of fossils that appear intermediate between fish and tetrapods

7. Ernst Mayr, *What Evolution Is* (New York: Basic Books, 2001), 202.
8. Even staunch Darwinist Ernst Mayr agreed on this point regarding the presence of gaps in the fossil record, writing in his last book that "the fossil record is one of discontinuities," *What Evolution Is*, 14.

have been found, including *Acanthostega* and *Ichthyostega*.[9] The gap between land-dwelling artiodactyls such as hippos and whales has also been narrowed with the finds of *Basilosaurus*, an early whale that dates to 38 million years ago, and *Rhodocetus*, an intermediate form that dates to 47 million years ago.[10] However, it is important to note that even with these impressive finds, significant gaps, some more than 10 million years in length, still remain in both of these lineages.

Other examples of recently found intermediate forms are the feathered dinosaurs such as *Microraptor gui* and *Sinornithosaurus millenii*.[11] These fossils have characteristics intermediary between dinosaurs and birds, bolstering the case that birds evolved from dinosaurs. Yet while these fossils do indeed exhibit characteristics between birds and dinosaurs, they are *younger* than the famous *Archaeopteryx*, which is considered to be the first bird. This indicates that transitional forms are still missing in this lineage.

One of the most studied discontinuities, which we examine in more detail in chapter 5, is the "Cambrian explosion." This refers to a short period (10 million years) during which most known metazoan (animal) phyla first appear in the fossil record. The complex metazoan phyla with bilateral symmetry such as chordates, mollusks, nematodes, and arthropods all appear for the first time in the fossil record during the Cambrian period. While some simple bilaterian fossils have been found in pre-Cambrian rocks as well as trace fossils suggestive of bilaterian animals,[12] a discontinuity remains.

While discontinuities exist in some lineages, other proposed transitions fare better with regard to the fossil record. For example, the transition between reptiles and mammals has left a relatively complete sequence in the fossil record, at least in terms of skull features.[13] Creationists, however, dispute that the similarities from one form to the next in these sequences represent any sort of evolutionary connection.

9. Jennifer Clack, *The Emergence of Early Tetrapods* (Bloomington: Indiana University Press, 2002).

10. P. D. Gingerich, M. ul Haq, I. S. Zalmout, I. H. Khan, and M. S. Malkini, "Origin of Whales from Early Artiodactyls: Hands and Feet of Eocene Protocetidae from Pakistan," *Science* 293 (2001): 2239–42.

11. Xing Xu et al., "Four-winged Dinosaurs from China," *Nature* 421 (2003): 335–40; Xing Xu, Zhi-Lu Tang, and Xiao-Lin Wang, "A Therizinosauroid Dinosaur with Integumentary Structures from China," *Nature* 399 (1999): 350–54.

12. Keith Miller, "The Precambrian to Cambrian Fossil Record and Transitional Forms," on the American Scientific Affiliation web site, www.asa3.org/ASA/topics/Evolution/PSCF12–97Miller.html#36; Eugenie Scott, *Evolution vs. Creationism: An Introduction* (Berkeley: University of California Press, 2005), 172–74.

13. Mayr, *What Evolution Is*, 222.

In summary, when looking at the data as a whole, it is clear that numerous intermediate forms exist and can be found in nearly every proposed lineage. At the same time, gaps remain and neat sequential progressions delineating the orderly gradual transition of one species into another species are relatively rare. More details regarding the nature of the fossil record are found in chapter 5.

Stasis in Most Fossil Forms

Another curious aspect of the fossil record data is the fact that most fossil forms remain stable for relatively long periods of time; that is, they do not tend to undergo any directional change. One of the most popular college textbooks on evolution puts it this way: "Perhaps the most prominent pattern in the history of life is that new morphospecies appear in the fossil record suddenly and then persist for millions of years without apparent change."[14] Stasis is the rule rather than the exception. In fact, the stability of fossil forms is a valuable tool for biostratigraphers who date rocks for the purpose of locating oil and minerals. Because fossil forms are stable, biostratigraphers date rocks by identifying a particular index fossil that is always found in a certain rock layer. They do not date rocks by looking for stages of evolution in specific species because these stages are more difficult to identify and are often absent.

Many researchers have inferred that stasis in the fossil record is an illusion based upon the limited amount of data that can be had from fossil remains, and that soft tissues that do not fossilize are changing over the time during which a species is considered to be in stasis. This is certainly possible, but the raw observational data are that most of the traits appearing in the fossil record are stable over vast expanses of time (millions of years).

Radiometric Dating

Radiometric dating is a tool that is based on well-known physical principles and is designed to assign an absolute age to a particular rock or rock layer. When it works well, it gives an absolute age for a particular rock, such as 135 million years, rather than a relative age, such as younger than a rock containing an *Archaeopteryx* fossil.

All radiometric dating techniques are based on observed properties of certain chemical elements, such as carbon (C), potassium (K), and argon (Ar). These and most elements appear in two or more forms, known as isotopes. Isotopes of an element have the same number of protons and

14. Scott Freeman and Jon Herron, *Evolutionary Analysis*, 3rd ed. (Upper Saddle River, NJ: Pearson-Prentice Hall, 2004), 522.

electrons, but differ in the number of neutrons. As a result, they differ in certain properties, such as their stability, that is, whether they decay through radioactivity to other elements. To be useful for dating purposes, an element must have at least one unstable isotope. In the case of carbon, the unstable isotope of interest is C-14, which decays over time to a stable isotope of nitrogen, N-14. It takes 5,730 years for half of the C-14 initially present in a sample to decay. This time is known as the half-life $(t_{1/2})$ of C-14. Because of its relatively short half-life, as compared to argon and potassium, carbon dating is typically used for dates in the range of a few hundred years to a maximum of about 50,000 years.[15] Other elements are employed for longer time spans, such as dating rocks of geological interest. The most commonly used isotope pairs, or parent-daughter pairs, for that kind of dating are potassium-argon (K-40; Ar-40, half-life 1.3 billion years), rubidium-strontium (Rb-87; Sr-87, half-life 47 billion years), thorium-lead (Th-232; Pb-208, half-life 14.1 billion years), uranium-lead (U-235; Pb-207, half-life 713 million years), and uranium-lead (U-238; Pb-206, half-life 4.5 billion years).

To perform a radiometric calculation, one first must know (or rather infer, since it generally cannot be known with absolute certainty) the amounts of the parent and daughter isotopes that were present in the rock at the time it was formed. For some isotope pairs such as K-40; Ar-40, this is straightforward: it is assumed that since Ar-40 is an inert noble gas, none was initially present in the rock. Other methods are used to determine this for the remaining isotope pairs, although all of these methods have limitations.[16] The next piece of information needed is the amount of both the parent and the daughter currently found in the rock. These values can be directly measured from the rock and are not controversial. They, along with the half-life, are then used to calculate the age of the rock according to the following formula:

$$age = 1.443\, t_{1/2} \ln \frac{original\ amount}{present\ amount}$$

A sample calculation using K-40; Ar-40 shows how radiometric dating is done. Suppose that you take a rock and, by direct measurement, determine it contains 0.075% K-40 and 0.025% Ar-40. Now you know the present amount of K-40 but still need to infer the original amount. Assuming as stated above that all the Ar-40 in the rock arose from the

15. Roughly speaking, the maximum span of time for which any given isotope pair is useful ranges from about 1/20 of the half-life to 20 times the half-life, though it may be less.
16. Roger Weins, "Radiometric Dating: A Christian Perspective," on the American Scientific Affiliation web site, www.asa3.org/ASA/resources/Wiens.html.

decay of K-40 originally present in the rock, the total original amount of K-40 in the rock must have been 0.1% (0.075% + 0.025%). Substituting the values 0.1% for the original amount, 0.075% for the present amount, and 1.3 billion years for the half-life into the formula yields:

$$age = 1.443 \times 1.3 \text{ billion} \times \ln \frac{0.1}{0.075} = 540 \text{ million years}$$

Given that the physics involved is well verified, the basic process of radiometric dating is relatively straightforward. However, the methods of inferring original isotope concentrations in the rock can be quite involved, depending upon the isotope pair used. As a result, problems often arise in determining the *original* amount of the element used in the dating. Assumptions about the rock must be made before the original amount of the element can be calculated. For example, in order for the calculation to be reliable, there must have been no contamination of the rock during its history, such as the parent or daughter being added to the rock or escaping from it after its formation. The absence of any such contamination is essential to the reliability of the method, for it guarantees that the only factor changing the amounts of K and Ar is the radioactive decay of K into Ar. Unfortunately, there are many ways in which contamination can occur. It is quite possible for water flow through a rock to deposit minerals or remove them. Even exposure to air can cause the loss of the gas Ar. Any such contamination would significantly alter the amounts of the parent-daughter pair and compromise the method. Since it is impossible to know for certain that any rock is contamination-free, the calculations will have some margin of error.

Given these difficulties, geologists often try to determine the age of rocks using different parent-daughter pairs, hoping that the different measurements converge upon a concordant date. In addition, an improved version of radiometric dating, known as the isochron method, is generally used today. The method is more protected from contamination because multiple samples are used. And because it uses *ratios* and not absolute amounts, it does *not* require that we know the quantity of the elements *originally present*.[17]

Another confounding factor is that of the three types of rocks—igneous, sedimentary, and metamorphic—only igneous rocks work well because these are the only rocks in which the minerals being analyzed were formed at the time the rock originated. (A common example of an igneous rock layer is that formed by lava flow.) In sedimentary rocks, the minerals form elsewhere and then are brought to the rock loca-

17. For further discussion see Chris Stassen, *Isochron Dating*, at www.talkorigins .org/faqs/isochron-dating.html#isochron.

tion at a later date, so mineral composition at the time of the rock's formation is difficult to determine. Unfortunately, almost all fossils are found in such rocks, so the dating of fossils must be correlated with an overlying or underlying layer of igneous rock. The third type of rock, metamorphic, is also difficult to date because the minerals form at different times during the history of the rock, thereby giving such rocks widely disparate ages.

Radiometric dating methods usually give rather great ages for rocks and fossils, and this causes grave problems for Creationists. They have, therefore, sought to impugn the methods by pointing out that they frequently give dates that are known by other means to be incorrect, often by large factors.[18] The methods, when applied to formations such as lavas of a known young age, can yield dates of millions of years.[19] Another problem is the progression of dates that have been given for many formations, as dating methods and techniques change.[20]

Many of these arguments posited to undermine the accuracy of radiometric dating techniques rely heavily on anecdotal, rather than systematic, evidence. There is nothing wrong with anecdotal evidence, and to be sure, it illustrates clearly the problems associated with radiometric dating. But a single valid date of a million years, or even just a hundred thousand years, is enough to demolish the Creationist timescale. And despite methodological shortcomings, it is unlikely that the method *never* yields accurate dates. Furthermore, the method is firmly based in physics and does give dates consistent enough that it cannot be rejected out of hand, as some Creationists concede.[21] For this reason, Creationists now challenge another assumption of all radiometric dating methods, namely, that radiometric decay has been constant over time. This strikes at the soundness of the physics underlying the technique rather than the methodology. Many Creationists now argue that radioactive isotopes decayed much faster in the past, and so, they argue (infer), the calculations described above, which assume that the present rate of radioactive decay was the same in the past, give misleading older dates. These arguments are discussed in chapter 6.

A summary of the foregoing observations along with the inferences made by the four schools can be found in table 3.2.

18. An easy-to-read overview of the problem as well as the Neo-Darwinian response is found in Scott, *Evolution vs. Creationism.*

19. John Woodmorappe, *The Mythology of Modern Dating Methods* (El Cajon, CA: Institute for Creation Research, 1999), 18.

20. Ibid., 7.

21. Andrew Snelling, "Geochemical Processes in the Mantle and Crust," in *Radioisotopes and the Age of the Earth*, ed. Larry Vardiman, Andrew Snelling, and Eugene Chaffin (El Cajon, CA: Institute for Creation Research, 2000), 273.

Table 3.2. Inferences based on radiometric dating and characteristics of the fossil record

Observation	Inference	Position of evolution schools			
		Neo-Darwinism	Meta-Darwinism	Intelligent Design	Creationism
1. Progression in the fossil record	Documentation of evolutionary history	✓	✓	✓	
	Documentation of stages of creation			✓	✓
2. Extinct creatures and massive extinctions	Results of evolutionary failures	✓	✓	✓	
	Result of global flood				✓
3. Discontinuous nature of the fossil record	Transitional forms that are missing will be uncovered with further exploration or are missing because they were few and unstable	✓	✓	✓	
	Missing transitional forms never existed and ones that do exist still leave gaps in the fossil record			✓	✓
4. Stasis	Result of stabilizing or normalizing selection	✓	✓	✓	✓
	Demonstrates the limits of organismal change			✓	✓
5. Radiometric dating	Earth is old; radiometric dates calibrate ages of strata	✓	✓	✓	
	Radiometric methods are defective or inapplicable; geologic column is attributable to global flood				✓

Variation and Mutability in Populations

High Amount of Genetic Variation in Species

In order for natural selection to act, there must be sufficient amounts of variability within populations. Although we can see that members of most species come in all shapes and sizes, we must determine if there is variability at the genetic level. Most research in this area has found that genetic variation is plentiful and much of it has an effect on the physiology of the individual.

Invertebrates are heterozygous[22] at approximately 6% of their genetic loci while vertebrates are heterozygous at approximately 15%.[23] In addi-

22. Heterozygous means having two different alleles, or versions of a gene, at the same genetic loci. For example, at a genetic loci responsible for eye color, one may have an allele for blue eyes as well as an allele for brown eyes.

23. J. F. McDonald, "The Molecular Basis of Adaptation: A Critical Review of Relevant Ideas and Observations," *Annual Review of Ecology and Systematics* 14 (1983): 77–102.

tion, it seems possible that most of these alleles have differential effects on survival under certain conditions. In fact, in nearly all cases in which heterozygote alleles have been studied, the proteins encoded by the different alleles (the different versions of the gene) have been found to have biochemical or regulatory differences. This strongly suggests that they do indeed have different effects on the fitness of an individual.

Such heterozygosity is seen even in humans. The international human HapMap Project is an attempt to map this genetic variability within the human population in order to better understand (among other things) how a person's genotype affects susceptibility to disease. This project is still in its infancy but it has already demonstrated that there is an immense amount of variability in the human genome. The extent to which this impressive variability affects the phenotype is currently being studied.

Most schools argue that this variability arises from the accumulation of new mutations within a species. Some of the mutations are deleterious and are removed by selection (conservative selection), while others that are neutral or advantageous may be retained. For the Creationists, the genetic variability was imparted to the species (or "kind") when it was created, and it can only deteriorate over time.

Differential Survival of Organism Variants

With so much variation found in living organisms, we would expect that some variants would fare better than others. One of the most cited examples of such differential survival of organisms in the wild deals with the alteration of beak length in finches on the Galapagos Islands (see fig. 3.7). Researchers have documented that the average beak length in a finch population changes following drought or periods of ample rain. During drought conditions, the finches with long beaks are better able to break through tough seeds that would otherwise be ignored during times of plenty. The long beak variant thus has a better chance for survival and tends to increase in the population over time.

In many cases, the gene involved in causing differential survival is known. The hemoglobin beta gene is an excellent example. This gene is involved in binding and transporting oxygen in red blood cells, and there are two major alleles, the wild type and a particular mutant. A person with two mutant alleles develops sickle-cell anemia; those who are heterozygous or who have two wild-type alleles do not. The story doesn't end there, however, because the mutant allele also can have a beneficial effect on fitness. In fact, heterozygotes are protected from malaria, while persons with two wild-type alleles are not. This is referred to as a "balanced polymorphism," a situation in which there is an advantage in having one copy of each allele. In this case, the mutant allele protects against malaria and the wild-type allele allows red blood cells to function normally.

Probably the most discussed example of the differential survival of organisms comes from bacteria and has to do with the acquisition of antibiotic resistance. Several genetic variations allow bacteria to become resistant to specific antibiotics. In all cases, these resistant variants fare better than normal bacteria in the presence of antibiotics. A few examples of such variants are listed below:

1. Restoration of the ability to make mRNA. Certain antibodies work by binding to an RNA polymerase and interfering with the activation of bacteria genes. Commonly occurring spontaneous mutations can result in a change to subunits of RNA polymerase, so that it can operate effectively even in the presence of antibiotics and still produce mRNA.

2. Efflux pumps. Some bacteria, including the common gut bacterium *Escherichia coli*, actually possess an efflux pump that expels multiple antibiotics from their cytoplasm. A spontaneous mutation in a regulatory protein results in the synthesis of large quantities of the pump proteins, which can then expel high concentrations of antibiotic agents from the bacterium.

3. Genetic deletions or shutting off genes. Resistance to erythromycin and chloramphenicol can arise following a small deletion in the bacteria's genetic material. In a similar vein, suppressing the expression of a certain gene provides resistance to the antibiotic actinonin in one of the most dangerous bacteria, *Staphylococcus aureus*.

4. Reduction of transport proteins or channels. Resistance to the antibiotic kanamycin results from the reduction in synthesis of a particular transport protein called a membrane porin. Because kanamycin enters the bacteria via this protein, reducing its level naturally inhibits it from entering bacterial cells.

5. Adaptation of cellular genes to degrade specific antibiotics. For the β-lactam antibiotics (penicillin, cephalosporins), certain resistant bacteria can degrade these antibiotics and render them ineffective by expressing an enzyme called a β-lactamase. The origin of the β-lactamase enzyme appears to be the result of an alteration of a cell wall biosynthesis enzyme, which in the presence of β-lactam antibiotics has evolved to take on a new function.[24]

The fact that these resistant variants exist is seen by Neo-Darwinians and others as an example of evolution in action. Selection has favored

24. A. A. Medeiros, "Evolution and Dissemination of Beta-Lactamases Accelerated by Generations of Beta-Lactam Antibiotics," *Clinical Infectious Diseases* 24 (1997): S19–S45.

these variants because they have an increased fitness in an environment swamped with antibiotics. Specifically, they see the acquisition of antibiotic resistance as an example of how random genetic mutations can lead to information gain. In this case, the information gained is that which is necessary to become resistant to a specific antibiotic. Creationists and some Intelligent Design proponents, however, argue that the origin of genetic variants resistant to antibiotics coincides with a *decrease* in information content. This dispute is discussed further in chapter 5.

Limits of Breeding Experiments: Genetic Homeostasis

For centuries, breeders have selectively bred livestock and plants to enrich them for various desirable traits such as size, milk production, or hardiness. Although the results have been impressive, there appears to be a limit to the amount of change that a population can undergo. While breeders can select for and breed cows that produce more milk or beef, they cannot breed cows the size of elephants. After a few generations the increase reaches a limit. Fruit flies normally live for four to six weeks, but selection for longevity can produce populations that survive almost twice as long. Although this is impressive, no amount of selection has resulted in fruit flies that survive for a year. The story is the same with plants: tomato plants that give large tomatoes can be selected for with much initial success, but once again a limit is soon reached. Tomato plants don't produce tomatoes the size of watermelons.

It is possible that new mutations might allow breeders to transcend such bounds in the future. However, present evidence points to the contrary, namely, that new mutations tend to disrupt what is called genetic homeostasis within the population. Genetic homeostasis implies that the genes within a population are balanced such that "all gene-controlled processes during [development] are so precisely adjusted to each other that even modest phenotypic deviations have disturbing effects." As a result, "each more pronounced deviation from the mean value [would be] correlated through a feedback mechanism with a reduced reproductive success,"[25] and therefore change may be resisted and homeostasis maintained.

Supporting this notion is the fact that animals that are bred for high milk production are usually less robust than wild-type varieties because production of excess amounts of milk disturbs other physiological processes. For example, high milk production leads to animals that tend to be more susceptible to disease and injury. As a result, breeding experiments usually produce only *quantitative* changes in structures and functions,

25. Ernst Mayr, "Selection and Directional Evolution," in *Evolution and the Diversity of Life* (Cambridge, MA: Belknap, 1976), 51.

such as milk production, that are already present in the population, and organisms that are, in general, less robust than their wild-type cousins. For Creationists, the inference from these observations is that genetic variability has both a specific function (protection of the species) and hard limits that make evolution impossible. For Neo-Darwinians, the inference is that evolution can proceed only very slowly, because the desirable mutations—those capable of major beneficial change—are quite rare.

Mutations Can Have Small or Large Effects on Organisms

It is clear that mutations can increase the amount of genetic and phenotypic variation within species. Of critical interest to the study of evolution, though, is the extent to which mutations can alter organisms, particularly for the better. Do mutations usually have minor effects on an organism or can they produce more drastic changes? In biology labs throughout the globe, researchers have developed mutagenesis protocols designed to genetically alter the organisms they are studying. This approach, coupled with the ability to examine variability between related species, has shown that mutations can have both small and large effects on the phenotype of the organisms.

Such experiments have identified many strange mutants in which a minor genetic change has resulted in a large-scale alteration of the organism. For example, mutations in Hox genes, which are involved in developmental regulation, lead to drastic phenotypes such as legs attached to the head of a fruit fly (see fig. 2.7). This change is the result of an alteration in just one gene. Similarly, a single amino acid change can lead to such large-scale effects as resistance to insecticides in insects.[26] These mutations represent discontinuous variations; that is, the insect either has insecticide resistance or it doesn't. The trait does not evolve gradually but emerges at once when a single mutation occurs.

While large-scale effects do occur, many mutations have only small effects on the phenotype, such as mutations that affect susceptibility to diseases such as epilepsy and heart disease. In these cases, susceptibility to disease is affected by a multitude of genes, and so mutations in one gene have only a minor effect on the overall phenotype. The same holds true for such traits as height, intelligence, and longevity. Traits that are affected by many genes are called quantitative traits and vary in a relatively continuous manner.

A summary of the foregoing observations along with the inferences made by the four schools can be found in table 3.3.

26. H. A. Orr, "The Evolutionary Genetics of Adaptation: A Simulation Study," *Genetical Research* 74 (1999): 207–14.

Table 3.3. Inferences based on variation and diversity in populations

Observation	Inference	Position of Evolution Schools			
		Neo-Darwinism	Meta-Darwinism	Intelligent Design	Creationism
1. High amount of genetic variation within species, including new mutations	Raw material used during evolution of new species	✓	✓	✓	
	Allows species to adapt (within limits) to environmental change			✓	✓
2. Differential survival of variants	Selects for fitter variants and eventually can lead to the emergence of novel structures and new information	✓	✓		
	Little data available showing that this leads to the emergence of novel structures		✓	✓	✓
	Natural selection in its conservative role, demonstrating care of Creator for creatures or the loss of genetic information			✓	✓
3. Genetic homeostasis	Constraint that must be overcome by addition of new mutations	✓	✓		
	Demonstrates limits of organismal change			✓	✓
4. Large and small effects of mutations	Demonstrates the effectiveness of mutations to fuel change either continuously or discontinuously	✓	✓	✓	
	Demonstrates the variability of mutations but does not demonstrate that they are capable of fueling any type of change			✓	✓

Physiological and Developmental Similarities

Physiological Similarities at the Cellular Level

Cells can be divided into two groups: those with and those without nuclei. Nucleated cells (eukaryotic cells) all have compartmentalized cellular structures that include but are not limited to nuclei, mitochondria, Golgi apparatuses, endoplasmic reticulums, and chloroplasts. While non-nucleated cells (prokaryotes) do not share these structures, they do share a common mechanism for using genetic information to make proteins, which consists of transcribing the DNA into RNA and then translating the RNA into a protein product. The plasma membranes are similar, as is the mechanism for inserting proteins into the membrane. Many metabolic pathways are shared, and a similar mechanism for producing

97

chemical energy in the form of adenosine triphosphate (ATP)[27] is found in both cell types. Although the eukaryotic cell is much more complex, the foundation of the eukaryotic cell can be found in prokaryotes. And so a common inference (by non-Creationists) is that the eukaryotic cell arose from a prokaryotic ancestor (common descent). Creationists attribute most similarities to a common design plan.

Physiological and Anatomical Similarities at the Organ-System Level

Similarities in structure and function are not only found at the cellular level but also abound at the level of organ systems. Bone physiology in vertebrates is a good example: the bones of all vertebrates are formed from the same basic substances (collagen, calcium, and phosphate) and are used for the same purpose (to protect and support the body and to help regulate the levels of calcium and phosphate ions).

The nervous systems in vertebrates also share many anatomical and physiological similarities. In fact, the structure and function of the main areas of the brain are similar, or conserved, in different vertebrates.[28] This is shown in figure 3.4, which displays the brains of a rat, a monkey, and a human. In all three cases, the brainstem region is closest to the spinal cord, the cerebellum protrudes from the back of the brainstem, and the whole structure is capped off by the cerebrum, better known as the cerebral cortex. As animals increase in behavioral and cognitive complexity, the cerebrum occupies a larger portion of the brain. This pattern is conserved in all vertebrates.

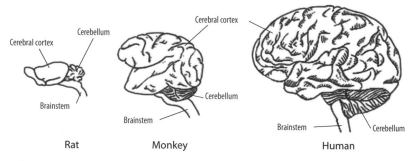

Figure 3.4. The organization of the brain in a rat, monkey, and human. The basic plan of the cerebellum on the dorsal aspect of the brainstem and the cerebral cortex on top is shared by all vertebrates. Figures are not drawn to scale.

27. ATP is the major energy carrier in the cell. It provides the energy needed for the cell to perform such energy-consuming activities as protein synthesis and cell division.

28. Conserved structures are shared structures that are found in all members of a specific group such as vertebrates. A vertebral column is an example of a conserved structure found in all vertebrates.

Similarities in Development of Related Organisms

Examining the development of different animals shows that even distantly related organisms pass through very similar stages. Starfish and humans both begin by going through a blastula stage (a hollow ball of cells) followed by the early gastrula stage (a hollow ball of cells with an invagination). Although it is easy to tell the two species apart, particularly at the gastrula stage, their early stages of development are similar. These similarities are even more pronounced for development within the vertebrates. As noted by the famous nineteenth-century embryologist Karl Ernst von Baer (1792–1876), the early developmental stages of all vertebrates are so similar that in the embryonic stage it can be difficult to tell them apart. These developmental similarities are used to infer common descent.

From this early starting point, though, the development of fishes, birds, and mammals all proceed along different paths. For example, all vertebrate embryos have a series of ridges called visceral clefts. These clefts differentiate into the gills of fish but in mammals they become structures such as the auditory canal that connects the ear and the oral cavity.[29]

Differences in Development of Related Organisms

Despite the fact that developing embryos often transiently pass through similar stages, the route they use to get there is highly divergent even among related organisms. Even the earliest stages of development—the cleavage patterns that divide up the fertilized egg—are not always alike:

> Vertebrate embryos that resemble each other at the pharyngula stage differ radically from each other during the cleavage and gastrulation stages that precede it. . . . Early embryogenesis in some frogs resembles that of reptiles and birds more than that of amphibians.[30]

Some frog species develop through a tadpole stage; many others develop directly into an adult without a tadpole stage. A similar phenomenon is seen with some related species of sea urchins that have drastic differences in gastrulation and cleavage patterns.[31] Creationists use this great diversity in development, even among related organisms, to argue against common descent.

29. Scott Gilbert, *Developmental Biology*, 4th ed. (Sunderland, MA: Sinauer Associates, 1994), 244.

30. Jonathan Wells, "Unseating Naturalism: Recent Insight from Developmental Biology," in *Mere Creation: Science, Faith and Intelligent Design*, ed. William Dembski (Downers Grove, IL: InterVarsity, 1998), 61.

31. Gilbert, *Developmental Biology*, 842.

Development of Homologous Structures

Two structures are considered homologous if they are derived from the same or equivalent structures during development in two different organisms. (In an evolutionary context, the two structures are believed to have evolved from a common ancestral structure via descent with modifications.) Two homologous structures therefore do not have to perform the same function; they only have to originate in the same developmental context. Examples of homologous structures abound in nature. As mentioned previously, during embryonic development both fish and humans exhibit visceral clefts. In fish these develop into the gills while in humans they go on to form such structures as the auditory tube and parathyroid gland.[32] Likewise, the limb buds in humans, moles, and bats all share a developmental pathway, but the results are entirely different structures (see chap. 2, fig. 2.3). Both examples demonstrate that similar developmental starting points can lead to vastly different endpoints. For Darwinists, this implies that natural selection has the ability to build a wide variety of complex structures from a simple structure that existed in a common ancestor.

A summary of the foregoing observations along with the inferences made by the four schools can be found in table 3.4.

Genetics and Genetic Code

Common Genetic Encoding Mechanism

In DNA molecules, all living organisms store the information needed to synthesize their requisite proteins, to create their physiological structures, and to reproduce themselves. DNA molecules are composed of a backbone supporting four bases: adenine (A), cytosine (C), guanine (G), and thymine (T), the order of which actually encodes the information. (In RNA molecules the thymine is replaced with uracil [U].) From bacteria to primates, the manner in which genetic information is used to make proteins is very similar. Each gene, regardless of the organism in which it is found, encodes the information to make a protein in virtually the same fashion. Every three bases (G, A, T, or C) within a gene code for a specific amino acid that will be incorporated into the protein. To use this information, the cell initially reads the sequence of the gene and makes an RNA molecule that is identical to the gene sequence except that U has replaced T in the RNA. The cell possesses the machinery to read this RNA molecule three bases at a time and build a protein. For every three

32. H. E. Lehman, *Chordate Development*, 3rd ed. (Winston-Salem, NC: Hunter Textbooks, 1987).

Table 3.4. Inferences based on physiological and developmental characteristics of organisms

Observation	Inference	Neo-Darwinism	Meta-Darwinism	Intelligent Design	Creationism
			Position of evolution schools		
1. Physiological similarities at the cellular level	Common descent	✓	✓	✓	
	Work of Intelligent Designer			✓	✓
2. Physiological similarities at the systems level	Common descent	✓	✓	✓	
	Results of evolutionary constraints		✓	✓	
	Work of Intelligent Designer			✓	✓
3. Similarities in development of unrelated organisms	Common descent	✓	✓	✓	
	Certain general developmental pathways are necessary to produce living organisms		✓	✓	✓
4. Differences in development of related organisms	Results of divergence and specialization of forms otherwise united by common descent	✓	✓	✓	
	Designer used unique pathways for different organisms			✓	✓
5. Presence of homologous structures	Descent from a common ancestor with subsequent modification	✓	✓	✓	
	"Shared developmental pathways" show considerable variation; if similar, it is merely a good design principle			✓	✓

bases it reads, it adds a specific amino acid to the newly synthesized protein. For example, if it reads GCU in the RNA sequence, it adds the amino acid alanine, while if it reads UAC it adds the amino acid tyrosine (fig. 3.5). While there is no necessity that they do so, all organisms use this three-base system, and although some organisms use certain codons preferentially (GCA rather than GCU for alanine), the same codon corresponds to the same amino acid in all organisms from bacteria to maple trees to humans.[33] This surprising unity at the molecular level is what makes it possible to put human genes in bacteria and fly genes in mice and have them function correctly. Such versatility has been the cornerstone of the biotech industry.

33. The main exceptions to this are found in the coding of genes in the mitochondrial genome. However, while mitochondria employ a few codons that differ from the nuclear genome, the majority of the codons are identical. Similar exceptions occur in single-celled protozoans known as ciliates.

		Second position					
		U	C	A	G		
First position	U	UUU } Phe UUC } UUA } Leu UUG }	UCU } UCC } UCA } Ser UCG }	UAU } Tyr UAC } UAA Stop UAG Stop	UGU } Cys UGC } UGA Stop UGG Trp	U C A G	Third position
	C	CUU } CUC } CUA } Leu CUG }	CCU } CCC } CCA } Pro CCG }	CAU } His CAC } CAA } Gln CAG }	CGU } CGC } CGA } Arg CGG }	U C A G	
	A	AUU } AUC } Ile AUA } AUG Met/Start	ACU } ACC } ACA } Thr ACG }	AAU } Asn AAC } AAA } Lys AAG }	AGU } Ser AGC } AGA } Arg AGG }	U C A G	
	G	GUU } GUC } GUA } Val GUG }	GCU } GCC } GCA } Ala GCG }	GAU } Asp GAC } GAA } Glu GAG }	GGU } GGC } GGA } Gly GGG }	U C A G	

Figure 3.5. The universal genetic code depicting the sixty-four possible codon sequences as well as the amino acids for which they encode. The first base is on the left, the second on the top, and the third is listed at the right. This table is the same for nearly all organisms.

Similarity in DNA Sequences

Given that the DNA encodes the proteins used to build and sustain the life processes of an organism, one would expect that similar organisms would share similar DNA sequences, given that they would need similar proteins. The field of comparative genomics, which has exploded in the past decade, has greatly facilitated our ability to address this issue. Rather than merely comparing a single gene or a handful of genes, comparative genomics compares entire chromosomes or entire genomes. What this field has discovered, for the most part, is that organisms within the same genus or family have very similar genomes as compared to organisms in different orders, phyla, or kingdoms. For example, recent DNA sequencing efforts have revealed that roughly 95% of the human genome can be aligned directly with the chimpanzee genome, and human and chimp genes are more than 98% identical.[34] In addition, the chromosomal

34. Chimpanzee Sequencing and Analysis Consortium, "Initial Sequence of the Chimpanzee Genome and Comparison with the Human Genome," *Nature* 437 (2005): 69–87;

organization between humans and chimps is highly similar, such that, for example, chimp chromosome 22 is very similar to human chromosome 21. A multitude of such examples can be cited for a variety of related or similar organisms.

Genetic similarities are also found across diverse groups of organisms. Many genes, which encode proteins that perform similar essential functions within cells, have a high degree of sequence similarity across diverse organisms. One example is the cytochrome proteins that are needed to perform cell respiration and/or photosynthesis, functions necessary for the organism's survival. When cytochrome sequences are compared between flies, yeast, worms, and humans, regions near the active site are found to be identical or nearly so.

These genetic similarities are so striking that in some cases genes from one organism can be used to rescue defects in other organisms. For example, genes involved in cell division in humans can rescue yeast that lack these genes.

While DNA sequence comparisons have yielded a wealth of information about the related nature of most genes, unfortunately sequence comparisons between taxonomically distant organisms have not always produced a consistent pattern. In fact, one often gets different results depending upon the gene that is chosen. For example, comparing ribosomal RNA sequences between various single-celled organisms gives a very different picture than comparing tRNA synthetase genes.[35] This variability makes it difficult to assess accurately the relationships and similarities between organisms based solely upon molecular data. However, observed similarities are great enough that most schools infer common descent from them. Creationists accept common descent only to a limited extent, and therefore they use this data to infer a common design plan; that is, the Creator used similar genes in similar species.

Gene and Chromosome Duplications

The comparative analysis of whole genomes has produced a wealth of information relevant to the study of evolution. For example, it has revealed the presence of gene families, which tend to expand as organisms become more complex. In other words, higher organisms tend to have multiple copies of genes that are found only in a single copy in lower organisms. For example, the primitive chordate *Amphioxus* has one copy of an insulin-like gene, while most mammalian genomes have

H. Watanabe et al., "DNA Sequence and Comparative Analysis of Chimpanzee Chromosome 22," *Nature* 249 (2004): 382–88.

35. E. Pennisi, "Is It Time to Uproot the Tree of Life?" *Science* 284 (1999): 1305–7.

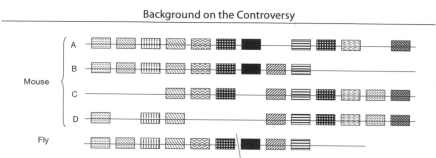

Figure 3.6. The four Hox clusters of the mouse are shown at the top, the single Hox cluster of the fruit fly *Drosophila* below. Homologous genes are aligned vertically. The fly Hox cluster is split between two regions of the genome, as indicated by the slash. The nine Hox genes in the fly are homologous to the first nine genes in the four mouse clusters. While there are a total of thirteen different Hox genes in the mouse, none of the four clusters contain all thirteen genes.

three copies.[36] The multiple copies in the mammalian genome are not identical, but rather are specialized for different tasks and are thought to have arisen via the duplication of the ancestral gene.

One of the most widely cited examples of gene duplication involves the myoglobin-hemoglobin gene family.[37] Primitive protozoans and flatworms have a single globin gene, while the more advanced jawless fish have two globin genes, a primitive hemoglobin and a primitive myoglobin. In mammals, the number of hemoglobin genes expands to at least seven. These multiple genes are needed at various stages of mammalian life, some being expressed in the adult while others are expressed in the embryo or the fetus. This pattern of expanding gene families as one proceeds from lower to higher organisms is repeated throughout the genome.

In addition to apparent duplications at the gene level, many organisms appear to have duplications involving large sections of chromosomes. In these cases, the duplicated regions have virtually the same genes in the same order as the original chromosomal segment. One of the most studied examples is the Hox cluster duplication. Hox genes are involved in regulating development. They act like switches to turn on or off other genes that are needed to form specific structures. Invertebrates have only a single Hox cluster, which consists of up to thirteen separate Hox genes. Mammals, like all higher vertebrates, have four Hox clusters, and although each cluster has a variable number of Hox genes, the Hox genes always are found in the same order, Hox 1 followed by Hox 2, and so on (see fig. 3.6). Such ordered duplication points to common descent, which is the immediate inference of most

36. S. J. Chan, Q. P. Cao, and D. F. Steiner, "Evolution of the Insulin Superfamily: Cloning of a Hybrid Insulin/Insulin-like Growth Factor cDNA from Amphioxus," *Proceedings of the National Academy of Sciences USA* 87 (1990): 9319–23.

37. Hemoglobin is a protein used to bind and transport oxygen within the body. Myoglobin is a related protein that is found in muscle and is used to store limited amounts of oxygen for the muscle to use when necessary.

schools. Creationists must once again fall back on a common design plan.

Synteny Blocks

Probably even more suggestive of common descent than the fact that different organisms have nearly identical Hox clusters is that the genes surrounding the Hox clusters are similar in different organisms. For example, in mammals, three of the four Hox clusters have a collagen and a calcium channel gene on one side and an integrin and a sodium channel gene on the other side of the cluster.[38] As there does not appear to be any functional reason that these genes should flank the Hox cluster (there is no obvious benefit to the organism—these genes could work just as well if found on other regions of the chromosome), the gene order is highly suggestive of a chromosomal duplication event in some distant ancestor that has been inherited by all mammals.

This situation, in which unrelated genes are found in the same order on chromosomes of different organisms, is quite common. These similar regions, called synteny blocks, can be found when comparing the genomes of mice and humans or even when comparing certain invertebrates and humans. A good example is human chromosome 9, which shares synteny blocks with, among other organisms, mice and zebrafish.[39] If one looks at human chromosome 9, one finds that it is largely a composite of regions found on mouse chromosomes 19, 4, 13, and 2. The short end of human chromosome 9 is similar to a region of mouse chromosome 19, the middle is similar to regions on mouse chromosomes 4 and 13, and the long end is similar to mouse chromosome 2. In comparative genomic studies, synteny blocks turn out to be the rule rather than the exception, reinforcing the inference of common descent by most schools. Creationists again invoke a common design plan to explain synteny, although there is no known functional reason why synteny blocks would be designed as they are.

Similar Genes: Different Functions

While it is true that organisms share many similar genes, two organisms may use the *same* gene for entirely *different* functions. This is generally not true of genes coding for enzymes involved in catalyzing reactions such as breaking down sugars, but it is true of most genes

38. N. W. Plummer and M. H. Meisler, "Evolution and Diversity of Mammalian Sodium Channel Genes," *Genomics* 57 (1999): 323–31.
39. J. H. Postlethwait et al., "Zebrafish Comparative Genomics and the Origins of Vertebrate Chromosomes," *Genome Research* 10 (2000): 1890–1902.

involved in regulating development. The Hox gene family described above provides a good example of this. In fruit flies, the Hox genes regulate the segmentation of the fly body, causing one segment to develop wings, another to develop legs, and so on. In mice, however, a similar set of Hox genes is used to regulate the development of different regions of the brain. In fish, Hox genes regulate the formation of fins, while in mice the corresponding Hox genes are used to regulate development of limbs. These regulatory genes, although they share similarities in sequence across organisms, tend to be used at widely disparate stages of development for vastly different purposes.

Pseudogenes

One of the most widely cited pieces of evidence used to infer evolution in general is the presence of pseudogenes within the genome of many organisms. Pseudogenes are stretches of DNA with sequences very similar to that of functioning genes; however, the information within the pseudogene is no longer transcribed to produce a protein or RNA product. These pseudogenes are comparable to vestigial traits at the molecular level.

While most pseudogenes are believed to be nonfunctional because they are no longer transcribed, some are involved in regulating the expression of other "normal" genes. For example, defects in one pseudogene, makorin1–p1, can cause a skeletal defect in mice. The absence of the pseudogene apparently reduces expression of another gene needed for normal bone formation.[40] This fact gives Creationists some hope when they dispute the inference of evolution from pseudogenes. But while some pseudogenes may have a function, the majority, especially processed pseudogenes in which the sequence has degenerated to a large degree, are almost certainly devoid of function.[41]

Large Amounts of Repetitive "Junk" DNA

The human genome consists of approximately 25,000 to 30,000 genes. If each gene coded for 10,000 base pairs, more than enough information to build a protein, it would require only 300 million bases to make up the human genome. In contrast, the actual human genome consists of 3 billion bases, at least ten times more DNA than is apparently needed. Much of this extra DNA consists of highly repetitive sequences, which

40. S. Hirotsune et al., "An Expressed Pseudogene Regulates the Messenger-RNA Stability of Its Homologous Coding Gene," *Nature* 423 (2003): 91–96.

41. Felix Friedberg and Allen Rhoads, "Calculation and Verification of the Age of Retroprocessed Pseudogenes," *Molecular Phylogenetics and Evolution* 16 (2000): 127–30.

vary in size and are repeated thousands of times throughout the genome. Humans have a repetitive sequence called the *Alu* sequence, which is about three hundred bases in length and is found at roughly a million sites in the human genome. This particular repetitive sequence, which does not encode any gene although it is 90% similar to the 7SL RNA gene,[42] rather surprisingly accounts for one-tenth of the human genome.

Because repetitive sequences are not used to encode genes (in fact, as little as 3% of the genome may be dedicated to encoding functional genes), the function of this noncoding repetitive DNA is still not understood. What is known, though, is that eukaryotic organisms have larger amounts of noncoding repetitive DNA than simple prokaryotic organisms. The amount, however, varies widely, with some plants having more noncoding DNA than primates. The reasons behind this are unclear, but the data does reveal a paradox termed the "C-factor paradox"—that most eukaryotic organisms appear to have more DNA than they need. For most schools of thought, the clear inference of this large quantity of junk DNA is that evolutionary processes have a lot of castoffs that, not doing any harm, do not disappear immediately. Creationists can only argue that the function of the junk DNA is not understood as yet but will be understood in the future, when its indispensable role will be found. Certainly some of this noncoding DNA is important in regulating gene expression and maintaining chromosome structure, but it appears that much of it is excess.

A summary of the genetic characteristics of organisms along with the inferences made by the four schools can be found in table 3.5.

The Nature of Living Species

Species in Neighboring Regions Are Often Very Similar

When one examines different species in locations that border one another or are separated by a small distance, one often finds it difficult to distinguish them. Darwin stressed this point throughout *The Origin of Species*, noting the similarities of organisms on islands with those on the neighboring continent. During his visit to the Galapagos Islands, Darwin found that each island had species that were quite similar to ones from neighboring islands as well as to those on the mainland of South

42. The sequences in the *Alu* segments are not identical but display much similarity to one another and to the 7SL RNA gene. This RNA molecule is involved in forming a complex that recognizes and binds to specific sequences on proteins that are destined to enter the endoplasmic reticulum.

Table 3.5. Inferences based on genetic characteristics of organisms

| Observation | Inference | Position of evolution schools | | | |
		Neo-Darwinism	Meta-Darwinism	Intelligent Design	Creationism
1. Common genetic encoding mechanism	Common descent	✓	✓	✓	
	Good design principle for storing and retrieving information			✓	✓
2. Similarities in DNA sequences in related organisms	Results from common descent with modification	✓	✓	✓	
	Similar species are built with similar blueprints			✓	✓
3. Duplications of genes and chromosomes	Results of random duplication events during the course of evolution	✓	✓	✓	
	Useful genes and chromosomes are used repeatedly by designer			✓	✓
4. Genes with sequence similarities yet different functions	Genes must have gradually acquired a new function	✓	✓	✓	
	Designer can use one gene for multiple functions much like a switch			✓	✓
5. Existence of pseudogenes	Remnants of once-functioning genes	✓	✓	✓	
	Function is not yet understood			✓	✓
6. Existence of large amounts of junk DNA	Artifact of replication process at some point in evolutionary past	✓	✓	✓	
	Function is not yet understood	✓	✓	✓	✓

America. Examples of this type of radiation (spreading) are not limited to the Galapagos. Darwin found that all the rodents of South America displayed a similar form that was distinct from European rodents. To name just one other example, Ernst Mayr observed a similar phenomenon with bird species in the Melanesia island region of the South Pacific. He found thirty-nine superspecies, groups of closely related species, in this region that are distinct from mainland continental species.[43] So geographic proximity often correlates with close biological kinship, and this fact is used to infer common descent.

Difficulty in Distinguishing True Species from Varieties

Distinguishing a population as a true species that is distinct and reproductively incompatible with other species is often difficult. The finch populations on the Galapagos Islands are a case in point. Each popula-

43. Ernst Mayr, *Evolution and the Diversity of Life* (Cambridge, MA: Belknap, 1976), 179.

tion has distinguishing characteristics—for example, beak size—but these characteristics often overlap with those of neighboring finch species (fig. 3.7). In addition, the different finch populations appear to be able to interbreed to some extent. Are these finch populations then to be considered separate species? Darwin spent a considerable portion of the second chapter of *The Origin of Species* cataloging cases in which the determination of whether two populations were distinct species, subspecies, or varieties seemed to be almost arbitrary. This is because in certain cases the characteristics of one population blended insensibly into those of another. Examples were drawn from organisms as diverse as birds, insects, and oak trees, and Darwin used this as evidence of natural selection being caught in the act of speciation. Since Creationists believe in speciation and common descent from a small number of animals following Noah's flood, they can agree with the inference of common descent via natural selection in some of these cases.

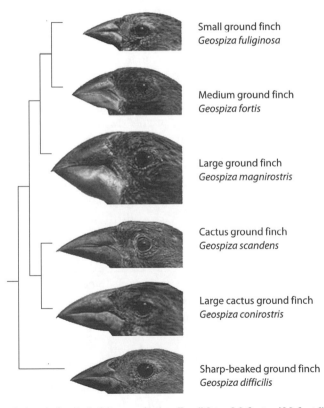

Small ground finch
Geospiza fuliginosa

Medium ground finch
Geospiza fortis

Large ground finch
Geospiza magnirostris

Cactus ground finch
Geospiza scandens

Large cactus ground finch
Geospiza conirostris

Sharp-beaked ground finch
Geospiza difficilis

Figure 3.7. Different finch species found in the Galapagos archipelago. (From K. Petren, B. R. Grant, and P. R. Grant. "A phylogeny of Darwin's finches, based on microsatellite DNA length variation." Proceedings of the Royal Society, Series B: Biological Sciences. Vol 266 (1999), 327, fig. 3; used with permission)

Modern biologists, naturally, have studied the phenomenon extensively. For example, when one examines tropical bird populations, one
often finds two isolated populations of birds that have *not* diverged from
each other even though they live on *different* islands. One also finds
populations in which small peripheral isolates *have* diverged to the point
that they would be considered subspecies or in fact two separate species. In addition to these two extremes, one can also find everything in
between, such as hybrid zones in which populations have acquired new
morphological characteristics but are not reproductively isolated, such
as the finches of the Galapagos. Such gradations can make it tedious to
determine whether two populations are of the same species, are of two
different subspecies, or are separate species.

Discontinuity of Living Species at Higher Taxonomical Levels

While different species within the same genera can in certain cases
be difficult to distinguish because of shared characteristics, the same
is not true when one looks at differences between higher taxonomical
levels. Living species are clustered in related groups, and these clusters
tend to be isolated and distinct from one another. For example, there are
no *living* forms that bridge the gap between humans and apes; likewise,
there are few that bridge the gap between reptiles and birds.

What this means is that while it is often difficult to determine if two
groups of elephants are members of the same species, there is no difficulty in distinguishing elephants from other related land mammals.
This is not to say that no intermediate forms are ever found. An example
of such an intermediate form is the egg-laying platypus, which occupies
a niche between reptiles and placental mammals. But even in this case
there is no difficulty in distinguishing a platypus from either reptiles or
placental mammals. In fact, there is no difficulty in distinguishing the
platypus from other egg-laying mammals. Darwin himself was cognizant
of this, and devoted a small section of *The Origin of Species*, titled "On
the Absence or Rarity of Transitional Varieties," to this issue, arguing
that such forms must be unstable and short-lived. This is still the inference of the Neo-Darwinian school. For Creationists, the inference of
course is that intermediate forms never existed and therefore evolution
did not happen.

Convergent Evolution?

Similar ecological environments often exist on different continents
that are obviously separated by unbridgeable distances. The plants and
animals that fill these niches on one continent are totally isolated from
those that fill a similar niche on a distant continent. Despite this separa-

tion, a multitude of cases exist in which organisms on both continents look remarkably similar in form and function. Good examples of this are the marsupials of Australia, which closely resemble placental mammals found in northern continents. Despite the obvious differences in the bearing of offspring, the marsupial Australian Tasmanian wolf is very similar to the northern wolf, and the marsupial Australian flying phalanger is nearly identical in form to the flying squirrel. Ernst Mayr points out that such cases are not rare: "Any knowledgeable zoologist would be able to list several pages of such cases."[44]

In addition to finding similar forms of organisms in disparate locations, certain complex traits are found in organisms that are totally unrelated. Echolocation, for example, is found in bats, two bird species, dolphins, and whales. Although the details of echolocation differ in each case, the basic principle is the same. The same can be said for photoreceptors, which in some form or another appear in roughly fifty different lineages. Six separate phyla have incorporated photoreceptors into complex eyes, two of which, the vertebrate eye and the cephalopod (squids and octopi) eye, are remarkably similar in appearance and function despite the fact that they develop from different tissues during embryogenesis.[45] Similar cases can be drawn from all classes of animals and plants.[46]

The observation that genetically unrelated animals have very similar forms or have independently developed similar complex characteristics, such as echolocation, is striking, and naturally the schools of thought have drawn different inferences from it. The Neo-Darwinian school has inferred that the power of natural selection coupled with random mutation to find good solutions is so robust that different lineages will converge upon the same solution if it is indeed a good principle. This they refer to as "convergent evolution." For Creationists, there is no convergent evolution. Rather, a creator endows all creatures with what they need to survive in their environment.

A summary of the nature of living species along with the inferences made by the four schools of thought can be found in table 3.6.

Conclusion

There is much empirical data relating to the history and evolution of life. Sorting through this data can be overwhelming, particularly given

44. Mayr, *What Evolution Is*, 222.
45. Stephen J. Gould, *The Structure of Evolutionary Theory* (Cambridge, MA: Belknap, 2002), 1126.
46. For more details, see Simon Conway Morris, *Life's Solution: Inevitable Humans in a Lonely Universe* (Cambridge: Cambridge University Press, 2003).

Table 3.6. Inferences based on characteristics of living species

Observation	Possible explanation	Neo-Darwinism	Meta-Darwinism	Intelligent Design	Creationism
		Position of evolution schools			
1. Geographic proximity often correlates with biological kinship	Species split from a common ancestor	✓	✓	✓	✓
	Similar design for species in neighboring environments			✓	✓
2. Difficulty in distinguishing species from varieties	Demonstrates the incremental species change needed for evolution to proceed	✓	✓	✓	
	Demonstrates species can change only to a limited extent (by degeneration)			✓	✓
3. Discontinuity of higher taxonomic levels	Intermediate forms were unstable and selected against	✓	✓		
	Intermediate forms never existed			✓	✓
4. Presence of similar forms in widely geographically separated species (convergent evolution)	Selection pressures must have been similar in both geographically separate environments	✓	✓	✓	
	Designer made similar forms for similar environments despite geographic separation			✓	✓

that what has been presented here is only part of the whole. Despite this limitation, the chapter does cover major areas that are in need of explanation by any scientific theory of evolution. When it comes to explaining this data, surprisingly, all four major schools of evolutionary thought seem undaunted, each confident that its solution is the best inference from all the relevant data. By imposing certain patterns on the data and by making particular assumptions, all four schools have been able to build a case for their position. These cases will be examined in chapters 5 through 8.

4

THE PRINCIPAL POINTS
IN DISPUTE

When it comes to the topic of evolution, nearly any point that can be disputed is disputed. Some of these points represent major issues that are worth discussing because their resolution could tip the debate in favor of one of the schools. Other disputed points have little bearing on the ultimate outcome of the debate. As a result, studying the rhetoric devoted to them is usually a waste of time. To help identify the issues in the debate that are worth investigating, we will list and describe in this chapter the major points in dispute, those that truly define the evolution controversy. It is important to keep in mind that in this chapter we are not trying to determine which side has the best arguments or is supported by the best science; we are only delineating the key points in dispute.

We have identified below six major points that define the evolution controversy. Four of them are empirical questions and go back to the observable facts discussed in chapter 3, since it is the inferences from those facts that are the basis for the theories advocated by the four schools. Obviously, the schools argue about what the fossil record tells us, what its gaps mean, what the significance of genetic commonalities may be, what the geologic column implies, what observable astronomical data say about the age of the universe, and many other questions. The last two points deal principally with philosophical issues or the philosophy of science. Nonetheless, points five and six have a great impact on the controversy since they affect how one interprets the scientific data that are at issue. As one can see from the list, the points are not entirely independent, but we believe the reader will find it useful to examine them separately:

113

1. Common descent of organisms from a single progenitor versus common design plan
2. The ability of random mutation to create new biological information
3. The efficacy of random mutation coupled with natural selection
4. The age and chronology of the earth and the universe
5. The scope of naturalistic explanations in science
6. What constitutes a bona fide scientific theory

To understand the issues at stake, we shall review each of these points, indicate how the schools differ, and explain how each point impacts a particular school's overall position.[1]

Common Descent of Organisms from a Single Progenitor versus Common Design Plan

The similar genetic makeup of organisms as diverse as *E. coli* and man and the shared morphology found in seemingly unrelated organisms virtually demands an explanation. To explain these observations, Darwinists and some in the Intelligent Design camp posit that organisms are related via common descent. The shared qualities that distinct species exhibit are assumed to be due to an early progenitor, which itself is descended from still earlier forms, ultimately going back to the first cell, discussed in chapter 3.[2]

The nearly universal genetic code is a prime example of common descent, as it is a characteristic shared by almost all life-forms and is believed to have originated in a universal common ancestor.

In addition to invoking a single progenitor of all living things at some distant time in the past, common descent also implies that any given or-

1. Let us note that one major point often cited as the focus of the evolution debate, namely the equivalence of microevolution and macroevolution, actually falls under the second and third points above and is better examined in those contexts.

2. There are other possible natural explanations for these commonalities, particularly at the morphological level. The same form could have been struck upon not because of recent common descent but because it was the most robust and beneficial form available to natural selection. This is the notion of convergent evolution, which could explain why the Australian flying phalanger is nearly identical in form to the flying squirrel of North America, despite the geographical separation. Note that this explanation does not refute the theory of common descent; it merely means that there is another mechanism by which forms may be related. For example, while Darwinists still believe that the Australian flying phalanger and the flying squirrel of North America do not have a recent common ancestor that explains their morphological similarity, they still believe that they are related by common descent to an ancestral mammal.

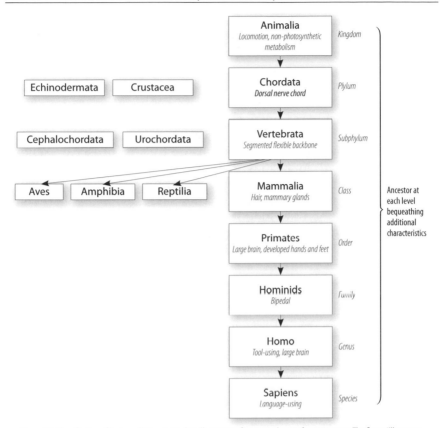

Figure 4.1. Classification of humans (*Homo sapiens*) as illustration of common descent from ancestors. The figure illustrates the connection of taxonomy with common descent, back to the kingdom level. An ancestor at each level contributed a key characteristic.

ganism alive today received its traits by descent from multiple ancestors. Thus common descent works in the following manner: characteristics at the deepest level are shared by all organisms, primarily the genetic code and the structures and systems that decode the information in the DNA to make proteins. Certain other structural characteristics of cells and metabolism fall more or less in this category, including the cell membrane and the enzymes used to break down sugars. As we examine the phylogenetic tree, we see that other characteristics are shared by many but not by all organisms. For example, a dorsal nerve, called a "notochord," is shared among many animals, who therefore are grouped under the phylum Chordata. These animals trace back to a progenitor who bequeathed this notochord to them (but not to other organisms). Beneath are those animals sharing the characteristic of a segmented and flexible backbone, and they are classified in the subphylum *vertebrata*.

Once again, an ancestor that developed the flexible backbone bequeathed it to its descendants. Reptiles and mammals form distinct classes under the vertebrata and so on. Thus the overall taxonomic scheme reflects a series of development and innovation steps. In this manner it is a reflection of common descent. The basic idea is shown in figure 4.1, with major innovations indicated at each level.[3]

Creationists dispute common descent from a single progenitor due in large part to their belief in the special creation of each "kind" or "type." They, together with some members of the Intelligent Design school, argue that observed similarities are the result of a common design plan, or set of plans, rather than common descent. According to this view, God created certain natural kinds or types, and these kinds share some features deemed good by God for the functioning of the world. Given their common design plan, each species then varies according to the specialization needed for its intended purpose in the natural order.

> The Scriptures are very clear in their teaching that God created all things as He wanted them to be, each with its own particular structure, according to His own sovereign purposes. The account of creation in Genesis 1, for example, indicates that at least ten major categories of organic life were specially created "after his kind." These categories are, in the plant kingdom: (1) grass; (2) herbs; (3) fruit trees. In the animal kingdom the specific categories mentioned are: (1) sea monsters; (2) other marine animals; (3) birds; (4) beasts of the earth; (5) cattle; (6) crawling animals. Finally, man "kind" was created as another completely separate category. . . . Even though there may be some uncertainty as to what is meant by "kind" (Hebrew *min*), it is obvious that the word does have a definite and fixed meaning. One "kind" could not transform itself into another "kind." There is certainly no thought here of an evolutionary continuity of all forms of life.[4]

While Creationists reject common descent at the higher levels (phyla and classes), they do accept common descent at *lower* taxonomic levels (species and genera) *from the created kinds*. (Their kinds, called "baramins," may not necessarily coincide with modern taxonomic levels.) Creationists need to posit common descent from a certain level downward in order to account for rapid repopulation of the world after the Genesis flood (see chap. 6).[5] In their view, the lower levels represent degeneration (loss

3. Based loosely on Lynn Margulis and Karlene Schwartz, *Five Kingdoms* (New York: Freeman, 1998). We have here retained the traditional breakdown into the phylum Chordata, rather than the newer one given in the book.

4. Henry M. Morris, *Scientific Creationism* (Green Forest, AR: Master Books, 1985), 216–17.

5. Glenn R. Morton, "Fossil Succession," *Creation Research Society Quarterly* 19, no. 2 (1982): 103–11.

of information) from the higher types, which were most rich in respect to the genetic information they were given.

The Ability of Random Mutation to Create New Biological Information

The Darwinian notion of common descent requires not just descent, but descent with modifications. Therefore, if new species and other taxa are to arise, then there must be some source of heritable new information, information that must enter the genetic code itself. Darwin recognized this, though of course he knew nothing of genetics. Though unclear on the matter, he seems to have assumed that some of this information was not random, but was somehow fed back to the organism from its environment. The Neo-Darwinian synthesis opted instead for a totally random source for the generation of new information—in their view, the only available option. This random source is the mutations that the genetic material, specifically DNA, continually undergoes due to uncorrected copying errors, radiation, chemical mutagens, and other sources. The idea is that while most of the changes introduced into the code by these sources are deleterious, once in a while a beneficial change arises, one that gives an organism a selective advantage. At this point, natural selection takes over, and the new genotype and phenotype become established in the population and transmitted to posterity. If sufficient random beneficial changes occur, a new species emerges (see below: "Natural Selection as the Filter for Random Mutations").

Without doubt, the efficacy of random mutation for the generation of new biological information is the most crucial issue for the evolution debate. The dominant school, Neo-Darwinism, absolutely requires random mutation as the sole engine of *new* information and thus the basis for all biological innovation. Other mechanisms exist that can (1) recombine existing information in hopes of creating new functionality, (2) duplicate existing information, or (3) reorder the expression of existing information. All of these processes can lead to changes in an organism's phenotype. In meiosis, for example, new combinations of genetic alleles are produced by the process of genetic recombination. By producing offspring with different assortments of alleles, some new phenotypes can be produced. But this is limited by the number and type of alleles that are present in the population. To evolve further, new alleles must be produced by mutation.

All the other schools accept random mutation as a reality but tend to be skeptical of claims that it can perform the role assigned to it by Neo-Darwinism. Any demonstration, whether theoretical or empirical, that

117

random mutation *cannot* supply the quantity and quality of new information needed to account for the history of life on earth will bring down Neo-Darwinism. Of course, any demonstration that it *can* do what its proponents (the Neo-Darwinians) claim will bring down the other schools.

Creationists flatly reject random mutations as having the potential to create new species or even new information. In their view, this is physically impossible, and they give arguments to that effect (discussed in chap. 6). As they see it, random mutation can lead only to degeneration or loss of information, never to its increase. The Intelligent Design and the Meta-Darwinian schools believe that the Neo-Darwinian picture, at the very least, cannot be the whole story. For the most part, both schools accept that some innovation may come through random mutation, but not nearly enough to account for the observed history of life-forms.[6] In addition, members of the Intelligent Design school argue that the complexity of some observed biological structures cannot be accounted for by any random processes, regardless of how long they may act.

Natural Selection as the Filter for Random Mutations

Natural selection is a well-known process that is independent of the Neo-Darwinian (or any other) theory of evolution. It was recognized by Blyth, Lyell, and others at least a quarter of a century before publication of *The Origin of Species*. Natural selection functions to optimize a population of organisms with respect to its current environment, a function that is not in dispute. Natural selection accomplishes this task by acting as a filter that takes raw material, in the form of the organisms produced in each generation, and selecting, on average, those that are most fit phenotypically. In practice, the filtering action works by virtue of the fact that those who are most fit (e.g., can gather the most food, can hide best from predators, etc.) will survive longer and hence will be able to reproduce at a higher rate, in a statistical sense. Natural selection will also choose those most fit genotypically, but due to the complex nature of the mapping from genotype to phenotype, the selective action is not so sharp with respect to the genotype.

This idea has been widely criticized because any reasonable expression of it turns into a tautology; that is, those that survive are the fittest,

6. Most members of the Intelligent Design school are extremely skeptical of natural selection's ability to create new information, but some believe it can lead to moderate amounts of information gain (see chap. 7). In contrast, members of the Meta-Darwinian school often accept it as being capable of supplying the necessary innovation, but place restrictions on it and the circumstances under which it may act. There are exceptions, though, as described in chap. 8.

and the fittest are those that survive. In fact, some have even claimed that its tautological nature was Darwin's great discovery. That problem, however, is not what is in dispute here.

Rather, the present dispute concerns the key contribution of the modern Neo-Darwinists, which was to combine the processes of random mutation and natural selection, and claim that together they could account for all of evolution. The basic idea is quite simple: random mutation occasionally produces a genotypic change that improves some aspect of an organism's performance and thus its fitness. This improvement is acted upon by natural selection—essentially survival of the fittest—which, on average, allows the most fit to breed and produce the next generation (see fig. 4.2). Over time, improved genes spread through the population, thereby causing an improvement in the average fitness of the population. Eventually many such changes accumulate and yield a new species.[7] Over still greater spans, new genera, families, and higher taxa emerge. This wonderfully elegant and powerful hypothesis, and its problems, will be discussed in detail in chapter 5. For now we will concentrate on the positions of the various schools.

Natural selection

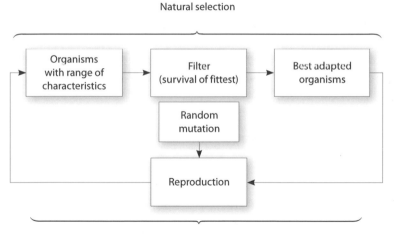

Random mutation and generation of variation

Figure 4.2. Operation of mutation and natural selection to produce increasing levels of fitness and complexity.

Creationists, naturally, reject the entire hypothesis as a mechanism capable of creating new higher taxonomic levels or even improved species. They argue that (1) such change would violate natural laws, especially the laws of thermodynamics; (2) it cannot work along these proposed

7. Exactly how many changes are needed is a matter of dispute.

lines because random mutation is incapable of generating the new information required for genetic improvement; and (3) there would not be enough time for the process to act. They do agree, however, that the process can produce new species through *degenerative* change, that is, through *loss* of information. Thus, from the original created *kinds*, new forms of life, new species, can arise as the original genetic richness, supplied by God, *deteriorates*. Obviously, this process is quite limited and will only lead to species similar to their progenitors, but that would be sufficient in their view to account for repopulation of the world after the Genesis flood.

The Intelligent Design school rejects the proposed mechanism as capable of creating new higher-level taxa, for many of the same reasons as the Creationists. They are especially concerned about the information creation problem and have focused on it with respect to the complexity of biological molecules, processes, and systems. They argue that any speciation event that requires the creation of what the Intelligent Design school calls an irreducibly complex system cannot be accounted for under the proposed mechanism.[8]

The Meta-Darwinian school does not reject the mechanism altogether, but believes that it can account for only some of the observed change and information creation. In effect, Meta-Darwinians accept the thrust of many of the criticisms of the Creationist and Intelligent Design schools about the feasibility of change coming about by this mechanism. But rather than turning to supernatural explanations, they contend that other natural processes can fill in the gaps.

Microevolution and Macroevolution

Evolution disputes are often phrased in terms of micro- and macroevolution. Microevolution is roughly synonymous with natural selection operating on an *existing* gene pool or species. Microevolutionary changes are those that result from selection of existing genes (actually genotypes) to better adapt a population to its environment. In most cases this deals with changes at or below the species level. Thus, skin color or beak length may change over several generations, as the appropriate characteristics are selected for in the presence of environmental stresses or changes, but no new information, and no innovation, is involved. These characteristics will revert to their "normal" condition if the environmental stresses or changes abate. Microevolution, as its name suggests, thus describes

8. In theory, it can be accounted for in at least some cases via the co-option argument; see chap. 7.

small-scale changes and is extremely efficacious in explaining such phenomena as the proliferation of different strains of HIV and at least some types of antibiotic resistance in bacteria. All schools of thought accept microevolution as an observable phenomenon of nature.

Macroevolution is the creation of new structures, characteristics, or features that result in the origination of higher taxa. That is, macroevolution, as its name suggests, involves large-scale changes. This, of course, requires *new information* or, at the very least, new uses of existing information. As described by John Rennie:

> Microevolution looks at changes *within* species over time—changes that may be preludes to speciation, the origin of new species. Macroevolution studies how taxonomic groups *above* the level of species change. Its evidence draws frequently from the fossil record and DNA comparisons to reconstruct how various organisms may be related.[9]

Let us consider the different points of view of the schools on micro- and macroevolution. The Creationists and the Intelligent Design camp claim that micro- and macroevolution are different processes, and therefore different evidence—much different than the passage quoted above suggests—must be adduced to support each. The question thus boils down to the *kind* of evidence needed for macroevolution. Those who support Neo-Darwinian evolution typically argue that, in reality, there is no meaningful qualitative or quantitative distinction between the two processes and that evidence for one is evidence for the other: "Evolutionists reason that if small changes can occur in a short time, large-scale changes can take place during the many millions of years of earth history."[10] Therefore there is no need for a separate set of evidence to support macroevolution. This, indeed, is why many proponents of evolution cite the cases of industrial melanism and Darwin's finches as examples of evolution in action, even though they are strictly speaking only microevolution at work. Critics of Neo-Darwinism naturally disagree, maintaining that direct evidence for the emergence of new species and higher taxa *by accumulation of microevolutionary changes* must be provided. In effect, the microevolution/macroevolution phrasing of the controversy combines two of the major issues in the dispute: (1) the ability of random mutation to create new biological information, and (2) the efficacy of random mutation coupled with natural selection. These two points focus directly on the *mechanisms* required to make

9. John Rennie, "15 Answers to Creationist Nonsense," *Scientific American* 287 (2000): 78–85; italics added.
10. Verne Grant, *The Origin of Adaptations* (New York: Columbia University Press, 1963), 36.

evolution work, rather than the *outcomes*, which tend to be the focus of micro- and macroevolution discussions. As we have noted, all sides accept microevolution. When it comes to macroevolution, things get more complicated. While Creationists believe that new species can form, they believe that this represents degeneration only. The Intelligent Design, Meta-Darwinian, and Darwinian schools all tend to believe that natural selection and random mutation can result in some level of innovation. What is bitterly disputed is exactly how complex these innovations can be.

Age of the Universe and Age of the Earth

Any form of Darwinian evolution requires long time spans to operate. Science has come to the conclusion that the earth is quite old (about 4.5 billion years), and the universe still older (about 15 billion years)—possibly long enough for evolution to work. A very young universe, or at least a young earth (thousands of years), would not provide adequate time for most naturalistic evolutionary mechanisms.

For all of the schools except Creationism, the age of the universe is reckoned according to the findings of astronomy and astrophysics, at about 15 billion years. The reader can find the evidence for this in any book on astronomy, and of course many galaxies visible to the naked eye or a backyard telescope are millions of light-years distant,[11] meaning that their light has taken millions of years to reach us. Since the Creationist school maintains that the earth and the heavens were created about 6,000 to 10,000 years ago, according to their literal interpretation of the first chapter of Genesis, this poses obvious problems. Their proposed theories to account for this difficulty are discussed in chapter 6. The stakes are quite high. On the one hand, if they can prove that the earth is young, all theories of evolution crash and burn, regardless of other evidence. On the other hand, any piece of evidence that shows the earth to be significantly older than 10,000 years will demolish the Creationist position.[12] The young earth Creationists are the only school to argue explicitly for a young earth; the other three schools either accept the astronomical and geological evidence for an

11. Technically, this is the *lookback distance*; the *present* or *comoving distance* would be greater due to the ongoing expansion of the universe. This distinction, however, does not affect the present argument in any way.

12. Note that in this discussion we are speaking only of the young earth Creationists, as they are the most vocal in the evolution dispute. "Old earth" Creationists are free to accept any age for the earth.

old earth or, in the case of the Intelligent Design school, do not make this an issue.

The Scope of Naturalistic Explanations in Science

The debate over the scope of naturalistic explanations in science is, in a sense, a battle for the soul of science: its contents, its place in the context of all human knowledge, and its limits. The Neo-Darwinian and Meta-Darwinian schools emphasize the need for scientific explanations always to be "pure"; that is, scientific explanations must exclude any notion of divine "intervention" or action. Only "natural" processes may figure in valid scientific explanations of phenomena. Such explanations are termed "naturalistic." All four schools are more or less in agreement that within the realm of science, naturalism reigns. Consequently, one cannot invoke nonmaterial causes (such as spiritual forces) to explain phenomena *when one is doing science*. For example, a scientist looking for the cause of epilepsy in a certain person must seek natural causes rather than invoking demons. Other causes may be proffered, but not within the context of scientific explanation. This is termed *methodological naturalism*.

The difference comes when a further step is taken, namely, when it is asserted that naturalistic explanations are capable of explaining *all* of nature, or, in other words, that *all* of nature can be explained scientifically. The scope of the "all" typically includes the origin of the universe and the origin of the first cell (abiogenesis), as well as mechanisms that account for the history of all flora and fauna on earth. Such is the doctrine of *naturalism*, sometimes referred to by the more loaded word, *materialism*. Naturalism, understood in this sense, has clear implications for other types of explanations (such as theological explanations): they are not needed.

On this point, the schools diverge sharply. The Neo-Darwinian and Meta-Darwinian schools accept naturalism, particularly when it comes to explaining the history of life.[13] They differ only in respect to the particular naturalistic mechanisms responsible for evolution. The Creationist school of course rejects naturalism as capable of explaining anything

13. Actually there would be some range of opinion within these schools on just how omnicompetent naturalism is; theistic evolutionists, in particular, would likely dispute any claims that there are naturalistic explanations for *all* phenomena and *all* of human experience; others would argue that philosophical knowledge, for instance, cannot be dispensed with, as it deals with the very question in dispute. But in general the Neo-Darwinian and Meta-Darwinian schools reject all claims that nonnaturalistic "intervention" in any form is required to explain any aspect of biological systems.

other than a very limited range of phenomena; the major events—namely creation of the world, creation of flora and fauna *kinds*, and sometimes geological phenomena such as radioisotope quantities—require direct creative action, that is, primary causality. Events such as the Genesis flood also indicate the need for explanations that go beyond naturalism. These explanations are usually termed "interventionist." The Intelligent Design school concentrates on other aspects of biological systems, which also indicate that naturalism is insufficient, in their view. They believe that the complexity of biological molecules and systems exceeds that which can reasonably be attributed to random processes, even when operating over the timescales associated with life on earth (about 4 billion years). Thus, they also claim that intervention is needed, though much less intervention than the Creationists claim is necessary. In general, the Creationist and Intelligent Design schools also reject naturalism's claim to be able to explain the origin of life, though that is a separate issue from evolution. Clearly, if the strong form of naturalism is false, science will encounter phenomena that it cannot explain. The Intelligent Design school, indeed, is heavily concentrated on finding just such phenomena and devising means to test for them.

Thus one may, in principle, view the issue as an empirical one: can science explain everything or not? But to answer this question one would need criteria to recognize a phenomenon that cannot be explained naturalistically, and agreement on those criteria will be difficult on account of the differing philosophical positions on naturalism. If one is convinced for philosophical reasons that all phenomena can be explained with naturalistic theories, then one is unlikely to agree to any meaningful set of criteria to classify a phenomenon as not explainable naturalistically. We say "philosophical reasons" because science itself is neutral on this issue; this is a metascientific question. The Creationist and Intelligent Design schools are of course committed to the existence of such criteria and to our ability to find them. The dispute over the scope of naturalism colors much of the evolution debate; even when not explicit, it is always lurking in the background.

To clarify the position of the Creationist and Intelligent Design schools regarding the limits of science, it is useful to distinguish two types or classes of explanations, rational and scientific. A *rational explanation* is one that uses some type of plausible reasoning from premises to explain an observed phenomenon or state of affairs. In the present context, we might have the following explanation: "God created the world. We observe that the world contains animals with similar body plans. Therefore God chose to create the animals with similar body plans." Such explanations are rarely subject to direct experimental test, as they refer to events that either cannot be repeated or are not capable of being tested at all. This does not mean that they are therefore inferior or incorrect, of course, nor does it mean that they are exempt from scrutiny.

Scientific explanations are a specific subset of rational explanations that are based upon the scientific method (discussed in the next section). Such explanations may take many forms, but in general, they require a hypothesis that is advanced to explain or generalize observed phenomena and which in turn is used to make predictions about what additional phenomena will be observed, and under what conditions. If these phenomena are indeed observed, the hypothesis is considered to be confirmed, but the hypothesis (which now may be referred to as a theory or a law) is subjected to further tests of this type on an ongoing basis. Both the ability to make testable predictions and the possibility of being refuted are considered essential to scientific explanation.

If the same event or phenomenon can be explained with both a nonscientific rational explanation and a scientific one, people typically choose the scientific explanation. However, the fact that an explanation is scientific does not mean that it is correct; it means only that it has been arrived at by a procedure that, empirically, has yielded explanations that meet a test of time and, in general, one of prediction.

In the case of evolution, Creationists and Intelligent Design advocates have advanced a mixture of rational explanations and scientific explanations for the appearance of flora and fauna, and their characteristics and history.[14] Evolutionists have what they claim are scientific explanations for all of these same things, explanations that the Creationist school argues are unscientific and incorrect.[15]

What Constitutes a Bona Fide Scientific Theory?

In theory, knowledge of proper scientific methodology and its application should help to resolve the evolution controversy. In practice, however, this has become yet another battleground. Questions about what science is and about the scientific method loom large, as all sides routinely exchange accusations that the others employ "unscientific methods," have "unscientific theories," or use an "unscientific approach." This is itself an indication of the highly political nature of the controversy. In today's environment, if one can tar one's enemies with the "unscientific" brush, they are effectively

14. For some phenomena, such as the distribution of fossils in geological strata, Creationists have also advanced what they claim are scientific explanations. However, they do not claim to have scientific explanations for all observed facts about the present nature, distribution, and characteristics of flora and fauna.

15. Actually, the Intelligent Design school and the Meta-Darwinian school also dispute the scientific status of some of these explanations.

discredited regardless of the merits of their case. As a result, one need not even acknowledge—much less respond to—their critiques.

The pejorative connotation associated with the label "unscientific" is the direct result of the fact that science *is* one of the great success stories of modern civilization. Much of that civilization, indeed, is based on engineering efforts stemming directly from scientific theories in physics, chemistry, and biology. Modern medicine, too, is largely based on science. Much of the success of science has been due to its increasingly mechanistic outlook, one that seeks completely naturalistic explanations of phenomena and eschews philosophical notions such as essentialism as well as all nonscientifically observable interventions.[16] Neo-Darwinian and Meta-Darwinian theories lay claim to being bona fide scientific theories in this tradition. The other two schools claim that their theories of evolution are scientific as well, even though they reject some or all of the mechanistic underpinnings of Neo-Darwinism. The schools, then, all agree that being "scientific" is important, and they even agree on many of the characteristics of true science. They disagree about which, among them, is in fact truly scientific. To help readers sort out this disagreement, we discuss the nature of and requirements for a bona fide scientific theory.

What Is a Scientific Theory?

The purpose of science is to investigate the natural order—natural phenomena—and explain what we have *already* observed as well as predict what we *will* observe and what we *won't*. A scientific *theory* is a set of hypotheses and definitions, together with certain rules of inference that, given some boundary conditions and empirical facts, can *explain* in a concise and compact manner many already-known natural phenomena. In addition, any theory should *predict* new natural phenomena, while at the same time *exclude* the possibility of others. Exclusion of some conceivable observations is extremely important, because it is this feature that guarantees that the theory will convey new information to us. A theory that can "explain" any conceivable observation does not explain anything at all—it is irrefutable, but at the price of imparting no real information.

While other branches of knowledge—theology, philosophy, and so on—make claims about explaining the world, science does so using a specific process, the scientific method. Exactly what constitutes the scientific method is a question that has been debated for many decades, but some aspects of it are clear enough for our present purposes. First, the

16. E. J. Dijksterhuis, *The Mechanisation of the World Picture*, trans. C. Dikshoorn (Oxford: Clarendon Press, 1961).

scientist is intrigued by some set of phenomena and wishes to explain them in a concise manner. That is, the scientist assumes that phenomena are the result of natural processes that act uniformly and therefore are capable of being recorded, measured, analyzed, and explained by the use of reason. Furthermore, he or she assumes that the regularity and uniformity of the processes can be captured in a suitable concise form, often mathematical. As Galileo said, "The great book of nature is written in mathematical characters."[17]

The key point is that the great variety of phenomena observed are the result of a small number of natural processes acting, and it is these processes that the scientist seeks to capture in the form of theories and laws. To find them, the scientist looks for regularities and uniformities and then formulates them in a suitable manner. For example, Newton looked for regularities in the motions of bodies under the influence of gravity and realized that the same type of "falling" that applies to apples dropping from trees also applies to the moon, which continually "falls" toward the earth in its orbit. This and other observations led Newton to develop his now-famous three laws of motion.[18]

Thus far, the work of the scientist and the work of the philosopher are perhaps not so different—and indeed for centuries what we now term "science" was referred to as "natural philosophy."[19] The full title of Newton's famous work, the *Principia*, is *Philosophia Naturalis Principia Mathematica* (1687). However, once the hypotheses have been formulated, the scientific method reveals its "scientific" character, because the scientist takes his or her formulation, or hypothesis, and uses it to *ask questions of nature*; these questions are *experiments*. But the questions are such that the scientist *has already formulated answers* (predictions); he or she only wishes to know if nature will *confirm* these answers (predictions). Thus science has a crucial *experimental* component, and we often use the term "experimental science." If the predictions are confirmed experimentally, the hypothesis receives a boost and further predictions and experimental tests are performed; if not, the hypothesis is either modified or discarded, and the process is restarted. A particular hypothesis could be confirmed for years, even centuries, before more accurate measurements, or measurements taken in a new area, reveal problems.

17. Galileo Galilei, *Il Saggiatore* [*The Assayer*], 1623, loose translation of famous passage from introductory pages of this work.
18. Shortened and simplified slightly; for more detailed treatment, see any physics text, such as Eugene Hecht, *Physics* (Pacific Grove, CA: Brooks/Cole, 1996), 116–192.
19. This, which might seem to be just a historical curiosity, is actually much more, because as we shall see, science is often called upon to fill the role of philosophy or metaphysics to tell us what is and is not real in the world, which is beyond its scope.

One central criterion of a scientific hypothesis is that the predictions it makes must be *falsifiable*; unfalsifiable predictions cannot be meaningfully verified by experimental tests, as no test can disprove them. In the case of evolution, for instance, Darwin told investigators to gather more fossils and they would find that most of the fossil record would consist of transitional forms. This is a perfectly valid type of experiment, even though it refers to events that happened millions of years earlier. The key point is that investigators *are told to look somewhere they have not looked before, for something they have not seen before*. It is this ongoing experimental verification and feedback that distinguishes science from philosophy and other types of knowledge. In a widely cited book,[20] Karl Popper has discussed the falsifiability criterion, arguing that it is a necessary condition for a theory to be a valid scientific theory. Popper then goes on to refine this notion, dividing theories into well-testable, hardly testable, and nontestable, and notes, "Those which are non-testable are of no interest to empirical scientists. They may be described as metaphysical."[21]

Criteria for a Valid Scientific Theory

Before presenting the criteria for a genuine scientific theory, let us review some definitions, which are taken directly from the National Academy of Sciences:[22]

Fact: In science, an observation that has been repeatedly confirmed and for all practical purposes is accepted as "true." Truth in science, however, is never final, and what is accepted as a fact today may be modified or even discarded tomorrow.

Hypothesis: A tentative statement about the natural world leading to deductions that can be tested. If the deductions are verified, the hypothesis is provisionally corroborated. If the deductions are incorrect, the original hypothesis is proved false and must be abandoned or modified. Hypotheses can be used to build more complex inferences and explanations.

Law: A descriptive generalization about how some aspect of the natural world behaves under stated circumstances.

20. Karl R. Popper, *Conjectures and Refutations: The Growth of Scientific Knowledge* (New York: Harper, 1968).
21. Ibid., 256–57.
22. National Academy of Sciences, *Teaching about Evolution* (Washington, DC: National Academy Press, 1998).

Theory: In science, a well-substantiated explanation of some aspect of the natural world that can incorporate facts, laws, inferences, and tested hypotheses.

Typically a scientist starts with a set of observations or *facts*, then formulates a *hypothesis* to explain those facts and to *predict* new ones, utilizing inferences based on the hypothesis. If the hypothesis looks promising, it is usually generalized or integrated into a larger explanatory framework, referred to as a *theory*. If a theory successfully predicts facts, and does so over a long period of testing, it may become known as a "law of science."[23]

Good theories always start from observed facts and ultimately return to them. As part of their role to tell us about reality, theories are general ways of capturing myriad facts and explaining them. Scientists want the best explanation of observed facts—the best theory—and they want to be able to predict new, as yet undiscovered facts. Ultimately, theories live or die on the basis of their ability to meet these goals. A single fact can bring down a well-attested theory and often has done so. Newtonian mechanics was felled by the discovery that the earth has no absolute orientation in space—a fact revealed by the famous Michelson-Morley experiment in 1887. Both facts and theories are essential to science and both have their important role.

Utilizing the foregoing definitions, we can formulate ten criteria that are intended to be very general, objective, and essential characteristics of any genuine scientific theory:

1. **Compactness.** The theory must consist of a relatively small number of hypotheses, expressed in general terms or equations, that can be combined with boundary conditions (specific empirical facts about nature) to yield meaningful statements about the world. A commonly cited but still excellent example is Newtonian mechanics.
2. **Simplicity.** The theory should be "simple," or "elegant"—that is, it should not require many hypotheses or ad hoc assumptions to accomplish its goals of explanation and prediction. As Einstein has said, a theory should be as simple as possible, but no simpler. Simplicity is often used as a criterion of a successful theory because of our underlying belief in the simplicity of nature. However,

23. Be aware of terminology problems. Not every well-established hypothesis is called a "law"; for example, the hypothesis of the special theory of relativity is called a "principle," namely, the "special principle of relativity." Nor is every "law" actually a scientific law; e.g., "Ohm's law" simply gives the relationship between current and voltage in certain types of materials.

simplicity is less in evidence in the realm of biology than in that of physics.

3. **Falsifiability.** The hypotheses (laws) must themselves be general statements about the world, capable of falsification. That is, they must convey *new* information, which means that they must express a choice among many alternatives. The hypotheses therefore can be neither tautologies (statements of the form *All A are A*, which convey no new information) nor self-contradictory statements, from which *any* conclusion or inference can be drawn. After a theory is well established, the hypotheses may be referred to as "laws"; thus we have "Newton's laws of motion."

4. **Verifiability.** The statements about the world must be capable of verification by experiment and observation, either directly, or through some inference based upon them. And the operational method to do this must be straightforward and not itself the subject of endless disputes. For example, it is quite easy to use Newton's second law, $F = ma$ (force equals mass times acceleration), to calculate the acceleration of an object subject to the force of gravity, and verify this calculated acceleration through measurement.

5. **Retrodiction.** When statements are about facts or observations already known, they must be in accordance with those facts or observations, and thus *explain* them. Again, Newton's laws allow for ready explanation of the phenomena of motion (at nonrelativistic speeds), and Mendel's laws explain the outcome of breeding experiments. The ability to explain a wide range of already-known phenomena in a concise and economical way provides great intellectual satisfaction and is taken as an indicator of a successful theory. This is especially true if those phenomena have received no prior satisfactory explanation.

6. **Prediction.** As we have already discussed, when the statements are about things that have not yet been observed, they are *predictions*, and these predictions must ultimately be borne out by observation and experimentation, that is, by *tests*, as Popper has noted. Predictions need not be about events or processes that will take place in the future; they can be about events or processes that occurred in the past but have not yet been observed, either because no one looked for them or because the means to observe them were absent. Darwin made several predictions based on his theory, even though the events referred to took place long ago. Astronomers and geologists similarly must, for the most part, make predictions about past events. There are two types of prediction: negative and positive. Negative predictions are of the

form, "You will never observe X." Such predictions are important but difficult to verify because we can never be sure that at some future time, if we look in the right place, the forbidden event will not be observed. Only higher and higher degrees of probability can be obtained. Better are positive predictions, which are of the form, "If you look in A, you will observe X." Such predictions are the true "gold standard" for science, as they provide unequivocal proof that the theory can tell us something new. Positive predictions are also the most dangerous for a theory, as they expose it to possible refutation if the predicted observations are not borne out.

7. **Exploration.** The theory must suggest new experiments, new avenues of approach to problems, and new ways of regarding accepted facts (otherwise it is too sterile to be a useful tool). For example, Einstein's special theory of relativity suggested many experiments related to time dilation, energy transformation, and appearance of objects traveling at high speeds.

8. **Repeatability.** The theory must be "publicly observable," which means that observations needed to verify it must be such that they can be carried out by any observer with the proper equipment, and observers repeating a given experiment should be able to obtain the same results. Anyone with modest equipment can repeat Newton's experiments or those of Mendel, for instance.

9. **Clarity.** There must be some common agreement among practitioners about what the theory says and the facts that it explains. If there is no such agreement, the theory is too ambiguous and too subject to individual interpretation to be of any use.

10. **Intuitiveness.** A theory must possess a more elusive quality, that of being intuitively satisfying, that is, of making us believe that it has penetrated into the nature of reality and truly does give us insight into it. This characteristic can be quite contentious, as the battles over quantum mechanics in the twentieth century clearly show. In some cases, it takes a generation or two before the world becomes comfortable with a new explanatory paradigm.

At the end of each chapter, we shall compare the four schools with respect to the criteria discussed in this section. In the meantime, bearing in mind the ten criteria should help expose the many bad arguments that are employed in the evolution debate. These substandard arguments are used by all partisans in the debate and are unfortunately all too common in partisan tracts. A few of the most common types of substandard arguments employed in the evolution literature are listed below:[24]

- *Tautology*: A statement that is necessarily true, such as "All *A* are *A*."
- *Circular reasoning*: Using *X* to prove *Y*, then using *Y* to prove *X*.
- *Analogy*: Drawing conclusions about one situation from a similar one.
- *Just-so stories*: Fanciful stories about the origin of some entity or characteristic.
- *Incredulity*: "I don't see how it could have happened; therefore it didn't."
- *Psychological plausibility*: "It sounds reasonable to me, so it probably happened."
- *Extrapolation*: Generalizing a small-scale change to a large-scale change.
- *Claiming liabilities as assets*: Better known as "putting a spin" on inconvenient facts.
- *Argument from authority*: "So-and-so says it, so it must be true."
- *Retreat into unknowability and untestability*: Postulating unreconstructible events.
- *Explaining by naming*: Using a name for a phenomenon as an explanation of it.

Although use of all these substandard arguments is widespread, we will avoid discussion of them in this book and will instead focus on solid reasoning techniques and arguments. All sides have a case to make, and we will focus on each school's best arguments. The evaluation of these arguments represents the crux of the book, and it is to this that we turn in the next four chapters.

Summary

The major disputed points, together with the positions of the four schools on each, are summarized in table 4.1, which is an expanded version of table 1.2.

---

24. Further details and examples of these arguments can be found at Thomas Fowler's web site: www.evolutionprimer.net.

Table 4.1. Summary of the principal points in dispute in the evolution controversy*

Disputed points		Position of evolution schools			
Point	Possible alternatives	Neo-Darwinism	Meta-Darwinism	Intelligent Design	Creationism
1. Common descent	All levels	✓	✓	✓	
	Some levels			✓	
	Lower levels only				✓
2. Genetic information and random mutation	New information created	✓	✓	✓	
	Information loss only			✓	✓
3. Adequacy of random mutation/ natural selection to account for change**	All	✓			
	Some		✓	✓	
	Degenerative only			✓	✓
4a. Age of universe	Old	✓	✓	✓	✓***
	Young				✓
4b. Age of earth	Old	✓	✓	✓	
	Young				✓
5. Scope of naturalistic explanations	Possible for all phenomena	✓	✓		
	Some gaps exist; limited intervention required			✓	
	Inadequate; significant intervention required				✓
6. Employs bona fide scientific methods and theory	Neo-Darwinism does	✓	✓		
	Meta-Darwinism does	✓	✓		
	Intelligent Design does			✓	
	Creationism does				✓

*Check marks in two or more adjacent vertical cells for a given school and issue indicate that the members of the school have divergent opinions on the issue.

**Includes related genetic mechanisms such as crossover, and also effects of control genes such as Hox.

***This represents the most sophisticated of the Creationist positions; however, many Creationists still insist on a young universe and all require a young earth.

PART 2

Discussion of the Major Schools of Thought

5

THE NEO-DARWINIAN SCHOOL

Currently the Neo-Darwinian school is the dominant player in the evolution controversy. Its advocates include most university science staff; major scientific organizations such as the National Academy of Sciences (NAS) and the American Association for the Advancement of Science (AAAS); the world's major natural history museums; and virtually all scientific journals, such as *Evolution, Paleobiology, Science*, and *Nature*.[1] Clearly, the Neo-Darwinian school has tremendous scientific prestige behind it, and what it says about the subject must be given careful consideration. In fact, the very organizations now supporting the Neo-Darwinian theory are those that are considered the repository of our hard-earned scientific knowledge.

Despite its ubiquity, the Neo-Darwinian school is by no means monolithic; there are constant disagreements among its members over the details of how evolution proceeded, the rate at which it proceeded, the factors that influence evolution, the classification of fossils, and myriad other issues, as one would expect in a fertile area of science. In fact, there is some degree of overlap between Neo-Darwinism and what we have called Meta-Darwinism (see chap. 8). In this chapter we will consider the Neo-Darwinian arguments, making reference to the Meta-Darwinian arguments where appropriate.[2]

1. These journals also publish articles by members of what we have classified as the Meta-Darwinian school, and at times the lines between these schools blur.
2. Among those who consider themselves Neo-Darwinists, there is at least one major division, that between the ultra-Darwinists and what we may term the ordinary members

What Is Neo-Darwinian Evolution?

As stated in chapter 3, any theory of evolution or creation has to account for an enormous range of empirical phenomena. The Neo-Darwinian theory claims to do this in a very economical way—certainly an apt illustration of Ockham's Razor: *Entia non sunt multiplicanda praeter necessitatem*, "entities should not be multiplied unnecessarily." In fact, the theory has only two essential hypotheses:

1. *Common descent*, which means that all organisms arose from a single ancestor at some remote point in the past.
2. *Gradual improvement* through occasional beneficial changes to genetic material. Such changes, arising from random mutations, are incorporated into the population's genetic reservoir by natural selection. Through this process, those organisms produced in any generation that are better suited to their environment than others have a greater chance of survival and propagation. Over the course of many generations this gradual improvement leads to the emergence of novel traits and structures. Note that this whole process, which is commonly referred to merely as "natural selection," actually involves more than just natural selection. In addition to selection, it also requires the occurrence of chance mutations and new genetic combinations.

These two hypotheses are often combined into a single phrase, "common descent with modifications," and forms the basis for the explanatory paradigm that we refer to as the "Neo-Darwinian theory." Neo-Darwinian evolution is thus a scientific theory, based on two hypotheses, that seeks to explain the history and characteristics of all life-forms on earth, including the origin of all taxonomic groups. The logical structure of the theory is depicted in figure 5.1.

A depiction of what Neo-Darwinian theory implies by the term "common descent with modifications" is shown in figure 5.2. In this figure, we see an organism, A, at time t_0 that is capable of producing offspring. Mutations can occur and accumulate in these offspring such that they become substantially different from their ancestor. In addition, populations can become split into isolated subpopulations. For example, by t_1 enough changes have accumulated in the subpopulations to create two

---

of the school. The ultra-Darwinists advocate an extreme form of Darwinism in which genes are the dominant players, and everything revolves around them. Other members of the school believe evolution centers on populations or individuals. For the majority of our discussion, this distinction is not important. What is important are the essential components of Neo-Darwinian theory, treated in this chapter.

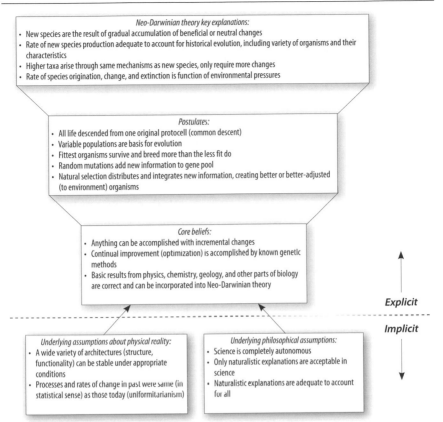

Figure 5.1. Logical structure of Neo-Darwinian theory.

new species (time has been collapsed in this figure). After many more generations (only a few of which are illustrated) still more changes have arisen, leading to two new genera at time t_2. By time t_3 these new genera contain a number of species, genera B having three species and genera C having two. The process may be envisioned as continuing indefinitely, gradually creating more variety and complexity. For example, if the progenitor is taken as the original cell from which all life descended, then eventually a point is reached when multi-celled organisms arise, which are able to occupy niches that their single-celled ancestors were unable to exploit, and the process continues and expands further.

Note that horizontal leaps are not postulated. Thus, a_1 in figure 5.2 cannot turn into a_2, and certainly not into b_1 or c_1. This is an extremely important point, because arguments against Darwinian evolution often start from the assumption that it sanctions such transitions—for example, a monkey turning into a man—and then go on to show that the odds against such a transition are astronomical, which indeed they are. No such transitions are envisioned or required, as they would entail many simultaneous or near-simultaneous changes, each of which has only a minute chance of occurring.

The gradual process that Neo-Darwinian theory relies upon to drive evolution implies that if one goes back far enough in the branching tree (fig. 5.2), one will find that any two species share a common ancestor. Thus, in figure 5.2, b_1, b_2, and b_3 share B; and c_1 and c_2 share C. Further back, all the a's, b's, and c's share a common ancestor, namely A. The common descent from a particular ancestor at various points explains the common features of organisms. As expected, those characteristics most widely shared (the most conserved) are the ones that must have arisen earliest, such as the genetic code and the mechanisms for decoding the information in the DNA in order to construct proteins. Subsequent evolutionary developments, such as the advent of genes that control development (of which the Hox gene clusters are perhaps the best known), are shared by many but not all organisms. Systems arising still later, such as the backbone of vertebrae, are naturally shared by fewer creatures.

Steven Rose has stated the core of Darwin's theory of natural selection very succinctly. First one can observe the following:

1. Like breeds like, with variations (random mutation).
2. Some of these varieties are more favorable than others.
3. All creatures produce more offspring than can survive to breed in their turn.[3]

In this first step, random mutations and novel genetic combinations occur via chance. Given this starting material, Darwin believed natural selection would drive evolution because of the following:

1. The more favored varieties are more likely to survive long enough to breed.
2. Hence there will be more of the favored variety in the next generation (natural selection).
3. Thus species will tend to evolve over time.

While nearly everyone agrees that such a mechanism can drive *some* evolutionary change, Darwin was convinced that it could account for virtually *all* evolutionary change. This does not follow directly, though, from the three statements above. To reach his conclusion, Darwin had to make three additional assumptions:

3. Steven Rose, *Lifelines: Biology beyond Determinism* (Oxford: Oxford University Press, 1998), 181.

140

1. There is a viable path of small variations connecting any two species (or other taxa).
2. Naturally occurring random processes can supply any new information needed for the variations necessary to traverse these paths.
3. It is possible to traverse the connecting path in a finite time.

For those who find these assumptions plausible, thc theory has great intuitive appeal. The notion that purely natural forces, primarily genetic

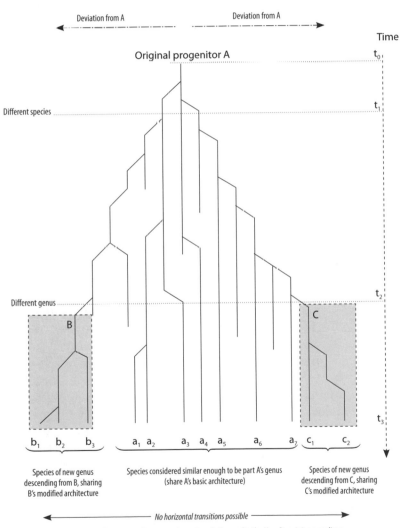

Figure 5.2. Illustration of common descent evolution under the Neo-Darwinian paradigm.

"accidents" and unplanned genetic modifications, could be harnessed to supply the driving force for the increasing diversity of evolution is startling and, for many, exhilarating. The ability to explain all of natural history with the two simple principles of common descent and gradual improvement through random mutations gives the theory unprecedented scope, and an inherent appeal due to its elegance and simplicity. The question, though, is whether common descent coupled with gradual improvement *can* explain all of evolutionary history. The answer to that question rests largely upon whether the assumptions listed above are correct. If they are not, Darwin's theory will turn out to be far more limited in scope than Darwin himself thought.

Establishing Neo-Darwinian Theory

In the introduction, we discussed the three tiers of evolution: historical, common descent, and strong Darwinian. If Neo-Darwinian theory is an adequate explanatory paradigm, all three tiers must be shown to be correct. This means establishing the truth of historical evolution as a fact, then showing that common descent explains key features of historical evolution, and finally showing that natural selection is the mechanism that can fully account for the emergence of new life-forms over time.

Historical Evolution

By utilizing the geologic column and dating methods from other sciences, Darwinists can assign dates to fossils and thus are able to date particular fossil organisms by their place in geological strata. In this way, a sequence of fossils can be established, showing development over time. As discussed in chapter 3, each geological stratum exhibits a representative group of fossils, with older (lower) layers having more primitive fossils than younger (higher) layers. In addition, these rock layers can be dated using radiometric dating techniques. These data have been used to establish historical evolution, the first and most basic tier. Since the arguments against historical evolution are presented in chapter 6, we will not deal with them here. In fact, since the basic physics of dating techniques is sound and the progression in the fossil record is clear, it is difficult to refute this level of evolution.

However, it is important to realize that, while the occurrence of historical evolution is necessary to establish Darwinian evolution, *it is not sufficient*. There could have been a historical progression of fossils over millions of years without common descent or natural selection.[4] Darwin's idea went beyond the mere historical progression of fossils to argue that all forms are related via common descent and that natural selection was the engine that drove the historical progression. Consequently, these two principles must be established to demonstrate the validity of Neo-Darwinian theory.

Common Descent

Common descent is the belief that all organisms descended from a common ancestor. If one goes far enough back in time, any two species, be they as unrelated as a dog and a flower, will share a common ancestor. While this is an essential component of Neo-Darwinian theory, it is not unique to it. For example, all Meta-Darwinists and some Intelligent Design advocates hold this position as well. Therefore, establishing universal common descent is not enough to establish the truth of Neo-Darwinian theory.

As pointed out in chapter 3, most organisms share an identical genetic code. A specific DNA triplet (codon), for example, codes for the same amino acid in a fruit fly as in a human. This fact makes it possible to put a human insulin gene in a bacterium and have the bacterium produce properly sequenced insulin protein. Given its ubiquity, it is presumed that the code evolved early on during life's history and then was inherited by all subsequent organisms. The universality of the genetic code is considered very strong evidence for common descent.[5] Hardwiring of the genetic code at an early point in evolutionary history made it virtually impossible for an organism to gradually evolve another code. Instead, organisms were left to use the original. Furthermore, studies have shown that the nearly universal code represents one of the best possible codes in terms of making organisms resistant to mutations.[6] It is likely that this advantage, which could have been selected for at an early stage of evolutionary history, may have kept the code from changing over time.

4. For example, a creator could have populated the earth at different times with different creatures.

5. Critics, however, have pointed out that the genetic code is not quite universal. Some organisms have slight variations in their genetic code, e.g., a stop codon such as UAA, which normally signals the termination of the protein, encodes for the amino acid glutamine. While these few exceptions demonstrate that the code is not universal, they tend to be minor variants that might represent slight tinkering with the code during evolutionary history.

6. S. J. Freeland, R. D. Knight, L. F. Landweber, and L. D. Hurst, "Early Fixation of an Optimal Genetic Code," *Molecular Biology and Evolution* 17 (2000): 511–18.

Creationist critics of common descent do not accept this position as the only possibility. They claim that the "universal" genetic code could also reflect a functional requirement for a robust code that is resistant to mutations. Given that all organisms need to be somewhat resistant to mutations, Creationists argue that a designer would have given all organisms the same robust genetic code. This is a rational but of course not a scientific argument. In addition, this counterargument rests upon the assumption that the genetic code is a near-optimal solution. Currently there is debate about that, although the evidence points toward the code being a near-optimal solution for minimizing the effect of random mutations.[7]

Beyond the genetic code, organisms possess other interesting similarities on the molecular level that suggest common descent. Organisms within the same genus or family have very similar genomes as compared to organisms in different orders, phyla, or kingdoms. For example, a recent whole genome comparison of humans and chimpanzees demonstrated that there is only a 1.23% sequence difference when one compares overlapping regions in these two organisms.[8] Bats have sequences more similar to that of other mammals than to that of birds. Similar examples can be seen between countless other species, and the argument is that the more similar the sequence, the closer the genealogical relationship. Two organisms that have diverged recently, such as humans and chimps, would have had less time for their sequences to diverge; therefore, they should have very similar sequences.[9]

Creationist critics once again fall back on a functionality requirement; similar sequences do not represent a close genealogical relationship but rather a common functional requirement. Since humans and chimps are

7. W. Zhu and S. Freeland, "The Standard Genetic Code Enhances Adaptive Evolution of Proteins," *Journal of Theoretical Biology* 239 (2006): 63–70.

8. There are also many cases in which a sequence in the human genome is not found in the chimp genome or vice versa. The deletions and insertions account for roughly 3% of the genome. This, coupled with the base pair changes, means that the two sequences are more than 95% identical. The Chimpanzee Sequencing and Analysis Consortium, "Initial Sequence of the Chimpanzee Genome and Comparison with the Human Genome," *Nature* 437 (2005): 69–87.

9. Some critics of Darwinian evolution have questioned the basic validity of DNA sequence comparisons by pointing out a few studies that seem to contradict the scientific consensus regarding common descent. (Gavin Naylor and Wesley Brown, "Amphioxus Mitochondrial DNA, Chordate Phylogeny and the Limits of Inference Based on Sequence Comparisons," *Systematic Biology* 47 [1998]: 61–76; Dan Graur, Laurent Duret, and Manolo Gouy, "Phylogenetic Position of the Order Lagomorpha [Rabbits, Hares and Allies]," *Nature* 379 [1996]: 333–35.) While this does occur on occasion, it is not as damning as it may appear. Since most sequence comparison studies use only a few genes rather than a complete genomic comparison, it is likely that they would occasionally produce lineages that are at odds with conventional evolutionary wisdom.

both social primates, they would be expected to have similar functional requirements for energy utilization, blood transport, rest, childbearing, and so on. As a result, one would expect that they would have similar sequences in much the same way as two trucks, a Ford and a Chevy, are built using similar blueprints.

In addition to DNA sequence similarities, morphological similarities have been used as evidence for common descent. Of particular importance are what we described in chapter 3 as homologous structures. Neo-Darwinists define these structures as those that were derived from the same or an equivalent feature in the nearest common ancestor.[10] The classic example is that of the forelimb of mammals, which consists of a single bone in the upper arm (humerus), two bones in the forearm (ulna and radius), the wrist bones (carpals), and then the bones of the hand. Forelimbs as functionally different as the wing of a bat, the leg of a horse, the fin of a dolphin, and the hand of a human all share the same basic bone structure (see fig. 2.3). Given that each animal has a different functional requirement (swimming, flying, walking, galloping), it is curious that they would all have the same bone structure. Neo-Darwinists argue that this bone structure first appeared in a common ancestor and was then modified to serve the necessary functions in these animals. Critics once again fall back on the functionality argument, stating that such a bone arrangement is optimal for the functioning of any type of mammalian limb.

While genetic and structural similarities are commonly cited as circumstantial evidence for common descent, one can see that this evidence can be interpreted differently by those opposed to common descent. Stronger evidence for common descent does exist, though, and it deals with similarities among organisms that cannot be explained by functionality requirements. There are three types worth discussing, the first of which is vestigial organs. As pointed out in chapter 3, vestigial organs are functionless structures that an organism possesses that serve an important function in other related species. At times it is difficult to determine if in fact a structure is truly vestigial; however, there are some particularly good examples of structures that bestow no functional advantage on their owner. For example, during development of the baleen whale, teeth appear and then are subsequently lost, while in certain snakes such as the rubber boa, degenerate hindlimbs form during development despite the fact that they serve no function.[11] There seems to be no reason for this, unless one could demonstrate, for example, that

10. Ernst Mayr, *What Evolution Is* (New York: Basic Books, 2001), 25–27.
11. Scott Freeman and Jon Herron, *Evolutionary Analysis*, 3rd ed. (Upper Saddle River, NJ: Pearson-Prentice Hall, 2004), 39–40.

there is a developmental necessity to form teeth in the toothless baleen whale, possibly to stabilize a certain stage of embryonic development. Currently though, the most parsimonious explanation, given the lack of functionality these structures exhibit, is that they were used in an organism's evolutionary past but are no longer needed and are therefore subsequently lost during development.

In addition to vestigial organs, there are vestigial traits at the molecular level called pseudogenes. These are "dead genes" that no longer encode for functional proteins or RNA products, and they can be found throughout the genome of humans and other organisms. Since these pseudogenes are no longer used, mutations accumulate in them at a much faster rate than in normal coding genes. As a result, one can infer the age of a pseudogene if one compares it to the sequence of a similar functioning gene. This has been done using a class of pseudogenes called processed pseudogenes, which are nonfunctioning genes that arise when an mRNA sequence is aberrantly inserted back into the genome by a process known as reverse transcription. Researchers have identified a number of processed pseudogenes in primates and have found that they comply with the expectations of common descent.[12] They found that the older the processed pseudogene, the more primate species that contain it. From the perspective of common descent, this makes sense because if the processed pseudogene originated early in the history of primates one would expect it to be widespread, while if it arose recently, one would expect it to be confined to only a few related species of primates. Again, this is exactly what investigators found.

The final piece of evidence comes from the recent explosion in comparative genomic studies. These studies have revealed that chromosomes consist of similar blocks of genes called *synteny blocks* that exist on different chromosomes in different organisms. These synteny blocks contain a specific set of unrelated genes that are found in the same order on the chromosome in various organisms. For example, a synteny block may contain a collagen gene followed by a sodium channel gene followed by a transcription factor gene. This block may be found on chromosome 2 in the human while it is found on chromosome 12 in the mouse. Though found on different chromosomes, the observed order of the genes in these synteny blocks is similar, despite the fact that there is no functional reason that this should occur. For example, there is no functional reason why a collagen gene and an epidermal growth factor gene should flank the Hox gene clusters in both mice and humans, but they do. In fact they flank all four of the Hox gene clus-

---

12. Felix Friedberg and Allen Rhoads, "Calculation and Verification of the Age of Retroprocessed Pseudogenes," *Molecular Phylogenetics and Evolution* 16 (2000): 127–30.

ters.[13] As it turns out, these synteny blocks are the rule rather than the exception in comparative genomic studies, providing strong evidence for common descent.

Note that all this evidence for common descent is circumstantial. In fact, it has to be, given the historical nature of evolutionary studies. Obviously, no one can observe humans and chimps descending from a common ancestor, but one can put together pieces of molecular and structural evidence left by this process to make a very good case for common descent. However, as we pointed out at the onset of this section, demonstrating common descent is not sufficient to establish the validity of Darwin's theory.

Neo-Darwinian Evolution

To establish the efficacy of Neo-Darwinian evolution, it is essential to provide support for the third tier of evolution, what we have called strong Darwinian evolution. This is the position that all of evolution can be explained by natural mechanisms. Specifically, it is absolutely essential for Neo-Darwinists to establish the capacity and efficacy of natural selection in conjunction with random mutation to innovate, to create new species and higher taxa. It must be the natural mechanism that drives common descent. If this cannot be demonstrated, the theory loses its explanatory power. It is at this level that most critics have attacked the theory because it is precisely here that the theory is most vulnerable. To understand why, let us look at three points one must demonstrate to validate Neo-Darwinian theory:

1. That random changes can yield improvements in organisms, as measured by overall organism fitness
2. That such improvements can become fixed in the population
3. That these small changes can accumulate to build any biological structure

The first two points are much easier to study than the third. In fact, one can directly verify these two hypotheses by means of either laboratory experiments or case studies from nature. For example, to investigate the first point, one could expose a laboratory population of flies to a mutagen that induces random changes. One could then see if the random changes that occur could in fact lead to some improvement in the organism. Alternatively, one could alter the environment in which a laboratory population is growing (exposing bacteria to antibiotics or

13. W. J. Bailey, "Phylogenetic Reconstruction of Vertebrate Hox Cluster Duplications," *Molecular Biology and Evolution* 14 (1997): 843–53.

exposing fruit flies to high temperatures) and determine if the organisms can improve their fitness relative to the new environment.

To demonstrate that these changes could become fixed within a population, one would have to demonstrate that the beneficial change (that which made the bacteria resistant to antibiotics) was able to spread throughout a specific population. This is best done using wild populations so as to avoid the influence the investigator can have on selecting laboratory populations. One could do this using case studies from nature in which a gene for body color or beak size, for instance, was followed through a number of generations to see if it increased in number to the point that it dominated the population. The field of population genetics, which mathematically models the ability of genes to spread through populations, would be helpful at this step to validate the various case studies.

Demonstrating the third point, however, is much more problematic. The reason behind this is clear. Demonstrating that the introduction of a predator to the habitat of a specific lizard population caused the lizards to evolve shorter tails does not lead to the conclusion that the entire lizard could have evolved via this same mechanism. Small changes may be possible, but it does not necessarily follow that complex structures can be built via natural selection. Such a conclusion would be a long-range extrapolation based upon the data at hand, and this is the crux of the problem. In other words, we cannot empirically demonstrate that eyes can evolve via natural selection in the same manner that we can demonstrate the evolution of antibiotic resistance. The former would occur on a scale that requires millions of years while the latter can occur in days or hours. This does not mean that Neo-Darwinian theory is incorrect, but it does mean that direct empirical verification is impossible.

How then does one establish point three? Absent empirical verification, one must hope to compile enough circumstantial evidence supporting the notion that small changes can accumulate to produce structures such as eyes and limbs. That is what Darwin attempted to do, and it is what most Neo-Darwinists continue to do today, taking advantage of DNA sequencing and fossil data that was unavailable to Darwin. Because this evidence is circumstantial, though, it is the subject of much debate both inside and outside Darwinian circles.

Neo-Darwinian Theory in Action

Evolutionary theory is often broken into two levels of analysis, microevolution and macroevolution. Microevolution focuses on changes in populations and species, while macroevolution deals with changes above the species level, such as the extinction or emergence of new phyla, the rate at which evolutionary change proceeds, and higher-level trends or patterns in lineages. While these distinctions are useful for different types of analysis, Neo-Darwinists argue that the processes that drive microevolutionary change (natural selection, genetic drift, migration, and mutation) are the same processes that drive macroevolutionary change. This point, however, is one of the most controversial topics in the debate and is one that goes to the heart of Neo-Darwinian theory. To investigate this topic, we will first examine the ability of Neo-Darwinian theory to explain microevolutionary change and its ability to model this change using population genetics. Then we will examine the ability of the theory to explain macroevolutionary change using these same processes.

Microevolution

The main reason that natural selection is so popular as a scientific explanation is the simple fact that it works, and nowhere is this more evident than at the level of microevolution. At this level, natural selection, along with other factors such as genetic drift and mutation, can explain how bacteria become resistant to antibiotics, how HIV is able to escape detection by the immune system, how the influenza virus changes its pathogenicity over time, and how the length of the tail in a lizard population can change when the population is faced with a new predator.

The cases of microevolution that we will investigate fall into two categories: (1) cases in which a species changes over time to adapt to a new environment, and (2) cases in which a species splits into two distinct species. Because the first type of change occurs relatively quickly, it can be directly observed in the laboratory or the wild, demonstrating irrefutably that natural selection does indeed occur. The second type requires more time but has been inferred from data collected in the wild.

One of the most studied examples of microevolution, both for theoretical and practical reasons, involves acquisition of antibiotic resistance by bacteria. Bacteria can develop resistance to antibiotics (discussed in chap. 3) through a wide variety of mechanisms. Some of these involve the loss of a gene; others involve the overexpression of a gene; and still others involve the modification of a gene. While it appears that some mutations conferring resistance to antibiotics existed before the introduction of antibiotics, natural selection has favored and tinkered with these chance mutations such that they have increased within bacterial populations over subsequent generations.[14]

14. Fernando Baquero and Jesus Blazquez, "Evolution of Antibiotic Resistance," *Trends in Ecology and Evolution* 12 (1997): 482–87.

One of the most impressive resistance mechanisms that has evolved is the adaptation of cellular genes to degrade specific antibiotics. In the case of the β-lactam antibiotics (penicillin, cephalosporins), certain resistant bacteria can cleave these antibiotics and render them ineffective by expressing an enzyme called a β-lactamase. The origin of the β-lactamase enzyme is interesting because it appears to be the result of an alteration of a cell wall biosynthesis enzyme, which in the presence of β-lactam antibiotics has evolved to take on a novel function.[15]

Another medically relevant example is the evolution of influenza type A strains, in particular the avian flu virus. The type A strains have caused global pandemics in the past, and there is concern that they will do so again, particularly as avian flu strains have begun to infect and kill humans in the past few years. The avian flu virus is constantly evolving as it moves from host to host and species to species, continually accumulating new mutations as it is subjected to different environmental pressures.Unfortunately, predicting exactly how this virus will evolve is difficult, given that the environmental pressures vary depending upon the defense mechanisms of the host organism, the density of the host population, and the migratory pattern of the host. Furthermore, avian flu virus can undergo more rapid change by exchanging genetic information with related viral strains.[16] Recently, strains have been found in which changes have occurred in the polymerase enzyme, which may make it better suited for replication in the upper respiratory tract, and in a hemoagglutination protein, which may make the virus better adapted at binding to human cells. Investigators are now trying to piece together how the forces of microevolution have caused these changes as well as which subsequent changes might transform the virus into something similar to the deadly 1918 pandemic strain.[17]

One of the advantages of studying genetic change at this level is that it can be modeled mathematically using theories and equations developed in the field of population genetics. This is claimed as one of the major successes of Neo-Darwinian theory and has allowed researchers to predict how organisms will adapt to various environmental perturbations.[18]

15. A. A. Medeiros, "Evolution and Dissemination of Beta-Lactamases Accelerated by Generations of Beta-Lactam Antibiotics," *Clinical Infectious Diseases* 24 (1997): S19–S45.

16. Centers for Disease Control and Prevention, "Avian Influenza," www.cdc.gov/flu/avian/gen-info/avian-influenza.htm.

17. Declan Butler, "Alarms Ring over Bird Flu Mutations," *Nature* 439 (2006): 248–49.

18. The ability of population genetics to model genetic changes in a population is not disputed by any of the schools. What is debated is the amount and type of change which can occur.

The ability to model genetic change is rooted in the Hardy-Weinberg equation, an equilibrium principle that mathematically demonstrates what would occur in a population that is randomly mating if there were no selection operating. Using the Hardy-Weinberg equilibrium, an investigator can study a population and determine if the allele frequency *is* changing from generation to generation, an indication that natural selection is possibly operating on the population. If selection is indeed operating, the Hardy-Weinberg equation can be used to determine how strong the selection happens to be. For example, if the frequency of the *aa* genotype increased in the population at the expense of the other two genotypes (*Aa* and *AA*), one might determine that the genotype *AA* is 20% less fit relative to *aa*. Knowing this, one can predict how selection will operate upon future generations and determine such things as how long the *A* allele will persist in the population.

While such mathematical models can determine whether a gene spreads within a population as well as the relative fitness of different genotypes, they do make some simplifying assumptions. The first assumption is that it treats genes as completely independent units and ignores possible nonlinear interactions between genes. These interactions between genes, or what is referred to as *epistasis*, are quite common and refer to the fact that the effects of any given gene on fitness are often dependent upon one or more other genes.[19] This can make it difficult to evaluate the effect a particular allele has on fitness. For example, the fitness benefit associated with the *aa* genotype in the above example may hold only in organisms that have the genotype *BB* at another locus. If an organism is *bb* at this second locus, it is possible that interactions between *b* and *a* would make the *aa* genotype less fit in this case. If this type of epistasis occurs between a number of genes, which it often does, then fitness becomes much more difficult, though not impossible, to model.

The second simplifying assumption in these models is that they ignore the difficulties that beneficial mutations have in initially establishing themselves within the population. Darwin himself argued that "natural selection is daily and hourly scrutinizing, throughout the world, the *slightest* variations, rejecting those that are bad, preserving and adding up all that are good."[20] What Darwin didn't recognize, which later population geneticists such as Sir Ronald Fisher did, was that most beneficial mutations would not be preserved because they would never be able to gain a foothold in the population. One may get the impression that any beneficial mutation, no matter how rare, would eventually make its way

19. Jason Moore, "A Global View of Epistasis," *Nature Genetics* 37 (2005): 13–14.
20. Charles Darwin, *The Origin of Species* (New York: Mentor Books, 1958), 91.

through the population. This is not the case. In fact, whether a beneficial mutation spreads through a population depends predominantly on three factors: first, whether it is dominant or recessive (dominant is better); second, the size of the population (the smaller the better); and third, the selective value of the mutation (the bigger the better).

The best case is when a mutation is dominant and its advantage is therefore expressed immediately. In this case, Fisher found that the probability of the beneficial mutation becoming established is approximately twice its selective value.[21] This means that if a dominant mutation appeared and it increased fitness by 10% (it had a selective value of 0.1), then there would be a 20% chance that it would spread through the population. Why is this so low? It is possible that the advantageous mutation would not make it into any offspring because each offspring had only a fifty-fifty chance of getting it. Furthermore, it is possible that although the new mutation raised the fitness of its owner, the owner may have had very poor fitness to begin with, such that even with the new mutation, the owner's fitness was still below average for the population. In the case of recessive beneficial mutations, the probabilities are even lower.[22]

Despite this, genetic changes can undoubtedly accumulate in a population, and Neo-Darwinian theory argues that if enough of these changes accumulate, speciation can occur. This might occur if the changes in gene frequency caused one population to become reproductively isolated from another population. Once this had occurred, the two populations could continue to diverge to the point that they are considered separate species. An excellent example of this occurring in nature is the adaptive radiation of fruit flies (*Drosophila*) on the Hawaiian Islands.

It is widely believed that *Drosophila* species initially colonized the older Hawaiian Islands such as Kauai and Niihau about 5 million years ago and then spread to the neighboring younger islands as these islands were formed.[23] There is evidence to support this notion, as DNA sequence

21. R. A. Fisher, *The Genetical Theory of Natural Selection*, 2nd rev. ed. (New York: Dover, 1958), 83–85.

22. It is much lower primarily because recessive mutations, when they first appear, are silent and have no effect on the phenotypes. As a result, recessive mutations must initially spread through the population such that individuals who are homozygous for the mutation arise. This initial step is random and is not favored by selection, which makes the probability of the recessive beneficial mutation becoming established much lower than twice its selective value. In fact, it is inversely related to the population size. Despite these caveats, the Hardy-Weinberg equation can be used effectively to model certain changes in gene frequency within a population and has been successful in doing so in countless cases, a fact that all four schools agree upon.

23. H. L. Carson, "Inversions in Hawaiian *Drosophila*," in *Drosophila Inversion Polymorphism*, ed. C. B. Krimbas and J. R. Powell (Ann Arbor, MI: CRC Press, 1992), 407–39.

comparisons of *Drosophila* species occupying a similar geographical niche on different islands have demonstrated this trend. In this study, older species are found to reside on older islands such as Molokai, while younger species are found to reside on the most recently formed island, Hawaii.[24] This data strongly supports the notion that the younger species originated when small founding populations from older islands colonized the younger islands. Since these founding populations were physically isolated from the parental population, they gradually diverged through random genetic drift and natural selection such that they became new species. As a result of this selection and drift occurring over millions of years, the Hawaiian drosophilids display an impressive amount of variability in lifestyle, courtship, wing size, body coloration, and food source. Note that all schools agree that this type of speciation can take place. They disagree only about whether it requires new information. Creationists believe that such speciation involves a loss of information.

The Hawaiian drosophilid radiation is not unique; many other examples of speciation can be found in nature.[25] These examples involve mammals, fish, plants, and crustaceans, to name just a few. Given the likelihood that these speciation events occurred via the accumulation of genetic differences, it would be useful to understand exactly the type and nature of the genetic changes that were involved in the speciation process. These changes can be inferred via a type of genetic analysis, called quantitative trait loci (OTL) analysis, a technique that can identify which genes contribute to the modification of a certain trait. Using this technique, much work has been done on *Drosophila* species to determine the amount and type of genetic change that exists between species. What these results have revealed is that there is no uniform pattern when it comes to species differences.[26] Some traits that vary between species do so because of changes in only one gene,[27] while others are the result of modifications at ten or more sites. Given that species vary in a number of traits, the genetic variation between species and the process by which these variations accumulate is difficult to model. Despite this problem, the DNA sequence similarities and the biogeography data strongly suggest that the *Drosophila* species in Hawaii evolved via natural selection to fit

24. R. DeSalle and V. Giddings, "Discordance of Nuclear and Mitochondrial DNA Phylogenies in Hawaiian Drosophila," *Proceedings of the National Academy of Sciences USA* 83 (1986): 6902–6.

25. See Freeman and Herron, *Evolutionary Analysis*, chap. 15, for many other interesting examples of speciation events.

26. H. A. Orr, "The Genetics of Species Differences," *Trends in Ecology and Evolution* 16 (2001): 343–50.

27. E. Sucena and D. L. Stern, "Divergence of Larval Morphology between Drosophila Sechellia and Its Sibling Species Caused by Cis-Regulatory Evolution of Ovo/Shaven-Baby," *Proceedings of the National Academy of Sciences USA* 97 (2000): 4530–34.

within different local environments. To summarize, microevolutionary changes can be observed on human timescales and inferred over longer time spans. Such change is accepted by all schools, though there remains some controversy about the degree of new genomic information these changes represent.

Macroevolution

The ability of natural selection and random mutation to effect small-scale change, such as the acquisition of antibiotic resistance or the adaptive radiation of *Drosophila* species on the Hawaiian Islands, is a point that few in the evolution debate dispute. The difficulty of identifying the nature of the genetic changes that separate species, however, indicates an important issue in the debate. Without a clear understanding of the nature of the genetic changes necessary to establish species differences or build novel traits, it is difficult to model exactly how (or whether) natural selection could produce such change.

This is not to say that natural selection cannot lead to speciation; the evidence suggests that it can.[28] However, if one has difficulty determining the pathway by which one *Drosophila* species evolved into another, the difficulties in determining the pathway by which the first eye or a specific metabolic process evolved becomes enormous. Without a clear understanding of how this could have occurred or what intermediates were important or critical in this process, it is difficult to demonstrate conclusively that natural selection had the ability to drive the evolution of the first eye.

What then is a biologist to do? Experimentally rerunning the evolution of the eye to observe how in fact it may have evolved is not possible. Without this direct proof, the evolutionary biologist is in the position of a detective trying to reconstruct what happened at a crime scene. Detectives do not have the luxury of going back in time to watch what happened when crimes were actually committed. All they can do is look at the evidence left at the crime scene as well as any other circumstantial evidence that can be acquired through careful investigation. If detectives believe they know who committed a crime, their hope is to assemble enough evidence to make the case beyond a reasonable doubt. This is analogous to what Neo-Darwinists must do with any macroevolutionary event: they must assemble enough circumstantial

28. For examples in which selection is driving speciation, refer to Freeman and Herron, *Evolutionary Analysis*, chap. 15. A number of examples are given, many of which refer to speciation being driven by food preference differences among members of a population. Recall that all four schools agree that speciation can and does occur. For Creationists, though, this always means a loss of information.

evidence to convince others that natural selection could have driven the transition in question.

What they should *not* do is posit that natural selection and random mutations can drive macroevolution by assuming macroevolution is nothing more than microevolution occurring over long time spans. Extrapolating from events that we can directly observe in our life span to events that occur over hundreds of millions of years is poor science. Despite this, arguments making this extrapolation are all too often put forward as proof that natural selection can explain macroevolutionary change. Take the following example:

> We now have direct observational evidence of evolutionary changes in successive generations of flies, corn plants, moths, hamsters and other animals and plants [i.e., microevolution]. Some of these changes have been made to happen in experiments; others have been witnessed in nature during the lifetime of one or more human observers. . . . The evolution hypothesis embraces, in addition to small evolutionary changes from one kind of moth or corn plant into another, also such far-reaching transformations as are involved in the differentiation of the carnivores, ungulates, primates, and bats from some common mammalian ancestor. *Evolutionists reason that if small changes can occur in a short time, large-scale changes can take place during the many millions of years of earth history.*[29]

This is certainly a possibility, but it is also possible that while natural selection can fuel some types of change, it is unable to bring about larger more complex transitions. To establish that natural selection is capable of macroevolutionary change, one must demonstrate that natural selection can build complex biological structures over time. In particular, one must demonstrate that there is a viable gradual path by which the macroevolutionary event could have occurred. Unfortunately, because of our lack of knowledge regarding how some transitions occurred, reconstruction of a viable path is not an easy undertaking, but one cannot therefore dismiss it:

> Given the state of our knowledge at any one time, we can reconstruct the evolutionary stages for some characters with some rigor . . . but not for others. For these cases, we can only guess at the possibilities because we cannot conduct a rigorous scientific investigation.[30]

In these cases, though, one needs to do more than just guess. What one needs to do is come up with a possible evolutionary scenario and then,

29. Verne Grant, *The Origin of Adaptations* (New York: Columbia University Press, 1963), 36; italics added.

30. Mark Ridley, *Evolution*, 2nd ed. (Oxford: Blackwell Science, 1996), 345–46.

like the detectives we described above, find evidence backing up the possibility. These hypothetical evolutionary scenarios are often pejoratively described as just-so stories, but they can be useful if they are used as a springboard into a serious investigation and are not offered as a substitute for it. If these stories are merely put forth as proof of the ability of natural selection, they are basically useless. With this in mind, we can now look at two examples of macroevolutionary events to see how effectively Neo-Darwinian theory deals with them.

Evolution of the Eye

Since it is one of the classic examples of macroevolution (one that was used by Darwin), the vertebrate eye is a useful adaptation to examine in light of the macroevolution discussion. The vertebrate eye is certainly a complex structure, one in which various components are arranged in a particular pattern that allows its owner to see a fairly accurate representation of the outside world. Darwin realized that the origin of such a structure posed a special problem to his theory, but he also realized the type of evidence that would have to be assembled in order to establish the possibility that natural selection could construct an organ "of extreme perfection."

> Reason tells me, that if numerous gradations from a simple and imperfect eye to one complex and perfect can be shown to exist, each grade being useful to its possessor, as is certainly the case; if further the eye ever varies and the variations be inherited, as is likewise the case; and if such variations should be useful to any animal under changing conditions of life, then the difficulty of believing that a perfect and complex eye could be formed by natural selection, though insuperable by our imagination, should not be considered as subversive of the theory.[31]

Darwin's account demonstrates that it is logically possible that natural selection could build the eye, but it does not establish whether it can or did do so during the course of natural history. Evidence must be proffered to demonstrate that there are simpler gradations of eyes that still afford their owner a selective advantage. If one could find a series of eyes ranging from complex to simple, one would have an initial piece of circumstantial evidence that natural selection could have gradually built the vertebrate eye.[32]

As it turns out, many types of eyes exist, and most are simpler in form than the vertebrate eye. Various mollusks, for example, have different

31. Charles Darwin, *The Origin of Species* (New York: Mentor, 1958), 172.
32. It is important to note that simply being able to arrange structures from simple to complex or from small to large does not imply an evolutionary relationship.

types of eyes, ranging from a simple pigmented spot to a complex lensed eye that is similar to the human eye (fig. 5.3). While it is apparent that various gradations of eyes do exist, it is important to realize that they do not represent an evolutionary history. Rather, they are all eyes that exist in living organisms. What they indicate is that simpler versions do exist and they are advantageous to their owners. The key question then for Neo-Darwinians is whether there are viable paths that could connect each of these intermediates and make them at each stage beneficial to their owners.

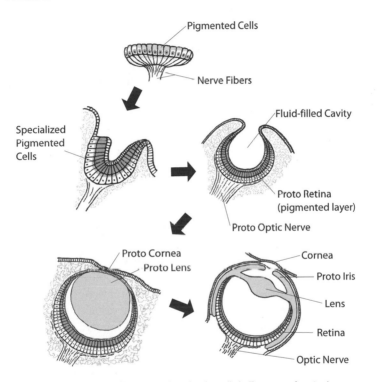

Figure 5.3. Drawings of different types of eyes found in mollusks. The eyes vary from simple light-sensitive spots to compound eyes similar to the human eye.

Absent the ability to re-create these forms, assessing this possibility can be difficult. At this stage, one must examine the structure of the eye to understand exactly how it works. Armed with this knowledge, a researcher may be able to demonstrate whether intermediate forms would be functional. As it turns out, given what we know about eye structure and development, this might indeed be possible.

To see this possibility, we will start with the relatively uncontroversial assumption that any change that would improve the spatial resolution of vision would be favored. Since mutations that would either cause the eye to invaginate or restrict the size of the aperture would increase spatial resolution, one would expect these changes to be favored (although this does not guarantee that they will spread through the population, as was explained in the above section on microevolution). Looking at figure 5.3, we see exactly this type of progression from simple to complex eyes, with invagination and a decrease in the size of the aperture.[33] A second reasonable assumption needs to be made as well, namely, that small changes that increase the refraction of light at the opening of the eye (this is what the lens does in the human eye) would be favored as they would also lead to increased resolution. Since many proteins could refract light if they were expressed in the anterior surface of the eye,[34] the origin of the lens could be explained by a mutation that caused any one of these particular proteins to be produced in high amounts by the cells in the anterior portion of the eye.

As a result, the evolutionary scenario for the eye is backed up by the existence of eyes of varying complexity and what we know about how spatial resolution can be increased in the eye. Taken together, this represents plausible circumstantial evidence that there is a viable path from point A, a simple eye, to point B, a complex eye. Since the mutations that are needed involve small changes in the folding of tissues and the misexpression of just one gene, it seems reasonable to assume that random mutations could supply the raw material needed for this transition.

At this point, one still has to determine if the necessary mutations can occur and accumulate in the available time. A viable path may exist between each of these eye types, but it may take too long to traverse the path. To address this question, researchers developed a computer model to estimate the time needed for an eye to evolve from a flat patch to a complex human eye. Their estimate, which they called conservative, was that it would take less than 364,000 years for this to occur.[35] While such a number represents a very short time span from an evolutionary

33. An increase in resolution, by itself, would be of no value unless the light-sensitive portion of eye, the retina, was able to take advantage of the increased resolution, the optic nerve to the brain could handle the additional traffic, and the brain could process this new information. This entails more light-sensitive receptors in the retina as well as more nerves to the brain and within the brain. Many of these changes could theoretically result from the environmental influences on development that would occur in organisms with increased resolution capacity. It is well known that visual input during development has a large effect on the wiring of the visual neurons in the brain, so the increased resolution capacity might drive developmental changes elsewhere in the visual pathway.

34. J. R. Ture and S. B. Carroll, "Gene Co-option in Physiological and Morphological Evolution," *Annual Review of Cell and Developmental Biology* 18 (2002): 53–80.

35. D. Nilsson and S. Pelger, "A Pessimistic Estimate of the Time Required for an Eye to Evolve," *Proceedings of the Society of London B* 256 (1994): 53–58.

standpoint, one must be careful when extrapolating from computer-generated data to the real world of organisms. Computer analysis of evolution can be misleading because many assumptions must be made (often these are buried deep within the computer code), many of which may not approximate what happens in the real world.

For example, this particular simulation assumed that in every generation there were individuals who had eyes that were significantly better at focusing light than the average. This assumption is true for the first few generations but does not take into account what happens in subsequent generations. To understand this issue, imagine you had a population of 100 individuals, all of whom were of varying heights. Suppose the average height was five feet, but a few individuals were close to six feet. If selection favored taller individuals, over a few generations the average height of the population would become closer to six feet. At this point evolution might stall for a while because no individuals in the population are taller than six feet. Evolution would have to wait for a mutation to occur that would increase size further. The eye simulation assumed that these mutations are always present. This is a gross oversimplification since (1) the occurrence of these new mutations is random, (2) such favorable mutations occur with a low probability, and (3) even when they occur there is no guarantee that they will spread through the population.

This objection may not be fatal, since no one knows how incorporating these conditions into the simulation would affect the outcome. However, it shows that caution is necessary when dealing with computer simulations of evolution. The simulations are only as good as the assumptions they make, and it can often be difficult to ascertain if the assumptions made by the programmers actually reflect the relevant biology.

There is another issue regarding the evolution of the eye that the simulation overlooks. The simulation starts with a relatively complex system, a flat eye spot that consists of photoreceptor cells surrounded by dark pigment and protected by a translucent skin covering. How did this structure originate? While this is a difficult question to answer, most biologists would agree that the first step is the origin of a molecule that responds to light, a photoreceptor molecule. Such molecules are found in bacteria, plants, and animals and have a wide variety of functions and uses.[36] In each case, though, the photoreceptor is connected to other signaling molecules that allow it to transmit information to the cell or organism about the amount, intensity, direction, and so on of the light. It is the construction of these systems, particularly integrated

36. A. Falciatore and C. Bowler, "The Evolution and Function of Blue and Red Light Photoreceptors," *Current Topics in Developmental Biology* 68 (2005): 317–50.

photoreceptor cells, that represents the largest hurdle in the evolution of the eye, and that is ignored in this simulation.

How then does natural selection produce an integrated photoreceptor cell? The first question to ask is what possibilities can be constructed based upon our knowledge of primitive light-sensing organisms? It has been speculated that photoreceptor cells evolved from the gradual adaptation of a photoreceptor organelle.[37] As cells within multicellular organisms became specialized, the photoreceptor organelles are thought to have become more prominent in certain cells to the point that the cells became dedicated photoreceptors. While this idea is plausible, it begs the question regarding the origin of the photoreceptor cell. How did natural selection assemble the photoreceptor organelle, which in many respects is as complex as a photoreceptor cell? In fact, the photoreceptor organelle in single-celled organisms contains "a cornea-like surface layer, a lens-like structure, a retina-like structure with stacked membranes (or microvilli), and a pigment cup."[38] In addition, these eye organelles are linked via signaling molecules to the flagellum, such that light levels can affect the movement of the organism. This structure is arguably more complex than a photoreceptor cell and little is known of its origin.

Because of this knowledge gap, many of the details of how the eye evolved, particularly at the earliest stages, are overlooked. Statements such as the following are common: "We propose the following scenario for the evolution of animal [photoreceptor cells] and eyes. Early metazoans possessed a single type of precursor [photoreceptor cell] that used an ancestral opsin for light detection and was involved in photoperiodicity control and possibly in phototaxis."[39] In this case, the photoreceptor cell is used as a starting point, while in other scenarios the photoreceptor organelle is used as the starting point. What seems to be the key point here is that, given a photoreceptor organelle or cell, there is evidence that an eye *could* have evolved via natural selection. However, data supporting the notion that natural selection could build the original photoreceptor is, at present, lacking.

While the ability of natural selection to build structures such as eyes has been debated since Darwin wrote *The Origin of Species*, recent critics of natural selection have questioned the efficacy of the theory's mechanisms at the biochemical level (see chap. 7 for more details). It is to this level that we turn for our second example of macroevolution.

37. W. J. Gehring, "New Perspectives on Eye Development and the Evolution of Eyes and Photoreceptors," *Journal of Heredity* 96 (2005): 171–84.

38. Ibid., 181.

39. D. Arendt, K. Tessmar-Raible, H. Snyman, A. Dorresteijn, and J. Wittbrodt, "Ciliary Photoreceptors with a Vertebrate-Type Opsin in an Invertebrate Brain," *Science* 306 (2004): 869–71.

Evolution of the TCA Cycle

The TCA cycle, or Krebs cycle, is a series of reactions cells use to extract energy from food molecules. Eight enzymes are necessary to complete the cycle which occurs in the mitochondria within our cells (see fig. 5.4). The TCA cycle plays a major role in metabolism in a wide array of organisms and is certainly an important macroevolutionary event that Neo-Darwinian theory must explain. It must do so by positing a gradual route by which the cycle could have been assembled because there is no realistic possibility that all eight enzymes appeared simultaneously to create the TCA cycle *de novo*. Further complicating matters, removal of any of the enzymes causes the TCA cycle to stop functioning.

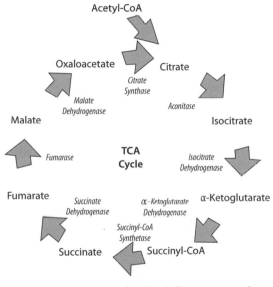

Figure 5.4. Enzymes and substrates of the TCA cycle. The enzymes are in italic. Most food molecules enter the cycle at the stage where oxaloacetate is converted to citrate via the addition of an acetyl group.

What this means is that a simpler TCA cycle, for example one using two or three enzymes, could not have existed because it would not have been functional. How then does natural selection build such a structure that lacks any obvious intermediates? In the case of the vertebrate eye, there were simpler versions that could perform the same basic function. Theoretically, these simpler versions could be built upon by natural selection to eventually achieve the complexity of the vertebrate eye. In the case of the TCA cycle, there are no simpler TCA cycles upon which to build.

The Neo-Darwinian answer to this dilemma is to argue that the TCA cycle was built gradually through the modification and adaptation of *other* existing pathways. This means that some of the TCA cycle enzymes initially were useful for some purpose apart from the TCA cycle. Once all the TCA enzymes accumulated in the cell, again because they were beneficial for some other purpose, they were modified to form the complete functioning TCA cycle. In this case, the simpler steps that natural selection built upon were not simpler TCA cycles but other simple systems that were co-opted for a new function. Along these lines, evolution is seen as a grand tinkerer, taking advantage of whatever material happens to be present in the cell by cobbling it together to make new pathways and molecular machines.

This scenario is certainly possible, but is it plausible? To establish the plausibility of the Neo-Darwinian explanation, researchers must be able to demonstrate three things:

1. The enzymes in the TCA cycle perform reactions that provide some benefit to the cell *in the absence of a functioning TCA cycle*.
2. The advent of the TCA cycle does not adversely affect the previous reactions that were useful to the cell.
3. The modifications that were needed could have occurred in a reasonable length of time.

It does appear that the first condition is true. All the enzymes in the TCA cycle except one have a role in other processes, making them beneficial to the cell for reasons that have nothing to do with the TCA cycle.[40] The first three enzymes in the cycle—citrate synthase, aconitase, and isocitrate dehydrogenase—are involved in the synthesis of the amino acid glutamate. Given that primitive cells would have had a need for this amino acid, it is likely that this pathway and the three enzymes listed above would have been present in the cell prior to the advent of the TCA cycle. The next enzyme in the pathway, α-ketoglutarate dehydrogenase, was likely necessary for the synthesis of succinyl-CoA. This molecule is a precursor needed for heme synthesis, which is a cofactor used in certain cellular reactions. Evolutionarily, these four enzymes, the three needed for glutamate synthesis and the one needed for succinyl-CoA and heme synthesis, are referred to as the oxidative branch of the TCA cycle.

To complete the TCA cycle, another branch, the reductive branch, was needed. This branch is thought to have been used initially by the cell

40. E. Melendez-Hevia, T. Waddell, and M. Cascante, "The Puzzle of the Krebs Citric Acid Cycle: Assembling the Pieces of Chemically Feasible Reactions, and Opportunism in the Design of Metabolic Pathways during Evolution," *Journal of Molecular Evolution* 43 (1996): 293–303.

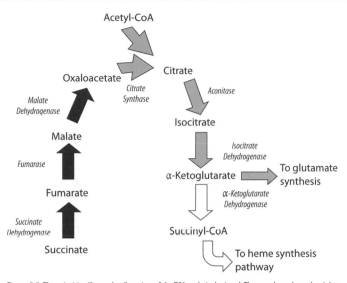

Figure 5.5. The primitive "horseshoe" version of the TCA cycle is depicted. The grey branch on the right consists of three enzymes involved in glutamate synthesis as well as one enzyme involved in heme synthesis (white branch). The black branch on the left side consists of three enzymes needed to maintain the reduction/oxidation balance in the cell. The reduction/oxidation balance is critical to the cell as many reactions are inhibited if the balance is disrupted.

to balance the NADH that accumulated in the oxidative branch.[41] The reductive branch involves three enzymes—succinate dehydrogenase, fumarase, and malate dehydrogenase—all of which initially worked in the opposite direction than that of the TCA cycle in order to maintain reduction/oxidation balance in the cell by removing the NADH produced by the oxidative branch. Such a primitive "horseshoe" version of the TCA cycle is found in several anaerobic bacteria and in some invertebrates (see fig. 5.5).[42] In these organisms, the NADH that accumulates in the oxidative branch during glutamate and succinyl-CoA synthesis is converted back into NAD+ by the reductive branch.[43]

41. A. Baughn and M. Malamy, "A Mitochondrial-like Aconitase in the Bacterium *Bacteroides fragilis*: Implications for the Evolution of the Mitochondrial Krebs Cycle," *Proceedings of the National Academy of Sciences USA* 99 (2002): 4662–67.

42. Melendez-Hevia, Waddell, and Cascante, "Puzzle of the Krebs Citric Acid Cycle," 293–303.

43. NADH and its counterpart NAD+ are molecules that are essential for specific reactions in the cell that involve reduction or oxidation. For example, NAD+ accepts electrons (it is reduced) during the breakdown of nutrients in the TCA cycle. Upon accepting the electrons, NAD+ is converted to NADH. The NADH is subsequently converted back into NAD+ (it is oxidized) by additional reactions which vary depending upon the cell type. Three steps within the TCA cycle require the presence of NAD+ to operate effectively.

Once these two branches were present, they needed only to be linked together to complete the TCA cycle. For this to occur, only one enzyme, succinyl-CoA synthetase, was needed. The emergence of this enzyme, as well as the reversal of the reductive pathway—hardly an insurmountable problem—would have led to a functioning TCA cycle. Here then is a possible route in which the TCA cycle could have gradually emerged from components that were already in the cell for other purposes. Furthermore, the emergence of the TCA cycle for cell metabolism would not have disrupted the old pathways such as glutamate synthesis. This is evident as our cells have the ability to either make glutamate or break down food molecules using the same TCA cycle enzymes. At this point, two of the three conditions have been satisfied, giving credence to the Neo-Darwinian explanation. The third condition is more difficult to ascertain, since the early history of the enzymes involved in the TCA cycle has yet to be reconstructed.

If the Neo-Darwinian scenario is correct, then the emergence of the TCA cycle occurred because evolution was able to tinker with other pathways that were already present in the cell: a pathway for the synthesis of glutamate (which in addition to the three TCA cycle enzymes mentioned above requires a fourth, called glutamate synthesis); the three enzymes of the reductive branch to balance this pathway; and the enzymes involved in heme synthesis that work with the TCA cycle enzyme α-ketoglutarate dehydrogenase to produce heme. (The heme biosynthesis pathway is quite complex and involves seven enzymes that are needed to convert succinyl-CoA into heme.[44]) The point in discussing these pathways is to demonstrate that explaining the emergence of one pathway (the TCA cycle) in terms of the co-option of other complex pathways often poses as many questions for Neo-Darwinian theory as it answers. For example, how might the heme synthesis pathway in all its complexity have evolved? How did the four enzymes necessary to make glutamate from the end product of glycolysis evolve? Did this glutamate pathway require the three enzymes in the reductive branch of the TCA cycle for redox balance? Are there simpler versions of all three of these pathways by which they could in turn be built?

Answering these questions in a definitive manner can be difficult, but the best place to start is by identifying and comparing organisms that have incomplete TCA cycles. This type of whole genome analysis has been done on some single-celled organisms, and it has been found that they possess a variety of incomplete TCA cycles.

44. Michael King and Sergio Marchesini, "Heme and Porphyrin Synthesis," University of Brescia web site, www.med.unibs.it/~marchesi/heme.html.

One bacterium has a "cycle" that only lacks the α-ketoglutarate dehydrogenase enzyme, while two archaea have half cycles that lack most of the enzymes in the oxidative branch.[45] While the presence of a variety of simpler systems in these organisms supports the notion that the TCA cycle could have evolved in stages from other distinct pathways, constructing a feasible pathway for building the TCA cycle from the ground up is still beyond our ability. Future work in comparing metabolism genes in primitive single-celled organisms may yield more clues regarding the feasibility of constructing this and other biochemical pathways.

In both cases described above, the vertebrate eye and the TCA cycle, there are some intriguing possibilities as to how natural selection may have been able to build such systems. Without denying the real potential of these possibilities, one must admit that we do not at present know whether or how natural selection actually could have built the systems in question. A recent review of eye evolution admits that while the evolution of the eye seems to have occurred "piecemeal, by myriad small steps, each an adaptive improvement over what went before, a detailed accounting of the steps is as yet beyond us."[46] In a similar manner, a recent article dealing with the evolution of the TCA cycle realizes that its origin remains an open question.[47]

The ability of the mechanisms proposed by Neo-Darwinian theory to explain how organisms change over time at both the micro and macro levels is by far the most important issue in the debate regarding the efficacy of Neo-Darwinian theory. Two possibilities exist: (1) natural selection can explain small-scale change, and given enough time, it can explain large-scale change as well, or (2) natural selection can adequately explain small-scale change, but it is deficient in explaining large-scale transitions. The discussion so far in this chapter should shed light on these two possibilities.

Those who argue that the second possibility is correct have come to a variety of different conclusions. Many have sought other natural causes to augment the ability of natural selection (these are discussed in chap. 8). Others have argued that Intelligent Design is the only explanation for some of these complex structures and pathways (see chap. 7). Still others have turned to Creationism (see chap. 6) for the explanation.

45. M. A. Huynen, T. Dandekar, and P. Bork, "Variation and Evolution of the Citric-Acid Cycle: A Genomic Perspective," *Trends in Microbiology* 7 (1999): 281–91.

46. T. Lacalli, "Light on Ancient Photoreceptors," *Nature* 432 (2005): 454–55.

47. Huynen, Dandekar, and Bork, "Variation and Evolution of the Citric-Acid Cycle," 281–91.

Other Important Issues in the Debate

While the efficacy of natural selection remains at the crux of the debate regarding Neo-Darwinian theory, other issues continue to surface. Some of these issues, such as the nature of the fossil record, are of central importance, while others are rather trivial. What we have attempted to do with the remainder of the chapter is to identify some major points of dispute that are raised against Neo-Darwinism and attempt to examine them in an evenhanded fashion. We do not claim to present an exhaustive list; rather, we focus on what we deem to be the most important issues, those that are of central importance to the theory of natural selection. At the end of this section we present a table summarizing the major arguments for and against the Neo-Darwinian theory, together with an evaluation of their impact.

The Fossil Record

The gradualistic paradigm for species change, or *phyletic gradualism*, which is an essential feature of the Neo-Darwinian theory, should fill the fossil record with not just a few, but innumerable transitional forms. According to this view, constant transition or constant change should be the main feature of the fossil record, since abrupt transitions or "saltations" are excluded from the theory because of their great improbability. A typical dictionary definition of "transition" is "movement, passage, or change from one position, state, stage, subject, concept, etc., to another." In the context of evolution, this means evolving from one species (or higher taxon) to another; it does *not* mean simply sharing characteristics of two different taxa, such as *Archaeopteryx*, which shares characteristics of birds and dinosaurs, as it has feathers and an opposable big toe (birds) as well as teeth and a rear neck attachment to the skull (dinosaurs). Taken alone, such sharing does not make *Archaeopteryx* transitional between dinosaurs and birds; at the very least it must also be found along a temporal lineage that links dinosaurs to birds. The fact that a fossil such as *Archaeopteryx* exists does indeed support the notion that birds and dinosaurs are related evolutionarily, but it must be found in the proper historical context to be deemed a transitional fossil.

The presence or absence of transitional fossils is probably the most debated aspect of Neo-Darwinian theory. The origins of the issue date back to Darwin, who realized the importance of it for his theory.

The number of intermediate varieties, which have formerly existed on the earth, [must] be truly enormous. Why then is not every geological formation and every stratum full of such intermediate links? Geology assuredly does not reveal any such finely-graduated organic chain; and this,

166

perhaps, is the most obvious and gravest objection which can be urged against my theory.[48]

As Darwin implies, the problem is not about any particular fossil, such as *Archaeopteryx*, and whether it is indeed "transitional"; rather, it is about the global question of the fossil record. As recently as 2001, Darwinist Ernst Mayr stated that lineages documenting a continual record of transition are rare and the fossil record is largely a record of discontinuities.[49] In fact, similar remarks have been made by other major figures.[50] Realizing the need for transition fossils within the Darwinian paradigm, Mayr argues, as Darwin did, that the nature of the fossil record will always be incomplete. Given the special conditions needed for fossilization and the difficulty of extracting the fossils that do form, such discontinuities may be expected. However, it is a situation that remains unsettling, given the hole that it opens up for critics of the theory.

To make matters more confusing, not all Darwinists agree that the fossil record is spotty. For example, the National Academy of Sciences has weighed in with a blanket assertion that there is no problem and that myriad transitional forms exist:

> So many intermediate forms have been discovered between fish and amphibians, between amphibians and reptiles, between reptiles and mammals, and along the primate lines of descent that it often is difficult to identify categorically when the transition occurs from one to another particular species. Actually, nearly all fossils can be regarded as intermediates in some sense; they are life-forms that come between the forms that preceded them and those that followed.[51]

What, then, are we to make of the fossil record, since even Darwinists seem to be unable to agree upon its status? Are there too many transitional forms or are there not enough? Perhaps the way to get a handle on this issue is to examine a few representative transitional lineages and see what the fossil evidence actually reveals.

The first transition of interest is the fish to land-dwelling tetrapod transition, which is believed to have occurred during the upper Devonian period, roughly 365 million years ago. The most likely candidate for the

48. Darwin, *The Origin of Species*, 293–94.
49. Mayr, *What Evolution Is*, 14.
50. Niles Eldredge, *Reinventing Darwin* (New York: Wiley, 1995), 95; R. A. Raff and T. C. Kaufman, *Embryos, Genes, and Evolution: The Developmental-Genetic Basis of Evolutionary Change* (Bloomington: Indiana University Press, 1991), 35.
51. National Academy of Sciences, *Science and Creationism*, 2nd ed. (Washington, DC: National Academy Press, 1999), 14.

ancestor of tetrapods is a suborder of lobe-finned fish called rhipidistians. These fish are found in the fossil record from 380 million to 375 million years ago. They share a number of skeletal features with the early fossil tetrapods, including uniquely folded teeth and a connection between the nasal and oral cavity to aid in breathing air. In addition, the fleshy lobe fins of the rhipidistians have a bone structure similar to the limbs of terrestrial vertebrates.[52]

The first hint of how these fish may have left the water appears in the fossil record 10 million years after rhipidistians disappear from the record. Two species of importance have been identified, *Acanthostega* and *Ichthyostega*, both of which display the early development of tetrapod limbs while retaining some fishlike features such as gills and tail fins (see fig. 5.6). These species, which date to 365 million years ago, are neither fish nor true tetrapods. While tetrapods have limbs with five digits, these fossils have extra digits, and unlike true tetrapods, the skeletal structure indicates that these intermediate forms were unable to support their body weight out of water.[53] The limbs most likely functioned more as paddles and were useful in scurrying through shallow creek beds that these species may have populated. While these species are not the only intermediate fossil forms from this period, they are the most complete.

Figure 5.6. A skeletal drawing of the Devonian tetrapod *Ichthyostega*. The fossil has six digits on its hind limb while the digits on the forelimb are missing.

After *Acanthostega* and *Ichthyostega* and their contemporaries, the next group of fossils in the tetrapod lineage forms do not appear until the Carboniferous period about 335 million years ago. These fossil forms are true tetrapods: they exhibit five digits on their limbs and were able to support their body weight out of the water. But the gap between *Acanthostega* and the true tetrapods spans about 30 million years and is so well known it has been named. It is called "Romer's gap" after the

52. Freeman and Herron, *Evolutionary Analysis*, 710.
53. Jennifer Clack, *The Emergence of Early Tetrapods* (Bloomington: Indiana University Press, 2002).

paleontologist Alfred Sherwood Romer.[54] Recently, a novel tetrapod fossil, *Pederpes finneyae*, has been found to help close this gap. This fossil dates to about 350 million years ago, smack in the middle of Romer's gap. This fossil appears to have five digits on its hind limb (its forelimbs are not complete) and is considered the earliest-known tetrapod to show the beginnings of terrestrial locomotion.[55] While this fossil helps close Romer's gap, it is a single fossil and there is still a 15-million-year gap between it and *Acanthostega* and friends.

This situation gives a good example of the status of the fossil record. At the times of major transitions, some intermediate forms do exist, although it is unclear which of these forms is on the transitional lineage between the ancestral and modern forms. Furthermore, there are often gaps between these intermediate forms, such as the gap of 15 million years between *Acanthostega* and *Pederpes finneyae*. This does not necessarily mean that fossils will not be found to close these gaps—*Pederpes finneyae* was discovered only a few years ago, for example—but at present gaps do exist, although they usually are not as severe as Creationist critics claim. Reconstructing exactly how and if Darwinian mechanisms could have driven the transition is made difficult by these gaps, as Jennifer Clack, the discoverer of *Pederpes finneyae*, points out: "Lacking a perfect fossil record or recourse to a time machine, we may never piece together the entire puzzle of tetrapod evolution."[56] As a result, there is usually debate within the scientific community as to how exactly any specific transition took place, and the fish-to-tetrapod transition is no exception. While this debate is seen as a sign of weakness by critics, it is also indicative of a fruitful area of research.

The fish-to-tetrapod example is particularly useful because it demonstrates three aspects that are characteristic of the fossil record in general. First, intermediate forms are often found in the proper temporal sequence. Second, gaps still remain in the fossil record. Third, recent fossil finds have been able to partially fill these gaps. Other transitions may fare better or worse than this one, but these three characteristics tend to define the fossil record.

In some cases, intermediate forms exist but they have not been found in the proper temporal sequence needed to be considered true transitional forms. The transition between dinosaurs and birds is an example. *Archaeopteryx*, the fossil with characteristics of both dinosaurs and birds,

54. Jennifer Clack, "An Early Tetrapod from Romer's Gap," *Nature* 418 (2002): 72–76.

55. Ibid.

56. Jennifer Clack, "Getting a Leg Up on Land," *Scientific American* (December 2005): 107.

has been found in late Jurassic rocks dating to 150 million years ago. Much has been made of the impressive fossil finds that have uncovered dinosaurs with feathers, such as *Microraptor gui* and *Sinornithosaurus millenii*.[57] While these feathered dinosaurs represent a possible transitional form leading up to *Archaeopteryx* and the first birds, they are found in the fossil record in early Cretaceous rocks dating to roughly 130 million years ago. This is 20 million years *after Archaeopteryx*, making it difficult to reconstruct a *temporal* lineage between dinosaurs and birds.[58] Such a sequence does not prove that the transition did not occur (as some Creationists argue), but since Neo-Darwinian theory predicts that the transition must have begun prior to 150 million years ago (before *Archaeopteryx*), one would expect to eventually find feathered dinosaurs in the Jurassic period before *Archaeopteryx*. Only time will tell if this prediction is borne out.

There is one more transition that we need to address because of its importance in the evolution debate. This transition involves the emergence of complex animal phyla from relatively simple animal phyla at the beginning of the Cambrian period, roughly 550 million years ago. The "Cambrian explosion," as the event has come to be known, is probably the most hotly debated aspect of the fossil record. Creationists (and others) have argued that the rapid appearance in the fossil record of complex animal forms such as chordates, arthropods, and mollusks is too much for Neo-Darwinian theory to handle, while Neo-Darwinists have been busy looking for transition fossils to explain this seminal event in evolutionary history via Darwinian mechanisms.

Prior to the Cambrian, the fossil record for animals is relatively sparse and consists largely of members of three simple animal phyla: Porifera (sponges), Cnidaria (jellyfish), and Ctenophora (comb-jellies). In contrast, the Cambrian rocks are filled with complex bilateral animal fossils including chordates, nematodes, arthropods, mollusks, and any number of strange phyla that have since gone extinct. Estimates vary, but a ballpark figure is that approximately thirty animal phyla made their appearance in the fossil record in the Cambrian period.[59]

How quick was this transition in the fossil record? The earliest Cambrian rocks date to about 530 million years ago, while the latest Precambrian rocks date to about 540 million years ago, leaving roughly

57. Xing Xu et al., "Four-Winged Dinosaurs from China," *Nature* 421 (2003): 335–40; Xing Xu, Z. Tang, and X. Wang, "A Therizinosauroid Dinosaur with Integumentary Structures from China," *Nature* 399 (1999): 350–54.

58. Peter Dodson, "Origin of Birds: The Final Solution?" *American Zoologist* 40 (2000): 504–12.

59. Freeman and Herron, *Evolutionary Analysis*, 667–77.

10 million years for this fossil transition to take place.[60] While this is a long time by human standards, it is but a blink of an eye in geological time, and so it is difficult to see how some thirty phyla would evolve via Darwinian selection in such a short period. Few if any Darwinists have argued that the transition occurred this fast; rather, they argue that transitional forms must exist in the Precambrian rocks. As a result, considerable effort has been made to scour the youngest Precambrian rocks, particularly those harboring the Ediacaran fauna (a well-preserved Precambrian ecosystem), in an attempt to close this fossil gap.

As more research has been done on the Ediacaran fossils, there is mounting evidence that some simple bilaterian phyla existed in the Precambrian period. There is evidence that annelid worms existed, given the many trace fossils of burrowing patterns as well as segmented fossils such as *Dickinsonia*.[61] Some of these trace fossils are found in rocks that date to 60 million years before the Cambrian, raising the possibility that some primitive bilateral animals existed long before the Cambrian period. In addition, a small mollusk named *Kimberella* has been found[62] as well as possible bilaterian embryos, although there is debate surrounding the nature of these embryos.[63]

Despite these finds, gaps remain. While some small simple bilaterans seem to have been present during the Precambrian period, there is still a striking change in morphology in the Cambrian fossils. As Stephen Jay Gould argues, this "absence of complex bilaterians before the Cambrian explosion rests upon extensive examination of appropriate sediments replete with other kinds of fossils, and located on all continents."[64] While it is true that these lineages likely were small, sparse, and soft bodied—characteristics that would have made fossilization difficult—the majority of Ediacaran fossils had the same characteristics, and they have been found in abundance.

The basic problem for the Neo-Darwinian theory is that the gradualistic paradigm upon which the theory rests calls for slow accumulation of changes. This implies that emergence of higher-order taxa takes longer.

60. R. Kerr, "Evolution's Big Bang Gets Even More Explosive," *Science* 261 (1993): 1274–75; S. A. Bowring et al., "Calibrating Rates of Early Cambrian Evolution," *Science* 261 (1993): 1293–98.

61. T. P. Crimes, "Changes in the Trace Fossil Biota across the Proterozoic-Phanerozoic Boundary," *Journal of the Geological Society* 149 (1992): 637–46.

62. Mikhail Fedonkin and Benjamin M. Waggoner, "The Late Precambrian Fossil Kimberella Is a Mollusc-like Bilaterian Organism," *Nature* 388 (1997): 868–71.

63. Freeman and Herron, *Evolutionary Analysis*, 676; S. Y. Xiao, Y. Zhang, and A. H. Knoll, "Three-Dimensional Preservation of Algae and Animal Embryos in a Neoproterozoic Phophorite," *Nature* 391 (1998): 553–58.

64. Stephen J. Gould, *The Structure of Evolutionary Theory* (Cambridge, MA: Belknap, 2002), 1157.

For example, emergence of a new genus would take longer than emergence of a new species; similarly, emergence of a new family would take longer than emergence of a new genus. Clearly, emergence of new phyla would take even longer as their emergence involves significant changes. The 10-million-year window of the Cambrian explosion appears far too short for this to have occurred via a gradual process.

Nonetheless, Neo-Darwinians could turn this difficulty on its head and argue that the rapid appearance of so many phyla actually confirms the theory. The reason for this has to do with the nature of the Darwinian evolutionary process as illustrated in figure 5.2, where we noted that horizontal changes at the bottom cannot occur. The implication is that one highly developed phylum (such as modern chordates) cannot change into another, because it corresponds to a highly complex architecture. But when life was simpler, architectures were simpler as well. Only at this stage could relatively minor changes result in what would later become vastly different organisms (upper part of the diagram). Once past this point, however, there was no turning back. So while acknowledging the difficulty that the 10-million-year window represents, Neo-Darwinists have argued that the first chordate, for example, could have been a worm with a stiff notochord on its dorsal side. If worms existed in the Precambrian rocks, adding a notochord in the Cambrian seems like a minor step.[65] Similar small changes, early in life's history, could have been the origin of the architectures we now refer to as phyla, but which, at the time, would have been barely noticeable.

There is, of course, much that is hypothetical here; critics will argue that these changes do not seem minor to them, and that this is just another case of the Neo-Darwinists using the argument from psychological plausibility. More research clearly needs to be done, on both sides.

Neo-Darwinists have also turned to sequence comparisons to bolster their case. Using sequences to determine divergence dates, studies have estimated that the complex bilateral forms that appear suddenly in Cambrian rocks must have diverged from each other about 1 billion years ago, rather than during the Cambrian explosion.[66] While these dates are subject to extensive debate, even if these phyla began to diverge at this early date, one still has to explain why in 10 million years the lineages exploded in size, number, and complexity and were suddenly fossilized at an enormous rate. For example, if the lineage leading to arthropods split from that leading to sponges 400 million

65. Alan Gishlick, "Icons of Evolution? Why Much of What Jonathan Wells Writes about Evolution Is Wrong," icon number 2: Darwin's "Tree of Life," National Center for Science Education, 2003, www.ncseweb.org/icons/icon2tol.html.

66. G. A. Wray, J. Levinton, and L. H. Shapiro, "Molecular Evidence for Deep Precambrian Divergences among Metazoan Phyla," *Science* 274 (1996): 568–73.

years before the Cambrian, this still does not explain the emergence of arthropod characteristics in the fossil record in such a short period. There are no obvious precursors in the Precambrian deposits that can be picked out as the likely forerunners of the arthropods. The hardened and segmented body plan appears only in the Cambrian, and this rapid morphological change needs to be explained. Not surprisingly, there is considerable debate on this subject.[67] This difficulty and the transitional form difficulties of the fossil record have caused many Neo-Darwinists to turn to the theory of punctuated equilibrium, discussed in chapter 8.

Before we leave the fossil record, it is important to point out what the fossil record can and cannot demonstrate. The fossil record can document only life-forms that existed and the order in which they existed. For example along the fish to tetrapod lineage, *Acanthostega* existed first, followed by *Pederpes finneyae*. While this is helpful in bolstering the case of Neo-Darwinian theory if transition forms are found in the proper temporal sequence, the fossil record cannot demonstrate whether the mechanisms proposed by Neo-Darwinism are adequate to explain the transition. Such a demonstration would require, at a minimum, an understanding of the molecular and developmental pathways involved as well as the selective pressures existing at the time. As a result, debate regarding the fossil record is unlikely to resolve the broader debate regarding Neo-Darwinian theory.

Information Creation

The entire history of evolution can be viewed as one continuum of information creation. Every evolutionary advance requires the presence or emergence of new amounts of information. Because of this, the question of whether information can be created by such naturalistic processes as natural selection has become an extremely contentious issue. Some critics of Neo-Darwinian theory have argued that there is no direct evidence for the creation of new information via natural selection, only for information loss. While information loss does occur in certain cases, the conclusion represents a narrow reading of the evidence at hand. It appears that new information can be created via natural selection, by at least three means:

1. Modification of an existing gene
2. Duplication and subsequent modification of an existing gene
3. Modification of a noncoding stretch of DNA

67. Gould, *Structure of Evolutionary Theory*, 1158.

Not all modifications of a gene create new information, but there are a number of well-documented cases in which this does happen. Consider the evolution of antibiotic resistance in bacteria to β-lactam antibiotics such as penicillin. As described previously, these antibiotics are destroyed by a β-lactamase enzyme that is expressed in many types of resistant bacteria. To overcome resistance, new classes of β-lactam antibiotics have been developed to better fight infections caused by these bacteria. The third generation of β-lactam antibiotics was successful in dealing with many strains that had the β-lactamase enzyme, because the enzyme was unable to efficiently deal with these newer antibiotics. Since their introduction in the late 1970s and early 1980s, though, a number of modifications have occurred in the bacteria's β-lactamase enzyme that have expanded its binding site to better fit these newer antibiotics.[68] This modification of an existing gene has provided the information necessary to afford resistance in bacteria to these third-generation drugs. The bacteria population now includes β-lactamase enzymes that have this information as well as those that do not and are still susceptible to the third-generation antibiotics. Information in the bacteria population as a whole has increased.

More significant increases in information can occur via the process of gene duplication. Gene duplication by itself does not create new information, but with proper modifications after duplication, new information may arise. Consider the case of potassium channels, which are found in abundance in excitable cells such as neurons and muscles. The channels act as holes in the membrane and control the flow of potassium by opening and closing. Most organisms contain a variety of different types of potassium channels. In fact, the human genome has a large variety of potassium channel genes, all of which serve slightly different functions.

Evolutionarily, these multiple potassium channels found in the human genome are thought to have arisen via serial duplications of an original potassium channel gene. These duplicated genes are free to mutate such that they may take on a novel expression pattern or a novel channel conductance or permeability (see fig. 5.7). For example, changing only a few DNA bases in a potassium channel gene can cause the channel to remain open longer or to allow more potassium through when it is open. New information has been generated in this case, and it is specifically encoded in the novel potassium channel that the organism might find useful.[69] The amount of information that has been generated, though,

68. Medeiros, "Evolution and Dissemination of Beta-Lactamases," S19–S45.
69. Most changes would be deleterious or neutral. As a result, duplicated genes would seldom take on a new function that the organism finds useful. When it does, though, the new gene can be preserved via natural selection.

may be quite small. In order for the duplicated gene to take on a novel function, it may have been necessary to change only a few bases of DNA. The *vast majority* of the information came already assembled in the form of the gene duplication. This point is critical to remember; although some new information is generated, the majority of the information already existed in the form of the original potassium channel.

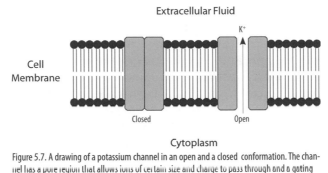

Extracellular Fluid

Cell Membrane

Closed   Open

Cytoplasm

Figure 5.7. A drawing of a potassium channel in an open and a closed conformation. The channel has a pore region that allows ions of certain size and charge to pass through and a gating mechanism that turns the channel on and off.

In addition to the duplication of existing genes, an even stranger example of information creation can be seen with the duplication of a noncoding section of DNA. This appears to be what has happened with the antifreeze glycoprotein (AFGP) from the Antarctic fish *Dissostichus*. The AFGP is a long peptide that is cleaved into forty-one identical pieces that inhibit ice crystal growth and prevent the fish's internal fluids from freezing. Interestingly, the sequence of this iterative segment is identical to a small sequence that spans part of a noncoding region in the *Dissostichus* trypsinogen gene, a gene used to cleave proteins.[70] Given this sequence similarity as well as the retention of sequences at both ends of the AFGP gene that are identical to the trypsinogen gene, it appears that (1) the trypsinogen gene was duplicated; (2) the noncoding region was modified to take on a novel function (inhibiting ice crystal growth); and (3) the noncoding region was duplicated to produce forty-one repeated segments. Together these changes have produced a functional protein from previously noncoding DNA. The information to inhibit ice crystal formation was not present originally, but emerged through a process of duplication and modification.

The AFGP and the potassium channel examples illustrate what natural selection and random mutation has been demonstrated to ac-

70. J. M. Logsdon and W. F. Doolittle, "Origin of Antifreeze Protein Genes: A Cool Tale in Molecular Evolution," *Proceedings of the National Academy of Sciences USA* 94 (1997): 3485–87.

complish. Given the duplication of a gene or a larger segment of DNA, these mechanisms can make alterations and create novel information. Despite this, the question of how the original potassium channel was assembled is still unanswered. To further complicate matters, it is very hard to pinpoint where the initial information actually originated. In the case of the fish antifreeze gene, which has a relatively simple function, the origin is relatively clear; in the case of the original potassium channel gene it is not. No one knows how the first potassium channel originated. In some cases this may be an intractable problem because "following the information trail for actual biological systems in the wild is rarely possible and depends on contingencies that may forever lie beyond the veil of history."[71] It is possible that the Neo-Darwinian mechanisms are up to the task of building the first potassium channel, but this is not self-evident. Just because mutation and selection can create new potassium channels given an initial one, or just because they can build antifreeze proteins from stretches of DNA, it does not necessarily follow that they can create all the complex information found in living organisms.

Given the difference in the amount of information that must be created in the scenarios described above, it would be helpful to have a way of quantifying the information that is needed at any stage of evolution. In the case of the alteration of the β-lactamase enzyme, only two amino acids had to be changed. In the case of the potassium channel duplication, only a handful of amino acids had to be changed (once the gene was duplicated). In both cases, only small amounts of information have been added at these steps. Such changes can be done quite readily by natural selection coupled with random mutations.[72]

In the final case, though, that of building the first potassium gene, much more information was most likely needed. The origin of information on this larger scale may require other processes than those described above. Rather than looking at all types of information creation as roughly equivalent, the varying magnitudes of information gain that have occurred during the course of natural history need to be acknowledged and studied empirically. Only then will we be able to fully understand the efficacy of Neo-Darwinian mechanisms to create new biological information.

71. W. Dembski, *No Free Lunch: Why Specified Complexity Cannot Be Purchased without Intelligence* (Rowman & Littlefield: Lanham, MD, 2002), 217.
72. Note that changing a gene does not necessarily mean an increase of information. In the case of a potassium channel duplication that takes on a novel function, though, information has certainly increased.

The Interrelatedness of Traits: Design Constraints

Before Darwin published *The Origin of Species*, the French comparative anatomist Georges Cuvier (1769–1832) argued against gradual evolution based on his functionalist understanding of the interrelatedness within organisms. As discussed in chapter 2, Cuvier viewed organisms as interconnected wholes, structures in which each part was dynamically balanced to remain in the proper relationship with the whole. For example, given the structure of the jawbone of a mammal, a certain type of jaw muscle followed by necessity. Given the jaw muscle, a certain type of skull bone followed by necessity as well, and so on through all the parts of the organism. Given this type of interdependence, any significant alteration to one part would send a disastrous ripple effect through the organism unless coordinated changes occurred in a number of other parts simultaneously. Thus, the gradual alteration of an organism, posited by Neo-Darwinian theory, would have been an anathema to Cuvier.

Cuvier's concern still resurfaces in the debate more than two hundred years after he first formulated it. The modern argument starts with the recognition that each species has a structural and functional organization that is hardly arbitrary. In fact, the specific anatomical arrangement—the species' architecture—allows the organism to survive and thrive within its environment. Architectures—or rather, good architectures—allow stable organisms to be constructed. Many desirable characteristics are combined in these stable organisms, such as adequate metabolism, ventilation, dexterity, and speed. As a result, architectures enable functionality but at the same time they impose performance and stability limitations because arbitrary combinations of characteristics are not possible. In other words, there are trade-offs, or what are called "design constraints."

To see this clearly, consider the architecture of *Homo sapiens*. Humans consist of a collection of interacting subsystems including the nervous, digestion, muscle, skeletal, circulatory, and immune systems. Each has an internal dynamic, which governs its behavior as a subsystem, and an external dynamic, which governs its interactions with other subsystems. The skeletal system, for example, is designed to support the weight of the body and protect it. The bones that compose the skeletal system are relatively light, considering their role in protection and support. Because of this they can be broken rather easily. Suppose one wanted to "correct" this situation and make the bones more rigid and compact and less likely to break. To do this, one could pack more collagen and calcium into the bones to make them thicker and denser. To compensate for such a "solution," though, the digestive system would have to absorb more calcium during digestion. The circulatory system would have to divert more blood to the bone tissues. The muscles would have to enlarge to move these heavier bones, and the joints would have to adapt in order to bear the excess weight of the new bones. Such larger bones may be less susceptible to breaking, but there will certainly be other less desirable effects on the body. For example, the new dense bone tissue would likely take longer to heal than normal bone tissue. The new heavier bones would probably make one less nimble and more prone to joint problems. The heart would be taxed to supply more blood to the bones, and the blood pressure might become elevated. Such trade-offs would limit the selective advantage of having denser, stronger bones.

A similar example can be seen with increased brain size. If evolution were to move toward a larger human brain, this pathway would be available only if other compensations were made simultaneously so that certain physical constraints—the narrow birth canal, the ability to cool the brain, the increase in metabolic demand—were dealt with. Without satisfying these physical constraints the large brain would not be stable and therefore would not be advantageous to its owner. That is, those with the larger brain would either get stuck in the birth canal, overheat, or die of poor nutrition. Multiple simultaneous changes appear to be needed.

The net effect of all this is to reduce the number of viable biological possibilities. While these constraints certainly put limitations upon evolution, it is unclear how drastically evolution would be affected. Obviously some possibilities like flying elephants and 3,000-pound arthropods are out of the question due to design constraints, but can an ape gradually transform into a human? Can the architecture of an apelike creature—small brain, large jaw, hunched posture—gradually transform into the architecture of a human—large brain, small jaw, upright posture—by passing through stages, each of which is viable? Certainly increasing the brain size puts extra demands upon other systems of the body, but it is possible that these other systems can compensate for the increased brain mass. If the brain mass increased gradually, then the metabolic rate and the blood flow might be able to adapt to track this change. In fact, the ability of blood flow to the brain and metabolism to adapt to this increase in brain size might not even require new mutations. As more is learned about development, it is clear that there is a certain amount of plasticity in development that allows organisms to compensate for a variety of circumstances. For example, the amount of capillaries that flow through a tissue is not fixed genetically. Rather it depends upon the size of the tissue as well as the metabolic demands put on the tissue. As a result, if a mutation increased the brain size modestly, the developmental process is robust enough to compensate for this by building more blood vessels to service the brain. In the case of blood vessel formation, this

178

process continues throughout life, such that blood vessel formation and regression matches the demands of the tissue.

This built-in buffering capacity is quite robust and can adapt to a wide variety of change. One of the most impressive examples is the brain's ability to compensate for the loss of the corpus callosum, the main connection between the right and left hemispheres of the cortex. There are many cases in which individuals are born without a corpus callosum. One might suppose that absent this structure, an individual would display obvious behavioral abnormalities, given that the two hemispheres of the brain are unable to communicate. Surprisingly, though, this is not always the case. What seems to happen is that during development, the brain strengthens other pathways that connect the two hemispheres of the brain, and these alternative pathways are able to compensate for the lack of a corpus callosum. In fact, many people who are born without a corpus callosum display no obvious behavioral defects. This flexibility indicates that there is not as tight a connection between form and function as Cuvier had argued. There is in fact room for adaptations to take place within a complex architecture, particularly given the robust nature of development. However, it remains to be seen exactly how pliable these developmental pathways actually are.

Recognizing this fact, critics of Neo-Darwinian theory have argued that certain changes are too drastic, certain chasms too wide, for natural selection and random mutations to have traversed. For example, some argue that the architecture of invertebrates such as arthropods is so drastically different from a vertebrate architecture that there is no way to change from one to another in a gradual fashion without traversing through unviable intermediate forms. The fact that vertebrates have an internal skeleton while invertebrates have an exoskeleton and the fact that invertebrates have two main nerve cords on the ventral side while a single nerve tube runs on the dorsal side of vertebrates are just two of the difficulties cited. Given the magnitude of this divide, it may well be that there is no way to traverse directly the design space separating vertebrates from invertebrates, but the argument misses a crucial point. Advocates of Neo-Darwinism do not argue that arthropods, for example, gave rise to vertebrates but rather that both of these body plans could have evolved from a common ancestor. While it is not clear what such an ancestor looked like, there has been speculation that both forms could have originated from a primitive flatworm body plan.[73] This scenario is largely speculative, but it is argued that this primitive ancestor might

73. D. K. Jacobs et al., "The History of Development through the Evolution of Molecules: Gene Trees, Hearts, Eyes, and Dorsoventral Inversion," in *Molecular Approaches to Ecology and Evolution*, ed. R. DeSalle and B. Schierwater (Basel: Birkhauser Verlag, 1998), 323–57.

have been more adaptable because it was simpler and therefore under less stringent design constraints. As a result, it could have given rise to vertebrates on the one hand and more complex invertebrates such as arthropods on the other.[74] While this scenario is logically possible, that does not mean that it did occur or even that it could have occurred. Much work remains to be done to determine the feasibility of such an evolutionary change.

Genetic Constraints: Pleiotropy

At the level of organ systems, it is clear that organisms are interconnected wholes in which changes to one system affects the other systems in the body. This interconnectivity is found not only at the level of the organ systems but also at the genetic level. Just as changes to one organ system can have wide-ranging effects on an organism, changes to one gene can also have wide-ranging effects, given that one gene can interact with multiple pathways and therefore affect multiple traits. Genes that behave in this manner are called "pleiotropic," and they are quite common. For example, tigers and Siamese cats have a gene that controls fur pigmentation but also influences connections between the cat's eyes and its brain—characteristics about as unrelated as one can imagine. As a result, a defect in this gene can cause both abnormal pigmentation and a cross-eyed condition. Another example is Marfan syndrome, in which one defect in a gene encoding for an elastic connective tissue protein called fibrillin results in, among other things, a weakened aorta, nearsightedness, and a malformed sternum.

The origin of pleiotropic genes can be explained easily by Neo-Darwinian theory: a gene with one function could gradually acquire a second function and then a third function and so on. A problem, though, occurs once a large number of pleiotropic genes exist within an organism. Research has demonstrated that pleiotropy leads to a situation in which little or no variation from a single, optimal form is possible.[75] The researchers who came to this conclusion were not investigating mechanisms of evolution; rather, they were seeking an answer to the question of why some structures are highly conserved, or shared, throughout the biological world, and presumably, throughout biological history. What these researchers found is that once a gene affects three or more characteristics, little or no variation in its product(s) is possible. The

74. Gould, *Structure of Evolutionary Theory*, 1151–61; D. K. Jacobs, "Selector Genes and the Cambrian Radiation of Bilateria," *Proceedings of the National Academy of Sciences USA* 87 (1990): 4406–10.

75. David Waxman and Joel Peck, "Pleiotropy and the Preservation of Perfection," *Science* 279 (1998): 1210–13.

reason, in qualitative terms, is not difficult to understand: any variation in the gene to improve one of the characteristics it controls will almost certainly reduce the effectiveness of the others. In more quantitative terms, the problem is that such linking of multiple traits constrains the ability of the population to explore new regions of its genetic space. The gene is basically sitting on the top of a fitness peak and there seems to be nowhere for the gene to go but downhill.

Such a situation poses an obvious problem for any evolutionary theory that relies on the accumulation of small changes for species improvement: the changes needed are inhibited due to pleiotropy. The problem is becoming all the more serious as research in molecular biology finds more and more pleiotropic genes. In fact, pleiotropy seems to be the rule rather than the exception. How, then, does an organism evolve in the face of this situation? The response of bacteria to antibiotics may provide a clue.

When bacteria acquire resistance to antibiotics, it often comes at a cost.[76] In many cases, the change that allows the bacteria to become resistant to an antibiotic adversely affects another trait such as protein synthesis. In fact, this is one manner by which bacteria become resistant to streptomycin. Streptomycin normally works by binding to the bacterial ribosome and inhibiting protein synthesis. If the ribosomal protein, rpsL, that streptomycin normally binds becomes altered via mutation, then streptomycin is unable to bind and the bacterium becomes resistant to the drug. However, there is an associated cost: once the bacteria ribosome is altered such that streptomycin cannot bind, the resistant bacteria are less effective at making proteins. As a result they are at a selective *disadvantage* when it comes to protein synthesis as compared to their streptomycin-sensitive brethren. The antibiotic mutation in this case has maladaptive pleiotropic effects.

At first glance, one might assume that this would severely limit changes in this gene. After all, one can only change the ribosomal gene so much before protein synthesis comes crashing down. Such a view ignores the possibility that compensatory mutations could occur in other genes that might alleviate the maladaptive effects streptomycin resistance has on protein synthesis. Indeed, researchers have found that such compensatory mutations do occur. If the streptomycin resistance strain is allowed to evolve in a laboratory setting, over time compensatory mutations occur that alleviate the adverse effect on protein synthesis.[77] Presumably these mutations occur in other ribosomal protein genes that are able to

76. Baquero and Blazquez, "Evolution of Antibiotic Resistance," 482–87.
77. S. J. Schrag, V. Perrot, and B. R. Levin, "Adaptation to the Fitness Costs of Antibiotic Resistance in Escherichia Coli," *Proceedings of the Royal Society of London*, Series B, 264 (1997): 1287–91.

compensate for the change in the rpsL protein yet allow the bacteria to maintain antibiotic resistance. Such compensatory mutations appear not only in bacteria but also in fruit flies and other organisms, and they seem to be able to overcome at least some of the limitations imposed by pleiotropy.[78]

The limitations imposed by pleiotropy may also be reduced via the robustness of genetic and biochemical networks. Theoretical research has demonstrated that if cellular networks are robust, that is, if they can adapt to the activity level of the network and the needs of the cell, then organisms will be buffered against small random changes.[79] For example, suppose there was a change in a pleiotropic gene that gives some benefit to the individual but at the same time it adversely affects the biochemical pathway for glucose synthesis. Rather than having a deleterious effect on glucose synthesis, the glucose synthesis pathway may be able to adapt to such a change so that the adverse effect is masked. Imagine that the change in the gene adversely affected one step in the glucose synthesis pathway, causing the enzyme catalyzing this step to have decreased efficiency. To overcome this, the pathway may be able to adapt by increasing the products and decreasing the reactants at this step. For example, if molecule C was converted to molecule D by the defective enzyme, other enzymes in the pathway can make more C or remove more D. This alteration in the levels of C and D (C going up and D going down) would make the reaction that converts C to D more favorable, a change that can compensate for the defect in the enzyme at this step.

This capability of networks to adjust to change involves large amounts of feedback loops within the pathway, such that it can adapt to the needs of the cell. Not surprisingly, such feedback loops are common in the biochemical and genetic networks found in living cells. While the fact that such networks exist is certainly suggestive, it does not prove that these networks can overcome the pleiotropy problem. Further studies, particularly genetic and biochemical studies, still are needed to determine exactly how robust cellular networks are and exactly how much change they can accommodate. If many cellular networks are found to be robust, though, it would demonstrate that organisms have "a larger parameter space in which to evolve and to adjust to environmental changes,"[80] despite the fact that they may contain large numbers of pleiotropic genes.

78. R. J. Kulathinal, B. R. Bettencourt, and D. L. Hartl, "Compensated Deleterious Mutations in Insect Genomes," *Science* 306 (2004): 1553–54.
79. N. Barkai and S. Leibler, "Robustness in Simple Biochemical Networks," *Nature* 387 (1997): 913–17.
80. Ibid., 916.

Probability and Mathematical Arguments

A ubiquitous form of argumentation in the evolution debate is the mathematical probability argument. To bolster their case for common descent, Neo-Darwinists point out the low probability that chance could account for a gene sequence in related organisms being so similar. Their critics attempt to calculate the probability that a certain structure evolved via natural selection in order to demonstrate the infeasibility of Neo-Darwinian theory. Sorting through all these arguments can be tedious, and their utility depends critically on the assumptions made—assumptions that are often unrealistic. Rather than sort through all such arguments, we will look at a few examples in order to demonstrate how these arguments are employed, and how they can be used by partisans on all sides of the debate.

The first example is an argument that is used by many Neo-Darwinists and it deals with the sequence similarity seen between related organisms. Comparing the sequences of the cytochrome c gene in various organisms shows that humans and chimps share an identical sequence at this gene.[81] This is interesting, given that more than 50% of the amino acids in the protein can be altered without knocking out the function. What this means is that many gene sequences would yield a functional cytochrome c protein. In fact, mutational analysis studies indicate that there are more than 2.3×10^{93} functional protein sequences for cytochrome c.[82] Given that any one of 2.3×10^{93} functional protein sequences would likely do the trick, the fact that humans and chimps have an identical sequence is strongly suggestive of common descent. Absent common descent there is only a 1 in 2.3×10^{93} possibility of chimps and humans having the same sequence.

Faced with such an argument, critics counter with the assertion that chimps and humans have the same cytochrome c sequence, not because of common descent, but because the specific sequence in question gives humans and chimps a functional advantage over the other possible sequences. While there may be many functional cytochrome c proteins, there might be only one that works optimally in humans and chimps. Thus, the sequence similarity is not the result of common descent, but of a functional constraint. It is, however, difficult to believe that none of the 2.3×10^{93} operational cytochrome c protein sequences would be functionally equivalent to the one humans and chimps possess. Even if only 0.0001% of the functional sequences are roughly equivalent to the

81. Douglas Theobald, "Molecular Sequence Evidence," on The Talk.Origins Archives web site, www.talkorigins.org/faqs/comdesc/section4.html.

82. Hubert Yockey, *Information Theory and Molecular Biology* (Cambridge: Cambridge University Press, 1992), 328.

human cytochrome c protein, that would still leave more than 1×10^{87} possible sequences.[83]

Yet this probability argument for common descent put forth by Neo-Darwinists can be modified by critics, particularly those in the Intelligent Design camp, to argue against natural selection. Given that there are 2.3×10^{93} functional variants of the cytochrome c protein sequences, one might want to know how many nonfunctional variants are there. Mutational analysis has determined that for every functional cytochrome c sequence, there are 5×10^{43} nonfunctional sequences.[84] As a result, the probability of building a functioning cytochrome c sequence from scratch is vanishingly small. There is no way that natural selection could sort through 5×10^{43} nonfunctional sequences in order to get to a functional one in the time available since the formation of the earth. Critics then come to the conclusion that natural selection and random mutation operating alone could not have built a cytochrome c sequence from scratch.

The problem with this approach is that the possibility of cumulative selection, that is, building the cytochrome c sequence gradually in pieces, is ignored. While the cytochrome c protein is approximately one hundred amino acids long, it may be possible that a ten-amino acid fragment of this protein could have provided some beneficial function to the cell. This would represent an intermediate step that natural selection would be able to build upon, such that it would not have to come up with a functioning cytochrome c sequence that is one hundred amino acids long, all in one fell swoop. It is also possible that many other intermediate versions of the cytochrome c protein could have existed, each representing a small improvement upon the initial small protein fragment. Such a gradual, step-by-step process is the basis of Neo-Darwinian theory, and it is usually ignored by those who perform such calculations. However, stating that such an event "may be possible" does mean that the event *is probable*, much less that it actually occurred; at the very least, though, it should be modeled or acknowledged in such calculations.

Ignoring the possibility of the gradual, step-by-step assemblage of a gene is problematic, but so too is the typical response of Neo-Darwinists. Rather than viewing cumulative selection and the presence of functional intermediates as a possibility, they view it as inevitable. In one of his well-known books, *The Blind Watchmaker*, Richard Dawkins presents the reader with an analogy in which a message of twenty-eight

83. While such numbers do suggest common descent, they do not *necessarily* support the belief that natural selection and random mutations were the mechanisms that drove it. As always, it is important to keep in mind the different levels of evolution.

84. Yockey, *Information Theory and Molecular Biology*, 328. The number (4.97×10^{43} ~ 5×10^{43}) is obtained by dividing two numbers in Yockey's text: the number of possible sequences (1.15×10^{137}) by the number of functional sequences (2.316×10^{93}).

characters, "METHINKS IT IS LIKE A WEASEL," has to be ascertained. In Dawkins's example, the process proceeds by starting with a random string of twenty-three letters, which of course is nowhere near the target sequence. At each step, a few variants of the random string are produced by randomly changing some letters. Of these strings, the one closest to the target is chosen and used in subsequent rounds of variation and selection. This is repeated until the target has been recovered, something that requires forty to sixty-five generations.[85]

The problem with this example, which Dawkins himself admits, is that it assumes that every intermediate stage is beneficial. This assumption, however, isn't true in his example and it isn't necessarily true during evolution. To see this, consider the following sequences *M*RDKWOLE DF SD *L*AZC B FHGEIO and *M*RDKWOLE DF SD *L*AZC B FHGEI*L*. The first one has two letters that are identical to the target; the second one has three (identified in bold and italics). In Dawkins's example, the second would be selected, despite the fact that it is just as nonfunctional as the first sequence. In either case, one cannot identify the target sequence from the letters written. You wouldn't be any more likely to guess the target sequence if you had the first sequence than if you had the second sequence, particularly since (if we want to simulate what occurs during evolution) there is no way you would know which letters were correct and which letters were incorrect.

The same problem holds true in evolution. If a sequence such as the cytochrome c could evolve only via cumulative selection, one has to demonstrate that functional intermediates exist. One cannot merely assume that they exist, as Dawkins's example incorrectly does.[86] One cannot assume that having a protein that has twenty amino acids identical to a functional 100 amino acid cytochrome c sequence is any better than a protein that has fifteen identical amino acids.

When it comes to probability arguments, both sides rely on tendentious arguments with problematic assumptions. Neo-Darwinists assume that there are numerous functional intermediates, since this is necessary for a step-by-step gradual evolutionary process to occur. Critics assume that no such intermediates exist. The truth may lie somewhere in between, but this requires more than probability and mathematical arguments; it requires biochemical and genetic research to identify and characterize possible intermediate forms. Much more is discussed

85. Richard Dawkins, *The Blind Watchmaker* (New York: W. W. Norton, 1996).

86. Dawkins claims that his example was merely to demonstrate the efficacy of cumulative selection and admits that it is incorrect in some of the details. There is no deliberate deception, but his example is, at the very least, misleading. When it is presented as an analogy of evolutionary change while missing a critical component—the existence of functional intermediates—it serves more to confuse than to illuminate.

regarding this matter in chapter 7, which deals with the theory of Intelligent Design.

At this point it is worth noting that there are other more involved mathematical challenges to Neo-Darwinism. Among those who have formulated such mathematical critiques are Lee Spetner,[87] Fred Hoyle,[88] I. L. Cohen,[89] and Herbert Yockey.[90] Their challenges center on the possibility of the transformations required for Neo-Darwinian theory to work, given its mechanisms. Another line of criticism uses a recent result from optimization theory, the "no free lunch theorem."[91] These arguments are quite technical, and they are contentious because they depend on the accuracy of the assumptions in the model. Nonetheless this may become a very important area for the evolution debate as the models are refined.[92]

Summary and Synthesis of the Neo-Darwinian Theory Paradigm

In this chapter we have attempted to give the reader an understanding of the basic principles of Neo-Darwinian theory as well as illustrations of how it works at both the level of macro- and microevolution. We have also attempted to address in a balanced fashion the main criticisms that have been leveled against Neo-Darwinism. The criticisms we have addressed do not represent an exhaustive list by any means. Rather, they are the issues that are most important in the debate. They are the issues that are written about most frequently or that are most critical for Neo-Darwinism to address in a sufficient fashion if it is to be a satisfactory explanation for natural history.

For those who are convinced that the history of life can be explained by natural mechanisms, most accept Darwin's theory as the *inference to the best explanation*. The following excerpt from the National Academy of Sciences sums up this predominant view within the biological community:

> The evolution of all the organisms that live on Earth today from ancestors that lived in the past is at the core of genetics, biochemistry, neurobiology,

87. Lee Spetner, *Not by Chance! Shattering the Modern Theory of Evolution* (New York: Judaica Press, 1998).

88. Fred Hoyle, *Mathematics of Evolution* (Memphis: Acorn, 1999).

89. I. L. Cohen, *Darwin Was Wrong—A Study in Probabilities* (Greenvale, NY: New Research Publications, 1985).

90. Yockey, *Information Theory and Molecular Biology*.

91. David Wolpert and William Macready, "No Free Lunch Theorems for Search," Santa Fe Institute Technical Report SFI-TR-95–02–010, 1995.

92. See www.evolutionprimer.net for details.

physiology, ecology, and other biological disciplines. It helps to explain the emergence of new infectious diseases, the development of antibiotic resistance in bacteria, the agricultural relationships among wild and domestic plants and animals, the composition of Earth's atmosphere, the molecular machinery of the cell, the similarities between human beings and other primates, and countless other features of the biological and physical world.[93]

The NAS is correct that common descent can explain many features of the world, particularly genetic similarities such as the cytochrome c gene, synteny blocks, and pseudogenes. Likewise, natural selection can explain many features of the world, particularly the types of change that lead to antibiotic resistance and mutation pattern in the flu virus. The question in dispute, though, is whether this mechanism is capable of explaining the whole tapestry of life. In particular, how does it fare in comparison with the theories described in the next three chapters? In addressing this question, one must be willing to take a critical look at the evidence. Unfortunately, many are not willing to take this critical look. Consider the following:

> Darwin's theory is now supported by all the available relevant evidence, and its truth is not doubted by any serious biologist. . . . I suggest that it may be possible to show that, *regardless of evidence*, Darwinian natural selection is the only force we know that could, in principle, do the job of explaining the existence of organized and adaptive complexity.[94]

According to the statement, any biologist who disagrees with Neo-Darwinian theory *cannot* be a serious scientist. Yet no scientist can take the position that Darwin's theory holds "regardless of evidence" and maintain his scientific dignity. Science needs predictions and objective evaluation of data in order to establish a theory, not bias. Darwin knew this, and in his time qualitative arguments were used to express predictions and evaluate the status of his theory. It should be no different today because Neo-Darwinian theory *does* make certain predictions that can be either falsified or verified. Below are just four examples:

1. Since evolution proceeds gradually, by incremental changes, there must be many stable transitional forms not only between species, but between all higher taxa as well. This is because whole populations of organisms must have had the intermediate characteristics as populations gradually changed. As it takes a consider-

93. National Academy of Sciences, *Science and Creationism*, viii.
94. Richard Dawkins, "The Necessity of Darwinism," *New Scientist* 15 (1982): 130; italics added.

able amount of time for such changes to become established in a population, the transitional forms should be evident in the fossil record. (This is one of Darwin's original predictions.)

2. Small beneficial changes (innovations) must be capable of becoming established in populations.

3. Random errors can increase the information borne by a DNA molecule, that is, the information in a genome.

4. Any structure, however complicated, can be created by natural selection and random mutation, given enough time, the correct environmental pressures, and a suitable population.

All of these predictions have been addressed in this chapter, yet none has been answered in a definitive manner, although it is clear that certain amounts of information can be created by natural mechanisms. A summary of the major arguments regarding Neo-Darwinian theory, both pro and con, are summarized in table 5.1. This table includes more information than has been addressed in the chapter, with some brief explanatory comments. None of the evidence cited in favor of the theory is marked as "Overwhelming" because none unequivocably establishes Neo-Darwinism over all of its competitors, especially the Meta-Darwinian school discussed in chapter 8. This is an important point to keep in mind. While evidence for historical and common descent may be strong, this is not sufficient to establish Neo-Darwinism. That requires another level of evidence. In addition, because of the historical nature of the evolution studies, an unequivocal reconstruction of what happened may well be beyond the reach of science. This does not mean that the theory cannot be verified or refuted, only that means other than historical studies will be required.

Table 5.1. Summary of arguments for and against Neo-Darwinian theory

Pro*	Value for theory	Comment
Commonalities in physiology, biochemistry, genetics (coding genes, synteny blocks, and pseudogenes)	High	Undoubtedly the strongest argument in favor of some type of naturalistic evolution, so not unique to Neo-Darwinian theory
Similarities of later forms of life to extinct forms in same area	High	Early but still forceful argument for common descent
Historical progression of life-forms from simple to complex	High	Important, though not as strong as above because of the many anomalies in the fossil record, such as the "living fossils" noted by Darwin
Sharing of innovations by subsequent life-forms	Medium	Important empirical fact, but may also be explainable by convergence rather than inheritance
Experimental demonstrations of the action of natural selection: Hawaiian drosophilids	Medium	Demonstrate speciation only; may not be able to extrapolate to macroevolutionary events
Offered micro- and macroevolution equivalence arguments	Low	Vacuous; long-range extrapolation; hard evidence is needed to support this assertion
Just-so stories	Low	Only useful as starting points for investigation (their only possible legitimate application)
Analogies from human-mediated change	Low	Intervention of human intelligence negates demonstrative value for purely naturalistic change

Con	Value as criticism	Comment
Information theory obstacles	High	Need to show empirically and theoretically that the complex amounts of information represented by specific proteins can come about by incremental processes.
Fossil record difficulties	High/ medium	The predictions of the theory often collide with what is observed. Much work will be required to show that observed patterns can be made consistent with the theory.
Architecture and system theory limitations	Medium	Though still only qualitative, represents a formidable difficulty in terms of how major changes can come about by means of gradual changes to a complex architecture
Pleiotropy restrictions	Medium	Since many characteristics of organisms are coupled, this represents a possible impediment to any major change
Mathematical arguments	Medium/ low	Difficult to demonstrate that these arguments take into account the relevant biology

* Much of the evidence cited in favor of Neo-Darwinian theory is also evidence for Meta-Darwinian theories.

Table 5.2. Neo-Darwinian school and the ten criteria of a genuine scientific theory

Criterion	Neo-Darwinian School
1. Compactness	Very compact, two major premises, 1) random mutations and 2) natural selection.
2. Simplicity	Very simple at its core.
3. Falsifiability	Difficult to falsify: explanations can be found for most observations.
4. Verifiability	Varies; difficult in some cases.
5. Retrodiction	Most data, once obtained, can be explained via retrodiction within the Darwinian framework.
6. Prediction	Predictions have been made, but if they are not borne out, often theory can still explain the data (retrodiction).
7. Exploration	Definitely suggests many avenues of exploration.
8. Repeatability	Normal scientific method applies.
9. Clarity	Very clear although explanations of certain individual evolutionary events can become muddled.
10. Intuitiveness	The mechanism at the heart of the theory is very intuitive.

6

THE CREATIONIST SCHOOL

With roots unabashedly in religious communities, the modern young earth Creationist (YEC) movement began in the early 1960s. Many consider that it traces specifically to the publication of *The Genesis Flood*, by John Whitcomb and Henry Morris, in 1961.[1] Building on earlier work by geologist George McCready Price in the 1930s and 1940s,[2] Whitcomb and Morris sought to explain the observed geology of the earth by means of the flood described in chapters 6 through 8 of Genesis. To further this work and develop a comprehensive biblically compatible explanation of natural phenomena, the Creation Research Society (CRS) was established in 1963. With a similar goal, the Institute for Creation Research (ICR) was founded in 1972, though it was an outgrowth of other organizations founded in the 1960s. Answers in Genesis (AiG), another major Creationist organization, was founded in 1994. The Center for Scientific Creation (CSC), which focuses primarily on physical processes to explain a young earth, dates to the 1980s.[3] The Geoscience

1. Further details on the history of the Creationist movement between 1925 and 1961 and thereafter may be found in Henry Morris, *History of Modern Creationism* (Santee, CA: Institute for Creation Research, 1993).

2. Ronald L. Numbers, *The Creationists* (Berkeley: University of California Press, 1992), 72–101.

3. Not all Creationists support the young earth theory. Some are content to argue that the strictly naturalistic mechanisms proposed by Neo-Darwinian theory are inadequate to account for observable facts about flora and fauna, and therefore direct creative action was necessary. Since these "old earth" theories accept all or nearly all of conventional geology and paleontology, they neither require nor postulate a fundamental rethinking of available

Research Institute (GRI) predates all of these, dating back to 1958, but has far less visibility.

All of these organizations argue for a "young earth" (6,000 to 10,000 years old), and postulate some type of cataclysmic flood event to account for observed geological features and fossil deposits. They fall into two camps, depending on their explanation of the floodwater and the cause of its movement: (1) those who advocate some type of rapid *plate tectonics* and/or *water vapor canopy* (mainly CRS, ICR, AiG), and (2) those who follow the *hydroplate* theory, which postulates that very large amounts of subterranean water were heated and escaped violently (CSC).[4]

Biblical fundamentalists, who currently make up the largest contingent of those commonly referred to as "Creationists," believe that science—or what they regard as bogus science—asserts positions and facts which they believe to be at variance with revealed truths of the Bible:

> The Biblical doctrine of origins, as contained in the book of Genesis, is foundation to all other doctrines of Scripture. Refute or undermine in any way the Biblical doctrine of origins, and the rest of the Bible is undermined. Every single Biblical doctrine of theology, directly or indirectly, ultimately has its basis in the book of Genesis.[5]

As a result of this situation, Creationists feel that it has become necessary to draw the line and stand up for what they believe. Accordingly, they have gone on the offensive, both in lectures and publications, claiming that the theory of evolution (at all levels) is not good science and is not supported by the facts. They have also undertaken research efforts to create a scientific basis for their beliefs; have issued numerous books, monographs, and videos; launched journals; and published numerous science textbooks aimed at primary and secondary schools.[6] This has provoked a somewhat belated response from the biological community and from other interested parties who were at first content to laugh off the challenge. The laughing stopped, however, when it became clear that

---

evidence. Their position is more that of interpreting biblical language to accommodate modern science. It is only the YECs who advocate an overthrow of key areas of modern science and who are therefore prominent in the evolution controversy. For this reason, we concentrate on their views.

4. Walt Brown, private communication to author (TF).

5. Ken Ham, "The Relevance of Creation," *Creation Ex Nihilo* 6, no. 2 (November 1983): special lift-out section.

6. Examples are the following: Alan Gillen, *Body by Design* (Green Forest, AR: Master Books, 2001); Jonathan Henry, *The Astronomy Book* (Green Forest, AR: Master Books, 1999); John Morris, *The Geology Book* (Green Forest, AR: Master Books, 2000); Michael Oard, *The Weather Book* (Green Forest, AR: Master Books, 1997); Walt Brown, *In the Beginning*, 7th ed. (Phoenix: Center for Scientific Creation, 2001).

the Creationists enjoyed widespread public support and were willing to use that support politically to leverage their position within academic circles, particularly at the secondary school level.

Young earth Creationists are often lumped together with Intelligent Design advocates.[7] However, this is a misguided oversimplification because the two share only a few ideas, primarily their critiques of Neo-Darwinism and their belief that naturalistic mechanisms cannot fully account for the history and development of life on earth.

In their critiques of Neo-Darwinism, Creationists are adamant that their explanations for the origin of life-forms fit the observed facts better and moreover that evolution, despite its near ubiquity in science books and classrooms, is indeed impossible. The principal argument of the Creationists is that the world is young, on the order of 6,000 to 10,000 years. If they can establish this time frame, they win hands down because all sides agree that large-scale evolution requires billions of years. If they cannot establish their time frame though, they lose everything, since their entire worldview, both theological and scientific, is predicated on a six-day creation and a consequent young earth. So the young age of the earth (and presumably the universe) is the make-or-break issue for Creationists.

In addition to their young earth arguments, Creationists put forth a number of critiques of evolutionary theory: (1) the theory of evolution is not scientific;[8] (2) evolution would violate established scientific laws, particularly the second law of thermodynamics;[9] (3) the proposed mechanism of Neo-Darwinian evolution (natural selection and random mutations) cannot account for the emergence of observed species and higher taxa (that is, evolution could not happen);[10] (4) the dating techniques used by geologists and evolutionary biologists are faulty; and (5) the fossil record does not support evolution via common descent.[11] Many of these critiques have already been examined, so in this chapter we shall concentrate on some of the main theories advanced by Cre-

---

7. Eugenie Scott, *Evolution vs. Creationism: An Introduction* (Berkeley: University of California Press, 2005), 57–67. Scott also attempts to smear the Creationists by putting them in a line with believers in a flat earth, geocentrists, and others. No serious Creationist has believed in any of these things for centuries. Readers must be on guard to reject such propaganda when evaluating the Creationists (or any other school).

8. Duane Gish, "It Is Either 'In the Beginning God'—or '. . . Hydrogen,'" *Christianity Today* 26, no. 16 (October 8, 1982): 29.

9. Henry Morris, "Evolution, Thermodynamics, and Entropy," in *Creation Acts, Facts, Impacts* (San Diego: Creation Life Publishers, 1974), 123–29.

10. Douglas Dewer, *Difficulties of the Evolution Theory* (London: Edward Arnold & Co., 1931), 93.

11. Sean O'Reilly, *Bioethics and the Limits of Science* (Front Royal, VA: Christendom Press, 1980), 55–60.

ationists to create a new vision of the world and explain observational data in biology, astronomy, geology, and paleontology, especially with reference to the young earth. Such work has recently become the focus of the Creationist movement,[12] and there are now much more sophisticated Creationist models and theories than existed even ten years ago. Increasing sophistication, however, is no guarantee that their theories will ever be capable of explaining observed facts well enough to displace existing theories in physics and biology for an old earth and universe and the development of life-forms.

The Creationist History of Life

The young earth Creationist worldview may be summarized briefly as follows: The universe was created 6,000 to 10,000 years ago (measured by earth time). The animals and humans were created as distinct "kinds," sometimes referred to as "baramin." All animals and plants existed then, but inhabited different ecosystems, so there was little or no mixing of dinosaurs with "modern" flora or fauna. Subsequently there was a great cataclysm, including a huge flood, triggered by plate tectonics or other forces. This led to the extinction of many plants and animals, including the dinosaurs, whose remains were buried together with other debris by the huge flood. The immense forces and high energy associated with the flood caused large and rapid changes to the earth, and thus led to the geological formations that we see today, including sedimentary rocks, canyons, the rise of mountain ranges, and also to the observed distribution of fossils. During this time, or prior to it, there were changes in physical constants, giving rise to an apparent old age for the earth. After the flood, conditions returned to "normal"—what we know today—and the animals preserved on Noah's ark then propagated and repopulated the world, with rapid speciation accounting for the great variety we see today.

In order to maintain this outlook, Creationists have two options. The first is the stereotypical classic Creationist view that merely proposes and fails to explain. It comes from Creationists who have abandoned science altogether (many claiming it is bogus) and who just postulate that in seven twenty-four-hour days, God created the universe and our world, complete with fossil record, functional and structural relationships (homologies) among flora and fauna, DNA that is at least 98% identical between humans and chimpanzees, and radioisotopes in proportions that suggest great

12. Lane P. Lester, "Mimicry," Impact number 18, Institute for Creation Research, September 1974.

age. Such an outlook is rejected by all serious Creationist researchers. The second option seeks to defend the Creationists' worldview by developing sophisticated theories, utilizing science, so as to give explanations that Creationists believe to be superior to an evolutionary worldview in accounting for observable facts. It is these theories and explanations that we shall examine in this chapter. As Kurt Wise, a Creationist geology professor from Bryan College, explains, 95% of young earth Creationists are "concerned consumers. . . . They just chose what feels good to them," or what they believe is morally important for their children. The remaining 5%, Wise believes, fall into two camps. In the first camp are those who are "crusaders" against evolution (4% of that 5%), well-meaning but often poorly trained, and who as a result give Creationism a bad name. "Because of their zealousness," Wise says, "they are the ones everyone sees when they look at Creationists." Wise claims that the "polarization between evolutionists and Creationists is primarily due to the action of these crusaders."[13] In the second camp (the last 1% of Creationists) are the scientifically trained "model builders" who use rigorous scientific methods to explore their belief in Creationism. In this chapter we shall explore the ideas proposed by these Creationists, who have formulated the Creationist movement's most scientifically advanced theories.[14]

To establish their position, or at least to make it credible, these model builders must deal with six major scientific issues:

1. Astronomical evidence of a very old universe, for example, light from distant stars and galaxies
2. Terrestrial geological evidence of a very old earth, for example, geological formations such as the Grand Canyon, the mid-Atlantic rift and plate tectonics, weathered craters, petrified rock, the history shown in rock formations, and the relative uniformity of the geologic column around the world
3. Radiometric dating techniques, which suggest very old ages for rocks
4. The fossil record, with its indication of successive rather than simultaneous flora and fauna (e.g., single-celled organisms, multicelled organisms, fish, dinosaurs, mammals, etc.) and long time spans that suggest that enormous numbers of creatures have lived in the past, too many to squeeze into 10,000 or so years

13. Quotes from "Creation-Evolution Debate Taking on a Less-Shrill Tone," *Washington Times*, January 17, 1996, 2A.

14. While it would be easy to criticize and poke fun at Creationists by quoting from the "crusaders" and other unsophisticated advocates, such an approach would not serve the purposes of this book.

5. Uniformity of the genetic code and the commonality of many genes, pseudogenes, and synteny blocks among organisms, which suggest common descent from a single ancestor
6. Similarities in physiological structure and function of organisms, also suggesting common descent from a single ancestor

This is, as the saying goes, "a tough row to hoe." To their credit, the model builder Creationists are not trying to paper over these difficulties but are aggressively seeking explanations of them. Nor do they deny or try to hide the fact that their belief system has its origin in the authority of the Bible. They do assert, however, that they can separate theology from science, and that they can argue against evolution and for creation strictly on nonbiblical grounds. As we have already mentioned, there is no guarantee that their efforts, however aggressive, will bear fruit in terms of viable explanations that can withstand the kind of rigorous scrutiny essential to science.

Major Players

Because of their obscurity to most scientists as well as the general public, we shall familiarize the reader with the major research arms of the modern Creationist movement, as they are responsible for most of the serious Creationist material published on evolution and related topics.

The Institute for Creation Research (ICR), in Santee, California, is probably the best-known Creationist organization. It was founded in 1972 by Dr. Henry Morris (1919–2006) and Dr. Duane Gish, its principal spokesmen. Both have written extensively on the subject, including books on the Genesis flood, refutations of Neo-Darwinism, and defenses of Creationist theories against attacks by their opponents; and both have participated in numerous debates over evolution. Morris was a professor of hydraulic engineering and author of a widely used textbook on the subject, so he speaks with some authority on matters related to hydraulics and thus the effects of the Genesis flood. (This does not mean that his views are correct; only that they cannot be dismissed out of hand.) The ICR facilities include the Museum of Creation and Earth History, a graduate school, and a library. The institute publishes and distributes a wide range of books (including textbooks), pamphlets, videos, and other materials related to Creationism. Its impact has been substantial, according to both Morris[15] and non-Creationists.[16]

15. Morris, *History of Modern Creationism*, 271.
16. Numbers, *Creationists*, 283.

The Creation Research Society (CRS), organized by Morris and other scientists in 1963, is focused on research and scholarly activities in creation-related areas.[17] To provide a publication outlet for creation-related research, it publishes a quarterly peer-reviewed journal, the *Creation Research Society Quarterly* (CRSQ), which first appeared in 1964 and is now in volume 41. (Peer review is by Creationists.) The CRS also publishes books on creation topics but does not generally engage in popularization of Creationist ideas or debates about it, leaving that function to ICR and AiG. It operates the Van Andel Creation Research Center, near Chino Valley, Arizona, and it sponsors scientific research related to Creationism. However, this research is but a tiny fraction of that done by scientists operating in mainstream science.[18]

Answers in Genesis (AiG), of Florence, Kentucky, is another well-known Creationist organization, founded in 1994 by Australian expatriate Ken Ham, who earlier was associated with ICR. Its agenda is a gloves-off, take-no-prisoners attack on evolution, accompanied by promotion of a young earth special creation (see fig. 6.1). This group is extremely ambitious; it has branches in the United States, Canada, the United Kingdom, Australia, Japan, and New Zealand. AiG publishes a popular magazine, *Answers*,[19] and until 2006, a more in-depth journal, *Creation Ex Nihilo Technical Journal*, with peer-reviewed articles, now in its twenty-first year.[20] Both the magazine and the journal are devoted to disseminating the antievolution, pro-young earth message. As in the case of ICR, the peer review is done by Creationists themselves. AiG has embarked on the construction of the $14 million, 95,000-square-foot Creation Museum, dedicated to explaining Creationist theories about the earth and its history.[21]

The ***Geoscience Research Institute*** (GRI), associated with Loma Linda University, is less well known than the other groups but has a distinguished history in Creationist research, with roots going back to 1957. It has a small staff of full-time scientists and publishes three periodicals, including *Origins*, a scholarly journal on creation issues. As a Seventh-day Adventist organization, it has steadfastly held to its beliefs in the recent creation of the earth, while still acknowledging that there

17. From web site: www.creationresearch.org/hisaims.htm.
18. Morris, *History of Modern Creationism*, 290–95.
19. Successor to *Creation*, published until 2006 by AiG and still published by Creation Ministries International in Australia, www.creationontheweb.com.
20. Also published by Creation Ministries International in Australia. This publication has been renamed *Journal of Creation*.
21. Answers in Genesis, www.answersingenesis.org/museum/overview.asp.

are problems to be resolved.[22] GRI tends to maintain a low profile but emphasizes careful research and quality work, in contrast to what they regard as the "sometimes slipshod presentations of Morris and his staff at the ICR."[23]

Center for Scientific Creation (CSC) is a small organization, dedicated to the promotion of physicist Walt Brown's hydroplate theory as an explanation of the physical features of the earth in a way compatible with recent creation. The hydroplate theory is integrated with other physical and biological theories so as to present a comprehensive scientific view of the world. The CSC publishes a widely used textbook on creation and evolution that expounds this view, a prominent feature of which is its series of thirty-one empirically testable predictions.[24] Many of these predictions have already been verified. CSC operates independently of other Creationist organizations, because the others reject the hydroplate theory as an explanation of the flood event.[25]

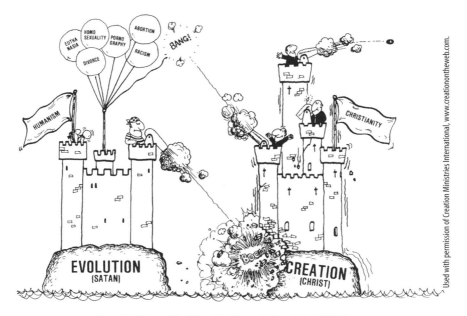

Figure 6.1. AiG view of the "Culture Wars." Cartoon by Steve Cardno, AiG/Brisbane.

Used with permission of Creation Ministries International, www.creationontheweb.com.

22. From the Geoscience Research Institute web site: www.grisda.org/about.htm.
23. Numbers, *Creationists*, 297.
24. In the context of scientific theories, the phrase "empirically testable" is generally synonymous with "refutable"; hence the value of such predictions.
25. Personal communication from Walt Brown to the author (TF).

Logical Structure of the Creationist Position

One can arrive at a Creationist position along two different paths. The first is to start with a belief in the inerrancy of the Bible, interpreted more or less literally, and then construct scientific theories compatible with that belief. This is the path that appears to be followed by most in the Creationist school. The other path emerges from dissatisfaction with standard scientific explanations of geological and biological phenomena (those discussed in chap. 3). Followers of this path construct theories that, in their view, better explain the phenomena but are based on (or point to) a young earth. A particular religious belief is not required for this path, though anyone following it is virtually certain to adopt such a belief if he or she is convinced of the truth of the theories. This is the path followed by those who are "refugees" from orthodox science, such as Russell Humphreys and Walt Brown.

Since everyone—Creationists as well as evolutionists—has access to the same observable data (the fossil record, geological formations, the geologic column, physiological and genetic data, as discussed in chap. 2), any differences must come either in immediate inferences from those data or in the theories proposed to explain the data and the inferences. A cursory examination suggests that the differences emerge from the immediate inferences. Indeed, some Creationists have claimed that the matter is really one of a "paradigm shift," along the lines of Thomas Kuhn,[26] so that the *same* evidence leads to *different* conclusions depending on the paradigm employed to understand it.[27] In general, the Creationist paradigm is based on rapid, high-energy processes rather than the slow, lower-energy processes characterizing conventional theories.

However, the mere existence of alternative paradigms—and hence alternative interpretations—does not mean that all are equally valid from a scientific standpoint. The true measure of a scientific theory is its ability to make predictions as well as explain convincingly all known facts. A standard criticism of Creationism is that it is untestable:

26. The phrase "paradigm shift" comes from Thomas Kuhn's famous book *The Structure of Scientific Revolutions* (Chicago: University of Chicago Press, 1962). The book emphasizes the fact that in periods of great theoretical change in science, people's whole way of thinking about the facts changes. Before, the collection of observed facts was perceived in one way; afterwards, the same facts are perceived differently. The classic example is perception of the motion of heavenly bodies, with respect to the geocentric and the heliocentric theories.

27. R. L. Wysong, *The Creation-Evolution Controversy* (East Lansing, MI: Inquiry Press, 1976), 365. Steven A. Austin, "Interpreting Strata of Grand Canyon," in *Grand Canyon, Monument to Catastrophe* (Santee, CA: Institute for Creation Research, 1994), 21.

> The reason that the ultimate statement of Creationism cannot be tested is simple: any action of an omnipotent Creator is compatible with any and all scientific explanations of the natural world. The methods of science cannot choose among the possible actions of an omnipotent Creator.[28]

While it is true that an omnipotent Creator could have created in any fashion that he saw fit, Creationists do not argue in this fashion. Rather, they posit that the Creator built the world in the manner described in the book of Genesis. This specific creation activity requires a specific physical history of the universe, the earth, and life. In fact, Creationists postulate a radically different sequence of events than that posited by evolutionists, one that must have left characteristic fingerprints in the cosmos and on the earth.[29] As we shall see in this chapter, they now correctly focus their efforts on the search for empirical evidence of these fingerprints, with theoretical work or projects planned or under way in nuclear physics, radiometrics, and astronomy. And as we have already noted, one of the branches of Creationist thought, the CSC, has made a series of refutable predictions based on its theory about creation. So while it is true that actions of a creator are "compatible with any and all scientific explanations of the natural world," only a limited set of scientific observations are compatible with the Creationist interpretation of the creator's actions. That is why the evolution debate must concentrate on empirical facts—they will ultimately tell the tale. We shall utilize these facts in our evaluation of Creationist theories in this chapter.

The general structure of all Creationist theories is shown in figure 6.2.

The Particular Scientific Theories of the Creationist School

To show that the science is consistent with a literal reading of the book of Genesis, Creationists need to explain a number of observations about the physical world that seem to contradict their ideas. Here we shall concentrate on theories proposed to explain directly observable facts, such as astronomical phenomena, geological features, and the

---

28. Scott, *Evolution vs. Creationism*, 19–20.

29. Science can and often does investigate whether explanations that invoke nonre-peatable events are viable *because they leave telltale signs*. Perhaps the best-known case is the big bang theory, which was accepted over the steady state theory after a key prediction and telltale sign, the cosmic background radiation, was discovered. Other examples include plate tectonics and biology's own "big bang," the Cambrian explosion.

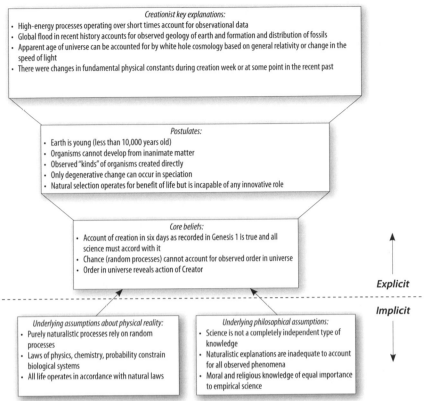

Figure 6.2. Logical structure of the Creationist theory.

fossil record, that seem to be at odds with the Creationist worldview. The theories proposed to support the Creationist view of the world include the following hypotheses:

1. The apparent age of the universe can be explained either in terms of gravitational time dilation, a decrease in the speed of light over time, or a distinction between local and cosmic time.
2. The apparent age of the earth (rocks and other materials), as disclosed by radiometric techniques, stems from a change in physical constants and/or erroneous or flawed radiometric data.
3. Observed geological formations and data, including the geologic column, can be explained by the Genesis flood, which in turn may be related to catastrophic processes such as volcanic activity, rapid plate tectonics, or the hydroplate theory. Fossils and the fossil record (historical evolution, in our terminology) can be explained by deposition occurring during the high-energy processes of the flood.

4. Flora and fauna in the fossil record all existed contemporaneously, or nearly so, inhabiting different ecosystems, rather than different time periods.

In addition to these four theories, Creationists have also posed two major challenges to the notions of common descent and strong Darwinian evolution:

5. Thermodynamics demonstrates that evolution is impossible.
6. Organisms (taxa) can change over time, but only in a degenerative manner.

Physicist Walt Brown has synthesized his version of these items into a consistent theory, albeit one that concentrates heavily on the physical processes involved in the first three of these hypotheses. Others tend to concentrate on one or a few of the areas. As a result of this heterogeneity, we do not claim that the six hypotheses listed represent any particular Creationist's position. There is disagreement among members of the school on many points, and they regard many of these subjects to be areas of active research. Limitations of space preclude discussion of all evidence cited by Creationists and of all theories they advance, and in general permit only a sketch of conventional science's critique of the Creationist theories. Readers who wish to pursue any of the topics further may readily find material in the bibliography and in the references cited.

The Creationists' Theories: Detailed Discussion

The Apparent Age of the Earth and the Universe

Because the age of the earth and the universe is so crucial to the evolution controversy, credibility requires Creationists to posit an explanation for the apparent age of both and reconcile this scientifically with the creation account in Genesis. Creationists do not dispute that the universe *appears* to be very old—after all, anyone with an inexpensive backyard telescope can view objects such as galaxies known to be millions of light-years distant, and whose light must therefore have taken millions of years to reach earth, based on the time–distance formula ($d = vt$, *distance equals velocity multiplied by time*) and the known speed of light.[30] Creationists recognize that this is a serious

30. This is because a light-year is the distance light travels in one year, and assuming the speed of light to be constant and ignoring possible time dilation effects and expan-

problem with their overall position, perhaps the most serious.[31] Over the years a number of solutions for this problem have been proposed, and some have become favorite targets of anti-Creationist groups.[32] Among the now discarded ideas is the "mature creation" or "created in transit" hypothesis, according to which the world was created with light waves in transit along their paths throughout the universe. This hypothesis implied that earthbound observers would "see" things that never actually happened, for example, supernova explosions in distant galaxies. Its untestability, combined with the disturbing implication that God purposely deceives mankind, gave ample ammunition to evolutionists and others who wish to ridicule Creationists and discredit them as serious scientists. For this reason, among others, new theories have been advanced. At present, at least three are on the table. Since the distance (d) between us and distant stars is not in dispute, either time (t) or the velocity of light ($v = c$) or both must have changed in the above formula.

A Change In the Speed of Light, C

This theory assumes that the speed of light has changed during the course of natural history.[33] Based on historical data back to the seventeenth century, and using extrapolation, Barry Setterfield has proposed that the speed of light has been decaying (see fig. 6.3).[34] The rate he determined was such that a mere 6,000 years ago, c was millions of times greater than today's value of 2.99792×10^8 meters/second. Such a speed change would effectively shrink the age of the universe from 15 billion years to around 15,000 years, because with increased speed, light from the most distant realms of the universe reaches earth much more quickly. It represents an ideal solution for

---

sion of the universe, the light from an object 1 million light-years away must have left it 1 million years ago.

31. D. Russell Humphreys, *Starlight and Time* (Green Forest, AR: Master Books, 1994), 9.

32. This discussion is based on Humphreys's book, *Starlight and Time*, Appendix A.

33. A variant on this idea has recently been advanced to resolve a problem of cosmology in the big bang theory, the so-called horizon problem, which has to do with the observed uniformity in the cosmic background radiation everywhere in the universe. The problem is that the present speed of light is too low to have permitted all of these parts of the universe to have communicated with one another and come to equilibrium. One solution—somewhat ad hoc—is the inflation theory proposed by Alan Guth in 1981. Another proposed solution is the decaying speed of light (also somewhat ad hoc); however, unlike the Setterfield hypothesis of a gentle slowing down to the present time, in this proposal all of the speed reduction occurred almost immediately after the big bang.

34. Paul Steidl, *The Earth, the Stars, and the Bible* (Grand Rapids: Baker Books, 1979); Barry Setterfield and Trevor Norman, *The Atomic Constants, Light, and Time*, privately published, 1980–1990, available at www.setterfield.org/report/report.html.

the Creationists because it has the capacity to make both the universe *and* the earth young, and may even be able to resolve problems such as the missing mass in the universe. Other Creationists have criticized this effort.[35] Recent observations do suggest that physical constants (of which the speed of light is one) may indeed change over time,[36] though by far less than the factor of a million required to shorten the age of the universe to 10,000 years. The signature effect of this change in c is a very marked time dilation,[37] which should be readily observable on astronomical scales, but is not seen. Another problem is that any change in physical constants threatens to destroy the stability of matter, which is very delicately balanced.[38] In particular, changes in the fine structure constant α, which would accompany a change in the speed of light, should be visible in the distant starlight, but they are not.[39] Finally, there is the problem in that stars and galaxies both appear to have life histories of their own. Stars appear to evolve in a certain pattern from their birth until their death. For most, this pattern, along what's known as the "main sequence," appears to be a viable inference from such standard astronomical tools as the Hertzsprung-Russell diagram, which suggests millions if not billions of years for stellar life—a problem not fixed by any speed of light change. If stars were created in their position along the main sequence, we are back to the issues that derailed the "created in transit" hypothesis. Observationally, stars (and galaxies) should look younger as distance from earth increases, if this theory is correct.[40]

35. Gerard Aardsma, "Has the Speed of Light Decayed Recently?" *Creation Research Society Quarterly* 25, no. 1 (1988): 36–40; D. Russell Humphreys, "Has the Speed of Light Decayed Recently?" *Creation Research Society Quarterly* 25, no. 1 (1988): 40–45; Roy D. Holt, "The Speed of Light and Pulsars," *Creation Research Society Quarterly* 25, no. 2 (1988): 84–86; R. H. Brown, "Statistical Analysis of *The Atomic Constants, Light, and Time*," *Creation Research Society Quarterly* 25, no. 2 (1988): 91–95; for a detailed mathematical analysis, see www.evolutionprimer.net.

36. Andreas Albrecht and João Magueijo, "Time Varying Speed of Light as a Solution to Cosmological Puzzles," *Phys. Rev. D* 59 (4) (February 15, 1999): 43516–19. Also João Magueijo, *Faster than the Speed of Light* (Cambridge, MA: Perseus Publishing, 2003).

37. In physics, "time dilation" means a slowing down of clocks, and thus of all physical processes. Both the special and the general theory of relativity showed that time does not have an absolute or "correct" value throughout the universe, but varies depending on the speed, acceleration, and gravity experienced by the observer.

38. John A. Peacock, *Cosmological Physics* (Cambridge: Cambridge University Press, 1999), sec. 3.5, "The Anthropic Principle."

39. The fine structure constant is a fundamental physical constant that characterizes the strength of the electromagnetic force. It is closely related to the speed of light and several other physical constants.

40. The theory and its problems are discussed in detail on Thomas Fowler's web site, www.evolutionprimer.net.

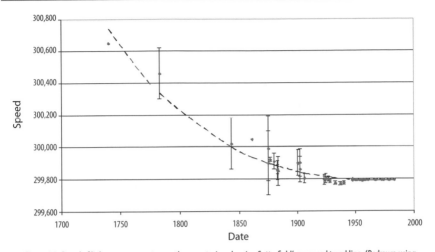

Figure 6.3. Speed of light measurements over three centuries, showing Setterfield's proposed trend line. (Redrawn using Table 11 data from Setterfield, *Atomic Constants, Light, and Time*.)

Gravitational Time Dilation

This second approach assumes that time (t) in the time–distance formula has varied; in particular, that time on earth slowed down relative to the rest of the universe, so that only some 6,000 years on earth passed when billions of years passed in distant parts of the universe. The theory utilizes standard physics from Einstein's general theory of relativity, but with a different set of boundary conditions. Russell Humphreys, the originator of this approach, examined the premises of the usual versions of the big bang theories and noticed that there is one premise that is not actually based on direct observations, but postulated as a "reasonable" assumption. It is the "cosmological" or "Copernican" principle: the universe is spatially homogeneous and thus has no boundary, that is, no edges and no center. According to this principle, the universe itself expands along with matter. It implies the general "sameness" of the universe, no matter where we look—something referred to as "isotropy." Humphreys argues that one does not have to start with the assumption of homogeneity that drives the standard big bang cosmology. The isotropy or sameness could be due to something else, namely, the assumption that "matter in the universe *has* a center and an edge (is bounded)."[41] In this case, isotropy follows not from homogeneity, but from the privileged place of earthbound observers, namely, near the center of the universe. The significance of this assumption is that it allows Creationists to solve the problem of the "old" universe within the context of general relativity. Roughly speaking, the argument is that when the universe was much smaller than at present, the density

41. Setterfield, *Atomic Constants, Light, and Time*.

of matter was so high that it would have to be in either a black hole[42] or a white hole. In such a hole, time slows down. A black hole is ruled out because it would cause the universe to collapse, so it must have been in a white hole (hence the name "white hole cosmology" for this hypothesis). As the universe expanded out of the white hole, the event horizon would have shrunk.[43] Due to gravitational time dilation, time slows down as one approaches an event horizon and effectively stands still at the horizon. If the event horizon crossed the earth's surface at the appropriate time, hypothesized to be the beginning of the fourth day of Creation Week (see Genesis 1:14–19), then billions of years of time would go by for the rest of the universe, while only a few hours transpired on earth. Thus the universe would appear to humans on the earth more or less as it does now, even though the earth itself was only a few days old (as measured by earth clocks). This is an elegant solution of the young earth problem, but the existence of white holes, though theoretically possible, has not been verified. Some dispute that they can exist with matter in them or claim that they would violate the laws of thermodynamics.[44] Moreover, this method would solve only the problem of the age of the universe, not the (apparent) age of the rest of the solar system, or the problem of its characteristics, such as the distribution of the planets and their moons. It has also been criticized because it does not yield enough local time dilation to achieve the desired result.[45] Variants on this model have been advanced, which also rely on some type of difference in earthbound clocks and those in other regions of space.[46] Observationally, objects more distant would look older than closer objects if this model is correct.

42. A black hole is an entity whose existence is predicted by the general theory of relativity. It is a place where the high density of matter creates such intense gravitational fields that even light cannot escape, thus the hole cannot be directly seen. However, matter in its vicinity is sucked into it, often reaching very high velocities in the process, and giving off considerable radiant energy. The "event horizon" of a black hole is the point of no return, so to speak; it is that boundary around the hole from which, once crossed, there is no escape. It is a sphere whose radius is known as the "Schwarzschild radius," which depends on the mass of matter within the hole. Due to quantum mechanical effects, British physicist Stephen Hawking has shown that black holes will eventually evaporate, as a result of what is known as "Hawking radiation." A white hole is a black hole running in reverse: nothing can enter it, and all light is reflected off of it, hence its name.

43. For a black hole, anything (including light) at or within the event horizon can never leave the hole. For a white hole, anything that approaches the event horizon from the inside will leave the hole forever.

44. Andrew Hamilton, "White Holes and Wormholes," http://casa.colorado.edu/~ajsh/schww.html.

45. J. G. Hartnett, "Look-Back Time in Our Galactic Neighborhood Leads to a New Cosmogeny," *Creation Ex Nihilo Technical Journal* 17, no. 1 (2003): 73–79.

46. J. G. Hartnett, "A New Cosmology: Solution to the Starlight Travel Problem," *Creation Ex Nihilo Technical Journal* 17, no. 2 (2003): 98–102.

Alternate Synchrony Convention

This third approach also utilizes changes in time (t) but in a different way, namely by distinguishing cosmic local versus cosmic universal time. Obviously biblical authors knew nothing about the size and structure of the universe as a whole, and they thought of time only in connection with the changes they saw around them. That is, the language of the Bible is phenomenological—it describes appearances only (cosmic local time). This convention is used even today by astronomers for some purposes. For example, supernova 1987a is so called because it was observed in 1987, even though, measured by cosmic universal time, it occurred 179,000 years ago. According to this hypothesis, stars and galaxies were made millions or billions of years prior to day four of Creation Week, but matters were arranged so that all the light, regardless of its origin, arrived at the earth on day four. Thus the time given in the Bible is a kind of "time stamp" for what was seen, but not what actually happened.[47] To the obvious objection that this theory implies that the stars were "really" created before day four, proponents argue that there is no "absolute" time, so local time is as good as any other time. This theory requires that stars and galaxies were created in shells from earth outward, as measured by cosmic universal time. So observationally the theory predicts that objects should look younger as distance from earth increases, with an age-distance relationship.[48]

If any of these theories were to be verified, it would shrink the age of the earth drastically and thereby demolish the Neo-Darwinian and Meta-Darwinian theories of evolution. However, if these theories fail, it bodes ill for the entire Creationist paradigm, presumably compelling it to fall back on some type of miraculous creation event, with many of the attendant problems of the "created in transit" hypothesis. The age-distance relation provides a possible experimental test of the theories presented here, so it appears possible to resolve this issue scientifically.[49] We leave it to the Creationists to decide if these theories are in fact compatible with a literal interpretation of the Bible.

Finally, Creationists often point out that the big bang has a light time-travel problem of its own. This is correct; it is the well-known "horizon problem," and curiously, one of the proposed solutions involves a change

47. Ibid.
48. This relationship should be of the general form $t(r) = t_{LSN} / (z+1)$, where $t(r)$ is elapsed time at distance r, t_{LSN} is Local Solar Neighborhood time, and z is redshift; based on a lecture by Jason Lisle in *Distant Starlight: Not a Problem for a Young Universe*, DVD (Petersburg, KY: Answers in Genesis, 2005).
49. Detailed technical discussions of these theories appear on Thomas Fowler's web site, www.evolutionprimer.net.

in the speed of light. However, the issue in question here is *not* between the big bang and Creationist theories, in the sense of either-or; one need not believe in the big bang at all to accept an old age for the universe. Rather, the dispute is about a fact: whether the universe and the earth are old or young.

Geology and the Apparent Age of the Earth

The earth is generally considered to be about 4.5 billion years old, based on geological evidence and radiometric dating methods. This poses a substantial problem for Creationists, who must reduce its age to 10,000 years or thereabouts. A summary of the different approaches of standard science and Creationism to the question of earth's history is shown in table 6.1. Creationists seek to substitute brief high-energy processes for low-energy processes acting over long time spans, the mode currently favored by most geologists.

Table 6.1. Comparison of standard and Creationist perspectives on historical geology*

Establishment [evolutionary] geologic paradigm (EGP)	Diluvial [Creationist] geologic paradigm (DGP)
The present is the key to the past.	The past is the key to the present.
Objective: Explain all geologic phenomena in terms of present processes, and, whenever possible, in terms of present rates. When this fails, invoke cyclical events or intermittent catastrophes.	**Objective:** Explain all geologic phenomena in terms of processes compatible with biblical events and chronology [as understood assuming a literal interpretation].
Primary variable: time	**Primary variable:** energy
Characterized by low-energy processes acting over immense periods of time. Evolutionist catastrophists see energetic processes acting intermittently through great spans of time.	Characterized by high-energy processes acting continuously over a limited period of time (principally Noah's flood). Other geologic events have been much smaller than the deluge, but have also tended to be brief, energetic events.

* Peter Klevberg, "The Philosophy of Sequence Stratigraphy Part II—Application to Stratigraphy," *Creation Research Society Quarterly* 37, no. 1 (June 2000): 36–46.

Given the widespread acceptance of the standard "uniformitarian" geological model (see "Laying the Foundations" in chap. 2), with its strata and timescale, there are two avenues open to Creationists. They can deny the validity of the usual stratigraphic order of rocks, or "geological column," as it is commonly known, asserting that it has no real chronological significance, either relative or absolute. Alternatively, they can accept the column's order as valid and of significance, that is, of relative value, but reject the usual absolute chronology associated with it. Increasingly, Creationists are turning away from the first approach, and adopting the second:

The creationist "myth" [is] that the geologic column is the product of an evolutionary/uniformitarian "conspiracy" and so essentially doesn't exist/isn't real. . . . However, the physical reality of the strata of the geologic column cannot be ignored, as they do exist. . . . It is time for the creationists to bury their "myth," face up to the reality of the geologic record (not the time scale imposed on it, of course), and tackle the exciting task of building the Flood model of earth history based on that record.[50]

(We note, however, that some Creationist researchers still reject the geologic column, regarding it as an artifact.[51]) Creationists have sought to clarify matters by pointing out that there are really three columns, not one. These are generally conflated in non-Creationist thought:

- Lithostratigraphic column (order of rocks found by geologists)
- Biostratigraphic column (order of fossils, commonly associated with layers or periods of the lithostratigraphic column)
- Chronostratigraphic column (dating system—age of the rocks)

Historically, these arose separately; for Creationists, this implies that they can be separated once again.[52] To support this position, they have zeroed in on what may be termed "discrepancies" in the standard geological model, owing to omissions or inversions in the usual stratigraphic layers, and also to the fact that the entire column is visible only in a very few places in the world. They take this as evidence that the layers do not represent a chronological sequence, but something else, namely, formations resulting from the Genesis flood. The discrepancies they allege are:[53]

- Paraconformities: there are instances of missing intermediate strata, for example, Devonian rocks resting on Cambrian rocks (Ordovician and Silurian missing).

50. Andrew Snelling, review of *Studies in Flood Geology*, by John Woodmorappe, *Creation Ex Nihilo Technical Journal* 9, no. 2 (1995): 162.
51. John K. Reed and Carl R. Froede, "Bible-Based Flood Geology: Two Different Approaches to Resolving Earth History—A Reply to Tyler and Garner," *Creation Research Society Quarterly* 37, no. 1 (June 2000): 61–66.
52. David J. Tyler and Paul Garner, "The Uniformitarian Column and Flood Geology: A Reply to Froede and Reed (1999, CRSQ 36:51–60)," *Creation Research Society Quarterly* 37, no. 1 (June 2000): 60–61.
53. This list follows John Woodmorappe, "An Anthology of Matters Significant to Creationism and Diluviology: Report I," in *Studies in Flood Geology* (El Cajon, CA: Institute for Creation Research, 1999), 140–42.

- Present sedimentation rates extrapolated backward yield erroneous (too large) thicknesses for formations.
- Placer deposits (glacial or alluvial deposits of sand or gravel that contain valuable minerals) appear mostly in Tertiary formations, rather than throughout the strata, as would be expected under the normal interpretation.
- Paucity of ecological relationships among ancient life as revealed in fossils.
- Orogenic (mountain-building) processes do not show expected punctuating effects within strata.
- Geologic deposition rates are inconsistent with paleontological evidence.
- Number of fossils known from fossil record is too low to be consistent with normal timescale of geological processes.

Of course, explanations for some of these problems can be found in conventional theories, and mainstream geology disputes that some are problems at all. The Creationists use these problems to criticize the conventional model and propose a new interpretation of the columns, namely that the strata were deposited rapidly under the violent conditions of the Genesis flood, a high-energy process giving rise to the geologic column.[54] Furthermore, the biostratigraphic column stems from the burial of various different ecosystems, all contemporaneous with the flood period (see below: "The Fossil Record: Simultaneous, Not Sequential").[55] As a result, the chronostratigraphic column is of relative value only; absolute dates associated with it are an artifact and can be explained by assuming one or more periods of accelerated radioactive decay (see below: "Radiometric dating").

While this could theoretically account for the geological column, to explain major geological features such as the ocean floor and mountains, the Creationists need still more help. For these features, some Creationists emphasize what they term "vertical tectonics," the large up and down movement of significant portions of the earth's surface, in contrast to plate tectonics, which deal more with lateral motion.[56] They

54. John Baumgartner, "Catastrophic Plate Tectonics: The Geophysical Context of the Genesis Flood," *Creation Ex Nihilo Technical Journal* 16, no. 1 (2002): 58–63, www.answersingenesis.org/home/area/magazines/tj/docs/16n1p58baumgardner.asp.

55. Robert Gentet, "The CCC Model and Its Geologic Implications," *Creation Research Society Quarterly* 37, no. 1 (June 2000): 10–21.

56. Of course, such lateral motion can lead to vertical motion, when plates collide, e.g., pushing up mountains.

believe that many if not most of these movements took place during the time of the Genesis flood, though they do not profess to understand the mechanisms. The hydroplate theory, discussed below, does propose mechanisms to explain these features.

Creationists use the vertical tectonics movements to resolve the obvious problem of the source of all the water needed to cover the entire earth.[57] The Creationist solution, based on Psalm 104:8–9, is that "the ocean did not have to rise to cover Mount Everest. The Himalaya Mountains rose out of the floodwaters."[58] This uplifting of course accounts for the deposition of marine fossils found there.

Erosion

According to Creationist theories, when the earth was covered with waters from the flood, several mechanisms[59] would cause large velocities in the water, including earthquakes, tidal effects, the uplifting of continents at the same time that ocean basins were sinking, and the rotation of the earth itself, which according to Creationists' modeling would cause currents up to eighty meters per second.[60] These large currents were the source of the erosion seen on the earth's surface.[61] The effect of this flow would be to cause sheet flow erosion, in which entire layers were removed almost instantly. Formations such as Devils Tower in Wyoming and Monument Valley in Utah and Arizona are cited as examples of sheet flow erosion stemming from the flood. Other evidence cited for the occurrence of sheet flow erosion includes large erosion surfaces—essentially smoothly planed rock. Such erosion surfaces can occur even on tilted sedimentary rocks, in which many layers of sediment are exposed at the surface (fig. 6.4a), all of which appear to have eroded evenly. According to Creationists, this suggests a recent cataclysm; otherwise, the different sediments would have begun to erode at different rates and the surface would cease to be flat (see fig. 6.4a and b).[62]

57. Bernard Ramm, *The Christian View of Science and Scripture* (Grand Rapids: Eerdmans, 1954).

58. Michael J. Oard, "Vertical Tectonics and the Drainage of Floodwater: A Model for the Middle and Late Diluvian Period—Part I," *Creation Research Society Quarterly* 38, no. 1 (June 2001): 3–17. Quote from p. 9.

59. Ibid.

60. D. Barnette and J. Baumgardner, "Patterns of Ocean Circulation over the Continents during Noah's Flood," in *Proceedings of the Third International Conference on Creationism, Technical Symposium Sessions* (Pittsburgh: Creation Science Fellowship, 1994), 77–86.

61. Oard, "Vertical Tectonics," 9–10.

62. Ibid.

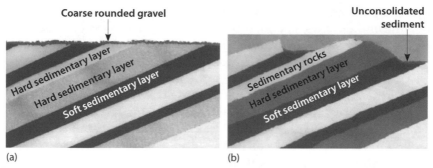

Figure 6.4. (a) Gravel-capped erosion surface on sedimentary rocks expected from recent erosion activity. (b) Similar surface eroded and weathered over a long period. (Figures 9 and 10 are from Michael J. Oard, "Vertical Tectonics and the Drainage of Floodwater: A Model for the Middle and Late Diluvian Period—Part I," *Creation Research Society Quarterly* 38:1 [June 2001]. Used by permission.)

Of course, one can agree that *some* geological features are the product of recent large floods without being committed to the Creationist position that *all* of them are so. In this connection, the reader needs to be on the lookout for the following reasoning: "If the Genesis flood occurred, then we would see sheet flow erosion. We see it, therefore there was the Genesis flood."[63] This does not follow because sheet flow erosion is the product of large-scale, rapid flooding, which occurs when natural dams break. Thus it can occur in the absence of the Genesis flood. As always, the reader has to be careful about what conclusions can legitimately be drawn from the evidence presented, and what alternate explanations there may be.

Sedimentation

The Creationist explanation of sedimentation, and the rapid formation of very thick layers of it, once again relies on the high-energy processes associated with the Genesis flood. Creationist geologists believe that such high-energy processes can account for observed stratification over very short periods and that strata form spontaneously when the right conditions are present, thus obviating the need for long time periods. They cite three sources of evidence to confirm their position. First is the work of French geologist Guy Berthault, who conducted experiments on the spontaneous formation of multiple laminations during sedimentation of heterogranular mixtures of sediments in air, in still water, and in running water.[64] Second, Berthault's work has been confirmed

63. Technically, the argument is an example of the logical fallacy of "affirming the consequent."
64. G. Berthault, "Experiments on Lamination of Sediments, Resulting from a Periodic Graded-Bedding Subsequent to Deposition—A Contribution to the Explanation of Lamination of Various Sediments and Sedimentary Rocks," *Compte Rendus Académie*

by other researchers, who discovered that spontaneous stratification into alternating layers of large and small grains takes place when the large grains have a larger angle of repose.[65] Their experimental application was to explain the observed stratification found in the snow of avalanches; however, the application to geology is immediate. Third, Creationists point to direct empirical evidence from nature, stemming from the eruptions of Mount St. Helens in the early 1980s (which in some ways has been a godsend for Creationists).[66] The formation in question is shown in figure 6.5. The same caveat applies here as in the previous section, however.

Figure 6.5. Deposits from the Mount St. Helens eruption, 1980–82. The 7.6-meter fine layering (center) was made by what Snelling describes as "hurricane velocity flows from the crater of the volcano on June 12, 1980. A mud flow layer from 1982 is above it, and air-fall debris from the May 18, 1980 eruption is below it. Photograph copyright 1991 by Geology Education Materials and used by permission of Dr. Andrew Snelling.

Fossilization

Fossilization is the process by which the remains of plants or animals are lithified—converted to stone—as a result of processes usually assumed to have taken long periods of time. Typically, it is assumed

des Sciences, Paris, 303 (Série II, no. 17) (1986): 1569–74; Berthault, "Sedimentation of a Heterogranular Mixture: Experimental Lamination in Still and Running Water," *Compte Rendus Académie Des Sciences,* Paris, 306 (Série II) (1988): 717–24.

65. H. A. Makse, S. Havlin, P. R. King, and H. E. Stanley, "Spontaneous Stratification in Granular Mixtures," *Nature* 386 (1997): 379–82.

66. Andrew Snelling, "Sedimentation Experiments: Nature Finally Catches Up!" *Creation Ex Nihilo Technical Journal* 11, no. 2 (1997): 125–26.

that a plant or animal dies, is covered by sediment, and the entire layer turns to sedimentary rock. Processes involved in the lithification include cementation, compaction and dessication, and crystallization. In cementation, "spaces between the individual particles of an unconsolidated deposit [the fossil] are filled up by some binding agent. . . . Apparently the cementing material is carried in solution by water that percolates through open spaces between the particles of the deposit."[67] Compaction occurs when the sediment is subjected to high pressures, as from overlaid sediments or earth movements; crystallization can occur when certain types of minerals are laid down in suitably pure form, though the nature of the mechanisms involved is not well understood.

The time required for lithification, and thus for fossilization, is not known because conditions vary widely, and there is uncertainty about the mechanisms involved. Creationists argue that, in fact, lithification can occur extremely rapidly, and point to numerous examples where this rapid lithification and fossilization has occurred. A petrified (lithified) hat is on display in a mining museum in New Zealand,[68] and petrified wood can form quickly as well. In fact, one process for making petrified wood has been patented. They also argue that fossilization by the methods ordinarily hypothesized could never occur because the animals or plants would disintegrate too quickly.[69] Other research (not done by Creationists) indicates that with only the right chemicals, opals and coal can form quickly, without heat or high pressure.[70] So it has been established that at least some types of fossilization can occur over short time periods. As before, the fact that fossilization *can* occur rapidly does not mean that it always *did* do so. It does not rule out the possibility that it can also occur over long time spans. Clearly, what is needed is more research on the mechanisms and timescales of fossilization.

Plate Tectonics and Vertical Tectonics

Another standard part of mainstream geology is plate tectonics, which has become a topic of spirited discussion within the Creationist community.[71] According to plate tectonics, the earth's surface is divided into

67. Don Leet, Sheldon Judson, and Marvin Kauffman, *Physical Geology*, 5th ed. (Englewood Cliffs, NJ: Prentice Hall, 1978), 99.
68. Carl Wieland, "The Earth: How Old Does It Look?" *Creation Ex Nihilo* 23, no. 1 (December 2000): 8–13.
69. Nearly all mainstream geologists agree that at least burial must occur quickly or fossilization will not take place.
70. R. Hayatsu, R. L. McBeth, R. G. Scott, R. E. Botto, and R. E. Winans, "Artificial Coalification Study: Preparation and Characterization of Synthetic Macerals," *Organic Geochemistry* 6 (1984): 463–71.
71. Michael Oard, "Is Catastrophic Plate Tectonics Part of Earth History?" *Creation Ex Nihilo Technical Journal* 16, no. 1 (2002): 64–68.

a number of large plates that slowly move, and their movements and collisions account for many of the earth's prominent geological features, such as mountain ranges. While the idea that continents move laterally in the earth's crust was first seen as an anathema, it appears that Creationist geologists are more and more accepting some form of it as a reality, and attempting to integrate it into their flood theories. Of course, they turn to high-energy events rather than slow and gradual movement of the plates.[72] By starting with all the continents in the primitive Pangean conformation (all the continents are united as one land mass) and assuming runaway subduction (very rapid movement of the plates), Creationist researchers have modeled a pattern of plate motions that resembles the inferred movement of the continental plates to their current positions.[73] What triggered the events leading to the hypothesized runaway subduction is not clear, and the theory has many critics. In fairness, it is worth mentioning that it is not clear what mechanisms triggered the traditional slow movement of the plates favored by mainstream geologists.[74] Regardless of the mechanisms, though, Creationists need to make predictions based on their work which differ from those of mainstream geologists. A very brief period for the tectonics, instead of hundreds of millions of years, must result in significant differences in some observable area leading to testable predictions.

Alternate Creationist Explanation of Earth's Geology: The Hydroplate Theory

The hydroplate theory of Walter Brown is an alternate explanation of the earth's geological features. The theory seeks to explain many of the problems facing mainstream geology, such as the source of all the pure limestone deposits around the world; the formation of metamorphic rock at great depths and pressures in the presence of water; the formation of the great plateaus of the world (Colorado, Tibetan); the formation of salt domes; the origin of numerous and often large submarine canyons; and the origin of the forces that move plates, among many others.

The theory rests on two postulates. First, approximately half of the water in the world's oceans today was once beneath the earth, in chambers that were interconnected.[75] According to the theory, the thickness of this water shell was a little more than a kilometer. A granite crust was above the water, and basaltic rock below it (see fig. 6.6). The second postulate is

---

72. John Baumgardner, "Catastrophic Plate Tectonics: The Geophysical Context of the Genesis Flood," *Creation Ex Nihilo Technical Journal* 16, no. 1 (2002): 58–63.

73. Ibid.

74. Richard Milton, *Shattering the Myths of Darwinism* (South Paris, ME: Park Street Press, 2000).

75. Walt Brown, *In the Beginning*. See www.creationscience.com.

that pressure in the subterranean water chamber continued to increase. Once the pressure reached the critical point, the crust began to rupture, and that rupture rapidly traveled around the earth (in 2–3 hours) and is visible to us today as the Mid-Atlantic Ridge and other such undersea ridges. This led to supersonic jets of water and sliding plates lubricated by the water underneath them (i.e., rapid plate tectonics).[76] The water forced out of the crack at high pressure led to worldwide torrential rains far exceeding anything before or since, leading to an enormous flood. Sediments (partly from the basalt below the water chambers) settled over the earth's surface, burying plants and animals and leading to the formation of most fossils. Carbon dioxide trapped in the subterranean waters under high pressure naturally escaped when the water exploded out of the chambers, resulting in the precipitation of limestone deposits seen around the world. Floods and violent conditions uprooted most vegetation and it was transported to locations where it accumulated. Some of it was later rapidly compressed and buried, forming coal and oil deposits. As the Mid-Atlantic Ridge began to rise, it caused the granite plates to slide away from it. Removal of this weight caused the floor to rise on the order of fifteen kilometers. Gravity thus caused the tectonic plates to move rapidly, lubricated by the water underneath, before they settled into their present position.[77]

Great earthquake activity followed, but it then slowed down and today's earthquakes are just part of a final "settling in" process. In the recovery phase of the flood, rain diminished, and most of the geological features we observe today were formed. The hydroplates sank into the basalt, which caused ocean floors to rise. As mountain ranges sank, plateaus adjacent to them rose (which is why plateaus are adjacent to mountain ranges). Some water remained trapped under the plates. Among the testable predictions of this theory are the following:

1. Large volumes of pooled saltwater, left over from the water in the chambers, should lie underneath major mountains.[78]
2. Salty water should be in cracks in granite eight to sixteen kilometers below the earth's surface, where surface water cannot penetrate.[79]
3. A sixteen-kilometer-thick layer of granite (a hydroplate) should exist one to two kilometers below the floor of the western Pacific.[80]

76. For an animation and quick overview of the theory, see www.thetaxpayerschannel .org/graphics/creation/fonte23.mov.

77. A good diagram of the tectonic plates can be found at the web site maintained by the USGS: http://geology.er.usgs.gov/eastern/plates.html.

78. Brown, *In the Beginning*, 105, prediction 1.

79. Ibid., prediction 2.

80. Ibid., 128, prediction 6.

4. Fossils of land animals, and not only shallow water animals, should exist near undersea trenches.[81]

5. Rotational speed of the inner core of the earth is decreasing.[82]

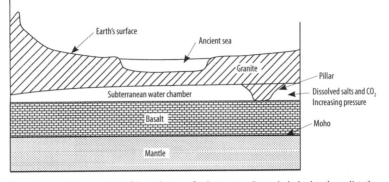

Figure 6.6. Representative geology of the earth prior to flood events according to the hydroplate theory. Note the water chamber underneath the granite.

Should they be confirmed by further research, Creationism, the flood theory, and the young earth hypothesis will all receive a boost. One prediction made in 1995—that crystalline rock under key locations such as Gibraltar, the Bosporus and Dardanelles, and the Golden Gate, should be found to be eroded in a V-shaped notch—was confirmed (at least for the Bosporus and Dardanelles) in 1998. According to the author's web site, prediction 1 above, first made in 1980, has been confirmed with the discovery of a 1.6-kilometer-thick layer of saltwater beneath the Tibetan Plateau.[83] However, Brown's theory does not address the problem of the age of the universe and of the earth. To resolve this problem, Brown currently favors the change in the speed of light hypothesis, which, as we saw, has a number of difficulties.

Some Problems with Creationist Geology

Creationists can point to evidence that stratification and sedimentary rock formation has occurred in a short period under violent conditions. They can also point to lithification and mineral formation that has occurred over short periods. But the fact that sedimentation, stratification, and lithification *can* occur in a short time does not imply that all such formations *did* originate this way. In particular, it is necessary to show that *all* types of sedimentary rock, including shale, sandstone,

81. Ibid., 129, prediction 7.
82. Ibid., 135, prediction 9.
83. From the web site www.creationscience.com.

and limestone, can form in a short time by means of the high-energy processes envisioned. For that conclusion, further evidence is required, especially since shale at least appears to require calm waters acting over long periods for its deposition.[84] Thus, what is needed is further scientific evidence that deposition and stratification, with the requisite fossil remains always in the proper strata, can occur as the result of the high-energy processes theorized for the Genesis flood.[85]

This poses an obvious difficulty because the large energy processes hypothesized in the flood geology paradigm would cause very large degrees of turbulence. At eighty meters per second, water is flowing extremely fast, and the flows would be laminar in few if any cases. The resulting turbulence would tend to churn anything in the water, and so fossils from the different ecosystems should be found mixed together in many places, rather than in separate layers. This is not observed; the fossils are almost always separate, even if inverted. Proof that such high degrees of turbulence can correctly sort organisms, as with the sand grains, is required, along with proof that all observed types of sedimentary rock can be formed by the hypothesized processes at the high speeds.

To help overcome the problem of sorting the fossils, Creationists put forth a number of hypotheses. First, they argue that the observed stratigraphic separation of fossils is an artifact of conditions of physical proximity. Creationists believe that the fossils found in different strata did not exist in close proximity to start with.[86] Furthermore, they claim that different fossils were deposited under specific tectonically differentiated conditions, thus giving rise to what they refer to as "tectonically associated biological provinces," presumed to be a principal cause of observed biostratigraphic differentiation.[87] However, it is unclear that these caveats can explain all of the significant features of the geologic column, such as the fact that small trilobites

84. Karen Bartelt, "A Visit to the Institute for Creation Research," part 4, www.talkorigins .org/faqs/icr-visit/bartelt4.html.

85. Actually limestone (calcium carbonate, $CaCO_3$) seems to pose some problems for mainstream geology as well, because the quantity of carbon in the world's sediments is greater than 6×10^{19} kg, whereas the total amount in the atmosphere, animals and plants (living and dead), coal and oil deposits, and as inorganic matter in the oceans totals only about 45×10^{15} kg. If the carbon in limestone were in the oceans and seas, it would make them far too toxic for life. In addition, the size of most limestone grains is too small to be explainable if they resulted from the death and disintegration of marine animals because wave action and predators cannot exert the forces required to break the shells into such tiny fragments.

86. John Woodmorappe, "A Diluviological Treatise on the Stratigraphic Separation of Fossils," in *Studies in Flood Geology*, 28.

87. Ibid., 44–53.

(low drag) are found together with large ones (higher drag), why flowering trees (which are large) should be found only in higher sediments, and why simple life-forms always fossilize in lower rock layers and more complex life-forms in the layers above. Argument is unlikely to resolve this issue, which seems to be an ideal case for empirical investigation, even if such investigation means recreating events with a large flood. It should be possible to determine whether the type of order needed in the sediments can occur as a result of rapid, violent processes and though the experiment would be large and difficult to mount, it is feasible.

Another sticky problem for many Creationists is the most recent ice age, usually ascribed to the period from 80,000 to 10,000 years ago. Creationists accept the reality of the ice age because there is abundant evidence of it.[88] But for them, the chronology obviously must be much different. They believe that the ice age occurred after the Genesis flood and lasted for only a few hundred years. But this does not appear to leave enough time for the glaciation that occurred, nor does it fit into the general historical record or reconstructed records of global temperatures. While Creationists can argue that an ice age is a likely or inevitable consequence of the flood, because it was a volcanic event with much soot and ash launched into the atmosphere,[89] this doesn't resolve the chronology problem. Walt Brown's hydroplate theory approach attempts to avoid the difficulty by arguing that when water from the subterranean chambers exploded into the atmosphere, it froze into cold hail, burying alive animals such as mammoths—which explains the condition in which they are commonly found.[90]

Radiometric Dating

The long time spans indicated by radiometric dating methods (discussed in chap. 3) are a direct challenge to Creationists' belief in a young earth. Even error factors of one thousand (often cited) are insufficient to solve the problem of reducing a 4-billion-year-old earth to just 6,000 to 10,000 years. In addition, other evidence suggests that much radiation has been emitted on and inside the earth—far too much to be explainable in a short earth lifetime, as Creationists concede.[91] Specifically,

88. "Was There an Ice Age?" www.answersingenesis.org/docs/218.asp.
89. "More on the Ice Age!" www.answersingenesis.org/docs/193.asp.
90. Brown, *In the Beginning*, 159–87.
91. D. Russell Humphreys, "Accelerated Nuclear Decay: A Viable Hypothesis?" in *Radioisotopes and the Age of the Earth*, ed. Larry Vardiman, Andrew Snelling, and Eugene Chaffin (El Cajon, CA: Institute for Creation Research, 2000), 335.

we observe a wide spectrum of nuclear decay effects: (1) daughter isotopes along the whole decay chain, (2) visible scars (halos) from α-decay, (3) the α-particles themselves (He nuclei), (4) visible tracks from decay by fission, and (5) the heat produced by nuclear decay. The most reasonable hypothesis is that all these products of nuclear decay were indeed produced by nuclear decay![92]

Therefore stronger medicine is needed. In order to save the young earth theory, Creationists are increasingly entertaining the possibility that radioactive decay has not been constant over time.

Many Creationists concede that available radiometric evidence points to hundreds of millions of years, at least, of radioactive decay. Under the assumption of a young earth of 6,000 years, and the 4.5-billion-year radiometric age found for earth, their conclusion is that there must have been an acceleration of radioactive decay by a factor of about 750,000, on average. However, this rate of radioactivity is far too high to permit life, and far higher than what is observed now, so the acceleration had to occur at a still higher rate at some time in the past, followed by a return to a normal rate. One possible scenario is that shown in figure 6.7, which postulates two periods of accelerated decay, one early in Creation Week and the other at the time of the Genesis flood. The problem, therefore, is twofold: (1) to find evidence that such accelerated decay did in fact occur, and then (2) to postulate mechanisms for it.

Creationists have devoted considerable effort to this subject. Indeed, a faculty group from the Institute for Creation Research has assembled a research initiative to investigate possible mechanisms that may be involved in the change of decay rates. The group, known as RATE (Radioisotopes and the Age of the Earth), completed its work in 2005 and has published its results.

Is there any evidence of accelerated decay in the past? Creationists point to the retention of He^4 given off by radioactive decay within zircons, a brown to colorless mineral that is often cut and polished to make gems that are embedded in crystals of biotite (black mica). The crystals in question are found in Precambrian granodiorite (granitic rock). According to Creationists, the high retention rates of He^4 in zircons are incompatible with the uniformitarian processes assumed by evolutionary (old earth) models, which call for steady low-rate production and diffusion out of the zircons leading to low concentrations. In their minds, the high retention levels fit better with a young-earth, accelerated-radioactive-decay scenario. No measurements have yet been done of diffusion of helium through biotite crystals, which does set the scene for an experimental test.[93] Figure 6.8

92. Ibid., 337.
93. Ibid., 346.

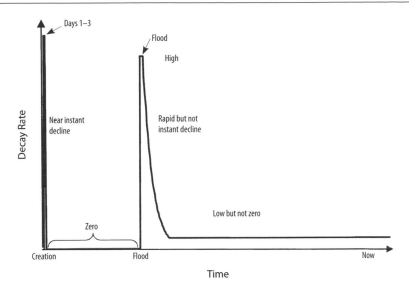

Figure 6.7. Decay rates hypothesized to account for observed radiation evidence and radiometric dates for age of the earth (vertical axis not to scale). (D. Russell Humphreys, "Accelerated Nuclear Decay: A Viable Hypothesis?" in *Radioisotopes and the Age of the Earth*, ed. Larry Vardiman, Andrew Snelling, and Eugene Chaffin [El Cajon, CA: Institute for Creation Research, 2000], 341.)

shows the calculations, which can be compared with experimental data. As yet, the rate of helium diffusion has not been measured, thus the value of a test. The mechanisms proposed by the RATE group for accelerated decay are rather technical and beyond the scope of this book.[94]

The Fossil Record: Simultaneous, Not Sequential

Directly observable to anyone who cares to dig through rocks are fossils, and in particular, their sequential arrangement, with simpler organisms generally found in lower strata and more complex organisms found in higher strata. The usual inference from this is that there is a *time series* involved; simpler organisms, such as single-celled organisms, precede the more complex ones, such as dinosaurs and mammals. This inferred time series, indeed, is the origin of the usual classification of geological time periods, from Precambrian to Tertiary. In conventional scientific theory, these periods are estimated to have lasted tens or hundreds of millions of years each. For the Creationists, of course, this poses a serious problem. The preferred solution is to make the geological periods correspond to different ecosystems that overlap somewhat in time and geographical extent, and that can shrink and grow. Unlike

94. Detailed technical discussions are on Thomas Fowler's web site, www.evolution primer.net.

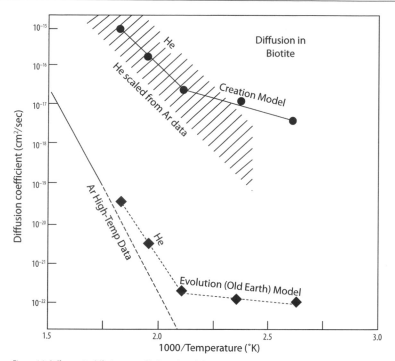

Figure 6.8. Difference in diffusion rates of helium through biotite needed to account for observed concentrations of He⁴. (Humphreys, "Accelerated Nuclear Decay," 348.)

the usual theory, then, these ecosystems were partly contemporaneous, rather than strictly sequential:

> Geologic "Periods" are, in reality, ecosystems each having characteristic life-forms. All "Periods" (ecosystems) were present *to some degree* on the earth immediately after Creation Week. Geologic events throughout pre-Flood times locally stratified changing ecosystems (later interpreted as "Periods" by early geologists). Man's life-sustaining ecosystem ("Cenozoic") is seen . . . as starting geographically very small and expanding during pre-Flood times allowing the growing human population to eventually fill the earth. Not all ecosystems ("Periods") are seen as surviving until the Flood, and few except those of the "Cenozoic" appear to be post-Flood.[95]

In this scheme, the flora and fauna inhabiting the different ecosystems generally do not mix, and all were buried separately in the Genesis flood, giving rise to observed fossil deposits.

The basis for this theory is that, according to Creationists, most species recorded in the fossil record still exist, many virtually unchanged,

95. Gentet, "CCC Model," 12.

especially at the higher taxonomic levels. They cite the insect families, 84% of which were alive in the Cretaceous period.[96] Similar statistics can be cited for terrestrial vertebrates. The inference is that all or most fossils have living counterparts, and thus the great extinctions of the past, such as that at the Cretaceous-Tertiary boundary 65 million years ago, which conventionally is assumed to have wiped out the dinosaurs, are illusory. In support of their position, Creationists cite an observation of paleontologist Niles Eldredge:

> There is just no rationale, no purpose to be served in giving different names to . . . virtually identical creatures just because they are separated by 3 million years of time. Yet that is the natural propensity of paleontologists: collections of otherwise similar, if not completely identical, fossils tend to get different names for no reason other than their supposedly significant age differences.[97]

Creationists argue that what appear to be distinct rock layers representing an evolutionary time sequence are merely different representations of contemporaneous ecosystems that have changed slightly over time. For example, Devonian rocks from 370 million years ago merely represent a semiaquatic ecosystem with many forms that still exist today. Slight changes to this ecosystem can be found in Carboniferous rocks that date to 330 million years ago. Using this reasoning, the Devonian and Carboniferous fossils are just two examples of the Paleozoic ecosystem (see table 6.2). Many Creationists posit three such contemporaneous ecosystems—the Paleozoic, the Mesozoic, and the Cenozoic—all of which show slight variations in the fossil record but all of which existed simultaneously.[98] In this framework, humans, part of the Cenozoic ecosystem, and dinosaurs, part of the Mesozoic ecosystem, were contemporaries.

The Creationists' approach requires that strata be associated with events and ecosystems during the relatively short life of the earth. For an example of the type of correlations utilized to accomplish this goal, see table 6.2. The table is based on a popular Creationist model, the creation/curse/catastrophe (CCC) model. In this model, it is postulated that the ecosystems indicated were all present at the end of the Creation Week, as given in Genesis 1. Typically Creationists place the Genesis flood at one of three boundaries: upper Paleozoic at the Pennsylvanian/Permian boundary, the Cretaceous/Tertiary boundary, or in the late Cenozoic, for example, before the Pleistocene epoch.

96. Carl Zimmer, "Insects Ascendant," *Discover* 14 (November 1993): 30, www.discover .com/issues/nov-93/departments/insectsascendant314.

97. Niles Eldredge, *Time Frames* (New York: Simon & Schuster, 1985), 21.

98. Gentet, "CCC Model," 12.

Table 6.2. The CCC Model interpretation of the geologic column (strata)*

Eon	Eras	Periods		Characteristic ecosystem
Phanerozoic	Cenozoic	Quaternary		This ecosystem is most characterized by the following types of life: human beings, most mammals, and flowering plants/grasses (angiosperms) in great profusion. It is the ecosystem wherein humans can thrive and characterizes what is biblically called "[Garden of] Eden." It now dominates the earth, whereas in the early earth it was only one of various ecosystems. Today it contains remnants of the earlier ecosystems in many life-forms termed "living fossils" because in old-earth evolutionary theories they have remained the same for long ages. The CCC model explains their continued existence as survivors from other created ecosystems, which in the early earth covered large areas.
		Tertiary	Neogene	
			Paleogene	
	Mesozoic	Cretaceous		This ecosystem is best known for the dinosaurs, the marine reptiles, and the flying reptiles. For most of the time, the dominant plant life consisted of cycads, conifers, ferns and other coarse herbage, and succulent water vegetation that supplied food for reptiles. In the Cretaceous strata, there is a noticeable change in vegetation from cycads, conifers, ferns, and so on to angiosperms. The CCC model interprets this change of vegetation as an indication that the surrounding Cenozoic ecosystem was expanding and beginning to invade the Mesozoic ecosystem. The result was a decline of the dinosaurs due to limited food supply and changing environmental conditions caused by geologic events of the preflood curse. Except for possible representatives from Noah's ark, the reptile preflood world may have been largely extinguished before the Genesis flood event.
		Jurassic		
		Triassic		
	Paleozoic	Permian		The Paleozoic ecosystem consists of many periods. Much Paleozoic strata is marine in origin. Much of the sea life characterized by this ecosystem is now extinct or thought to be extinct. The trilobites that especially characterize the Cambrian period are one obvious example. All of these life-forms were created during Creation Week, but some forms dominated the early earth and later died out or greatly diminished in the ecosystem as the food chain matured. Biblically, this is the result of life heeding God's command to "fill the earth." In the preflood earth, as the various ecosystems changed due to geologic events, pressure from surrounding ecosystems, and changes within the Genesis "kinds," and so on, the ecologic successions were preserved in the geologic record and interpreted as immense time periods by early geologists using uniformitarian theory.
		Pennsylvanian		
		Mississippian		
		Devonian		
		Silurian		
		Ordovician		
		Cambrian		
Precambrian Eon				Life-forms are relatively few and simple (algae, fungi, worms, etc.). Probably represents Creation Week/earliest postcreation/curse events.

* Gentet, "CCC Model," 13. Reproduced by permission.

Since the Creationist theory requires that three basic ecosystems were present simultaneously on earth (Paleozoic, Mesozoic, and Cenozoic), we might expect to find abundant Cenozoic fossils in lower rock formations. The reason we do not, according to Creationists, is that the Paleozoic ecosystem dominated early on and therefore occupies most of the lower layers. But if the ecosystems were contemporary there should have been some mixing, and indeed the fossil record does exhibit a small degree of mixing in places. The question is whether it can be attributed to contamination from normal geologic processes or if it does indeed reveal that the disparate ecosystems were contemporary.

Another problem is the belief that these ecosystems, especially the Paleozoic and the Mesozoic, were buried at the same time in the great flood. If this occurred, why is the Mesozoic ecosystem almost invariably buried on top of the Paleozoic, especially if they were "isolated ecosystems"? Such an observation argues for a time sequence that Creationists, for the most part, are unwilling to grant.

The hydroplate theory, however, does make specific predictions about the fossil record, predictions that are quite different from those of conventional geology. Among them is the following: "One should never find marine fossils, layered strata, oil, coal seams, or limestone directly beneath undisturbed rock ice or frozen mammoth carcasses."[99] Such predictions, if borne out, would give much-needed support to the Creationist interpretation of geology.

Creationists also believe that the fossil record, especially the Cambrian explosion, supports their view of multiple origins of life-forms (polyphyly) far better than the standard view, monophyly.[100] Further experimental work in this area might provide the key to deciding between these theories.

Thermodynamics, Information Theory, and Evolution

Thermodynamics is in some ways an ideal venue for arguing possibility and impossibility, because it allows us to draw conclusions about a broad range of cases with minimal information. Often it can be invoked if we know only the before and after states of a process; knowledge of operational details—how to get from one state to the other—is unnecessary. For example, we can determine if an inventor's claims about a new type of engine are credible if we know only a few facts about heat intake and exhaust, and nothing about the internals of the engine. Creationists, therefore, have sought to use thermodynamics to demonstrate that evolution is impossible, because if applicable, it could strike down evolution as a possible change in state *regardless of mechanisms proposed*. No knowledge of such mechanisms would even be necessary—we could be certain that none could possibly work based on the thermodynamics alone.

Some Creationists argue that evolution violates the second law of thermodynamics because it posits an increase in order and complexity. Since the second law states that entropy (disorder) will increase in any closed system over time, the increase in order and complexity during evolution seems to be in violation of this law.

99. Brown, *In the Beginning*, 175.
100. L. James Gibson, "Polyphyly and the Cambrian Explosion," *Origins* 52 (2001): 3–6.

The problem with this argument is that the site of evolution, the earth, is not a closed system. Rather, energy is constantly being added to this system in the form of sunlight. Furthermore, the kinds of processes required by natural selection involve relatively small changes in the key thermodynamic quantity, entropy, which can easily be supplied by the energy of the sunlight striking our planet. Therefore the process of evolution is not forbidden by thermodynamics alone, though there may be other problems that limit evolution, problems that fall outside the scope of thermodynamics. Thus thermodynamics alone cannot be used one way or the other in the evolution debate because the changes in the key quantity of entropy are too small to be significant on the timescale envisioned (billions of years).[101]

Can New Species and Higher Taxa Arise?

Contrary to popular belief, Creationists believe that speciation is possible, but only through a degenerative process in which information in the collective genome of a population is *lost*. This entails that creation of new higher taxa (genus, family, up through phyla) *cannot* arise through natural processes. Many Creationists, indeed, prefer to speak of "baramins" rather than species. A baramin[102] is an original created kind of organism, from which arose, by speciation, groups of animals or plants that we know. ("Baramin" comes from two Hebrew words, *bare*, to create, and *min*, kind.[103] Creationism does not use the concept of fixed and immutable species, which traces back to Linnaeus and Aristotle.[104]) The baramins were originally created by God; all plants and animals descended from them, and this descent can be down through several taxonomic levels, including family, genus, and species. Creationists are therefore quite comfortable with speciation and indeed require it for their model of the current world ecosystem:

> Poorly-informed anti-creationist scoffers occasionally think they will "floor" creation apologists with examples of "new species forming" in nature.

101. A technical discussion of this issue appears on Thomas Fowler's web site: www.evolutionprimer.net.

102. Other terms used by Creationists include "holobaramin": a group comprising only the organisms related by common descent, including all who descended from the original pair; "monobaramin": same as holobaramin except that all descendants need not be present.

103. Frank L. March, *Fundamental Biology* (Lincoln, NE: March Publications, 1941), 100.

104. Kenneth B. Cumming, "On the Changing Definition of the Term 'Species,'" Impact number 211 (Santee, CA: Institute for Creation Research, January, 1991).

They are often surprised at the reaction they get from the better-informed creationists, namely that the creation model depends heavily on specia-tion. . . . It seems clear that some of the groupings above species (for ex-ample, genera, and sometimes higher up the hierarchy) are almost certainly linked by common ancestry, that is, are the descendants of one created ancestral population (the created kind, or baramin). Virtually all creation theorists assume that Noah did not have with him pairs of dingoes, wolves and coyotes, for example, but a pair of creatures which were ancestral to all these species, and probably to a number of other present-day species representative of the "dog kind." . . . Demonstrating that speciation can happen in nature, especially where it can be shown to have happened rapidly, is thus a positive for creation theorists.[105]

The key point for Creationists is that no new genetic information enters the biosphere; what is there (or was there) was put there by the Creator. Mutations can only *decrease* the information. Thus, despite a belief in rapid speciation under certain circumstances, the Creationist is not adopting a large-scale evolutionary viewpoint.[106] The question of whether new information can be generated by the processes hypothesized by Neo-Darwinism has been discussed at length in chapter 5. Suffice it to say that on this matter, the Creationist and Intelligent Design schools tend to be in agreement (and a minority of Meta-Darwinists concur): no new information, or not enough new information, can come from random mutations alone.

Creationists thus accept many of the speciation pathways proposed by mainstream geneticists, such as allopatry (physical isolation of popu-lations) and hybridization. They also accept that speciation can occur even without physical separation, such as when different sets of the same species start to look for different types of food—the case of Darwin's finches.[107] Some problems for Creationists remain, however, such as the problem of the amount of genetic material that had to be present in the kinds on Noah's ark. If a single dog kind was on the ark and from it de-scended all present-day dogs, wolves, dingos, and so on, then a massive amount of genetic variability must have been present in this original pair. Even assuming maximum heterozygosity (genetic differences), it is not clear that this matter can be solved in the Creationists' favor. Some Creationists are proposing that yet-unknown mechanisms will be discov-ered; junk DNA has been proposed.[108] Given the time from Noah's ark,

105. Carl Wieland, "Speciation Conference Brings Good News for Creationists," *Creation Ex Nihilo Technical Journal* 11, no. 2 (1997): 135–36.
106. Ibid.
107. Carl Wieland, "Darwin's Finches Evidence Supporting Rapid Post-Flood Adapta-tion," *Creation Ex Nihilo* 14, no. 3 (1992): 22–23.
108. Wieland, "Speciation Conference Brings Good News," 135–36.

it should be possible to estimate mutation and speciation rates required to give known distributions of species worldwide, and to compare this with observed rates of change. Such empirical studies should be able to test the feasibility of Creationist speciation models.[109]

Other Problems with Creationism

At this point, it is worth listing and briefly describing several other problems with Creationism that must be addressed before it can claim to be a comprehensive explanatory paradigm. The most serious of these problems is the genetic evidence that strongly suggests common descent. As mentioned earlier in the book, related organisms such as chimps and humans have very similar DNA sequences with many genes that are nearly identical. Creationists believe these sequence similarities do not imply common descent but rather a common structural or functional requirement:

> We know that DNA in cells contains much of the information necessary for the development of an organism. In other words, if two organisms look similar, we would expect there to be some similarity also in their DNA. . . . Humans and apes have a lot of morphological similarities, so we would expect there would be similarities in their DNA. . . . Certain biochemical capacities are common to all living things, so there is even a degree of similarity between the DNA of yeast, for example, and that of humans. Because human cells can do many of the things that yeast can do, we share similarities in the DNA sequences that code for the enzymes that do the same jobs in both types of cells.[110]

Given the similar structural and functional requirements that related (and sometimes not-so-related) organisms share, the Creator would be expected to endow them with similar genes.

109. Creationists have made some efforts to justify their claims about created kinds. With respect to the dog kind, they observe that Canus familiaris, Canus lupus, and Canus latrans can interbreed in the wild. (Dennis Cheek, "The Creationist and Neo-Darwinian Views Concerning the Origin of the Order Primates Compared and Contrasted: A Preliminary Analysis," *CRSQ* 18:1 (September, 1981): 95–96. They have also observed that most of today's dog breeds have come from selective breeding over a relatively few centuries. (David Woetzel, "Understanding Transitional Forms," *Creation Matters* 6:2 (2001): 2. Other studies have been done on variations in the dental structures of dogs as a way of assessing the degree of variation possible in kinds or baramin. (Celedonio García-Pozuelo-Ramos, "Dental variability in the domestic dog: Implications for the Variability of Primates," *CRSQ* 35:2 (September, 1998): 66–75. None of this research, however, proves that the modern-day species cited actually did stem from a primitive kind or pair on Noah's ark.

110. Don Batten, "Human/Chimp DNA Similarity: Evidence for Evolutionary Relationship," *Creation Magazine* 19, no. 1 (December 1996): 21–22, www.answersingenesis .org/creation/v19/i1/dna.asp#f1.

While this is a possible explanation for the sequence similarities, there is other genetic evidence that is harder to dismiss. As discussed in prior chapters, researchers have found that chromosomes consist of conserved blocks of genes called synteny blocks. These synteny blocks contain a specific set of unrelated genes that are found in the same order on the chromosome in various organisms. Since there is no functional or structural requirement that these genes be next to each other on the chromosome (e.g., there is no reason a sodium channel gene must be next to a collagen gene), the obvious question is, why are certain blocks of unrelated genes often found together? The most parsimonious explanation is that these genes were initially in this order in a common ancestor and have been inherited via common descent. Creationists have yet to offer a convincing counterexplanation for synteny blocks. A similar situation exists with specific pseudogenes that serve no function but are found in related organisms.[111] Both of these issues are discussed in chapters 2 and 5 and remain stumbling blocks for Creationist theories.

The remaining issues we will discuss all concern matters of the relatively recent past that conflict with the dates and circumstances hypothesized by Creationists for major events. Such matters are ideal for resolving the evolution debate because their recent vintage makes them much more amenable to investigation than events hundreds of millions or billions of years ago.

First, there are a number of historical issues. Given that the Genesis flood is supposed to be so recent (around 2300 BC) and was of the great violence discussed above, it is difficult to reconcile this with the survival of any artifacts of earlier civilizations, such as the Sumerian or Egyptian, both of which predate it. The Step Pyramid at Saqqara is dated to 2630 BC and the Great Pyramid of Giza to 2560 BC. The interior of the Great Pyramid does not appear to have suffered water damage, nor did nearby subterranean chambers, artifacts from which are now on display in Egyptian and other museums. Many Sumerian artifacts dated to the fourth millennium BC have come down to us, including pottery and wooden musical instruments.[112] It is difficult to see how such delicate items could survive the type of flood envisioned.

Second, there is the problem of populating the Western Hemisphere, starting from the landing of the ark in about 2300 BC. It is known from both linguistic evidence, physiology, and genetic data that the peoples of

111. Felix Friedberg and Allen Rhoads, "Calculation and Verification of the Age of Retroprocessed Pseudogenes," *Molecular Phylogenetics and Evolution* 16 (2000): 127–30.

112. A wooden harp, dated to about 2650 BC, is in the collections of the University of Pennsylvania; see http://ccat.sas.upenn.edu/arth/hp/sumer/harp.gif.

the New World originated in East Asia, with the standard initial migration date estimated at about 30,000 years ago.[113] If they come to East Asia from Mt. Ararat in eastern Turkey, the raw distance across Asia, along the Alaskan coast, through North America, and on down to the southern tip of South America is on the order of 15,000 to 20,000 miles, depending on the route taken. Included is the most varied terrain, from nearly Mediterranean climate to steppes, mountains, deserts, extremely cold regions, and extremely hot regions. Fairly good evidence suggests that this entire area was populated no later than 2,000 years ago, probably much earlier. Under the Creationist model, that leaves about 2,000 years for this migration to take place, or about 7 to 10 miles per year.[114] While this rate may not sound so daunting, such a migration would cover regions with enormously different climates and topography, and thus would require completely different survival skills. It is extremely difficult to see how these skills could be learned—or rather invented—nearly fast enough to permit the indicated migration rate.

Third, we may point to purely linguistic evidence. The existence of language families is well established, and Indo-European is perhaps the best known. Creationists acknowledge this, along with virtually everyone else.[115] The problem this poses is that the individual languages in the families could not have diverged as rapidly as required to fit the compressed chronology of 2300 BC for the repopulation of the world as well as the Tower of Babel scattering, which is assumed to have been postflood, roughly about 2000 BC. For example, Linear B is a very early form of Greek. As it dates at least to the fourteenth century BC, Greek was already a well-developed language by that date. Hittite inscriptions dating to the late fifteenth century BC have also been deciphered.[116] It has to be assumed that the other major Indo-European language groups were well developed by then, including the Italic, Celtic, Balto-Slavic, Hittite, Persian, and Indian groups. The Indo-European proto language itself has also been partially reconstructed. The known rates of language change suggest that Indo-European was spoken in about 4000 BC, at which point the daughter languages began to diverge, most likely as a result of migrations.[117] This date, however, obviously conflicts with

113. Luigi Cavalli-Sforza, *Genes, Peoples, and Languages* (New York: North Point Press, 2000), 136; Merritt Ruhlen, *The Origin of Language* (New York: Wiley, 1994), 166.

114. Assuming the flood occurred about 5,000 years ago.

115. Carl Wieland, "Towering Change," *Creation Ex Nihilo* 22, no. 1 (December 1999–February 2000): 22–26.

116. Thomas V. Gamkrelidze and V. V. Ivanov, "The Early History of Indo-European Languages," *Scientific American* 262, no. 3 (March, 1990): 110–16.

117. J. P. Mallory and D. Q. Adams, *The Oxford Introduction to Proto-Indo-European and the Proto-Indo-European World* (Oxford: Oxford University Press, 2006), 100.

the Creationist chronology, and even though it is not too far off, this is clearly a case where "close only counts in horseshoes."

What is especially ironic about the linguistic question is that linguists have been steadily linking larger and larger numbers of language families, with the ultimate goal of perhaps an original protolanguage from which all languages derive—a return of the time before the Tower of Babel, though not, obviously, associated with Babylon and Nimrod (Genesis 11). The evidence often cited for such a protolanguage is the existence of "global etymologies," words with the same or nearly the same meaning in all language families.[118] Early steps in this process were the linking of several large families, including Indo-European, into the Eurasiatic or Nostratic superfamily.[119] Rough estimates suggest that if Nostratic was spoken, it would have been 15,000 to 20,000 years ago, with higher groupings perhaps pointing to a common language 100,000 years ago. As of yet, this is quite speculative, but further verification of such dates would pose a serious problem for Creationists, who, not surprisingly, dispute the superfamilies idea.[120]

Mainstream Science Critiques of Creationism

Many critiques of Creationism have been made by mainstream scientists. There is little point in reviewing them all here because they usually come down to claims about the key disputed points we have discussed throughout, such as the age of the earth, macro-/microevolution, the existence of transitional forms, or evidence of new species. Well known are books by Berra, Eldredge, Futuyma, Godfrey, and Kitcher, among others.[121] Even the National Academy of Sciences has weighed in with a brief tract.[122] Most of these books are out of date, because Creationists have developed far more sophisticated theories than those criticized.

118. Merritt Ruhlen, *A Guide to the World's Languages* (Stanford, CA: Stanford University Press, 1991), 260.

119. Ibid., 261.

120. Wieland, "Towering Change," 22–26.

121. Tim M. Berra, *Evolution and the Myth of Creationism* (Stanford, CA: Stanford University Press, 1990); Niles Eldredge, *The Monkey Business* (New York: Washington Square Press, 1982); Niles Eldredge, *The Triumph of Evolution and the Failure of Creationism* (New York: W. H. Freeman, 2000); Douglas Futuyma, *Science on Trial* (New York: Pantheon Books, 1983); Laurie R. Godfrey, *Scientists Confront Creationism* (New York: W. W. Norton, 1983); Philip Kitcher, *Abusing Science: The Case against Creationism* (Cambridge, MA: MIT Press, 1982).

122. National Academy of Sciences, *Science and Creationism* (Washington, DC: National Academy Press, 1999).

A recent catalog (2002) appears in *Scientific American*, and comprises fifteen arguments on the major points,[123] most of which have been covered in this chapter or earlier chapters of the book.[124] Creationists have responded point by point along lines that will not surprise readers who have come this far.[125] Like most creation/evolution debates, these discussions are generally too superficial and exhibit too little understanding of Creationist positions to be of much value, but they do show how mainstream science attempts to refute Creationist theories. Reading a detailed study of original source material (on both sides of the dispute) is unquestionably more profitable. This will allow the reader to make up his or her own mind based on primary, not secondary, knowledge of the issues. To attempt to learn about one side's views from the opposite side's presentations is never a good course.

Let us, however, examine what is perhaps the most commonly made objection to all things Creationist, namely, that Creationism is not science and therefore not worthy of consideration, debate, or refutation. Defining what science is and isn't has proven to be a difficult problem (see chap. 4). This does not mean, however, that no judgments can be made. There is little dispute that geology, chemistry, physics, and astronomy belong in the camp of science, whereas astrology, phrenology, and telekinetics, for example, are excluded. It should be clear to anyone who has read this chapter that Creationists make scientific claims; they posit specific hypotheses that can be tested scientifically. This does not mean they are correct or even that their ideas are viable, but it does mean they can be evaluated *scientifically*. Despite this, many critics simply assume Creationists have no scientifically testable hypotheses and seek to dismiss them summarily—presumably so that no effort will be needed to address the specific criticisms of evolutionary theory put forth by Creationists:

> The truth is that the existence in nature of evolution, natural selection, microevolution, and macroevolution have been well-established by scientists for over 100 years, there is no legitimate scientific objection to them, and no reason to debate or "dialogue" about them.[126]

123. John Rennie, "15 Answers to Creationist Nonsense," *Scientific American* 287, no. 1 (July 2002): 78–85, available at www.sciam.com/article.cfm?articleID=000D4FE C-7D5B-1D078E49809EC588EEDF&pageNumber=1&catID=2.

124. Some of the material deals with issues outside the scope of this book, such as abiogenesis and anthropology.

125. AiG web site: www.answersingenesis.org/news/scientific_american.asp.

126. Steven Schafersman, "The Challenge of the Fossil Record," www.freeinquiry .com/challenge.html.

This writer then goes on to claim that there is no "genuine controversy . . . within science about the truth of evolution" and that no debate with Creationists should be undertaken because such debates by themselves enhance the Creationists' stature.

But there is something fundamentally wrong with the refusal of evolutionists to debate Creationists. Nothing in our modern society is beyond question; everything can be and is questioned, and everything has to be defended. Neither evolutionists nor anyone else can hide behind the mantle of authority. Authority still stands for something, but only if it can defend its role and accomplishments. If a significant number of people in our society dispute the truth of some purported body of knowledge, its defenders have an obligation to come forth and speak out, regardless of how convinced *they* may be of the truth of their ideas. Many were the physicists convinced of the truth of the deterministic paradigm of scientific explanation, before the quantum revolution and Aspect's experiment.[127] And even though we live in a democracy, and democratic capitalism appears triumphant on the world stage, no one would suggest for a moment that we should scorn challenges to debate by representatives of communism, fascism, anarchy, socialism, or other competing systems. No astronomer would hesitate for a minute to debate the flat earth or any other hypothesis if a substantial number of people subscribed to the theory and that it was threatening established science.

The most common concern voiced by evolutionists regarding debating or discussing issues with Creationists is that debates between Creationists and evolutionists give Creationists some badly needed credibility that they otherwise would not be able to obtain. Alas, the problem is precisely the opposite: Creationists already *have* a great deal of credibility—not with most scientists, obviously, but with tens of millions of Americans who naturally incline to their point of view. Whether this is good, bad, or indifferent, it *is* a fact. Evolutionists, by refusing to defend their views in a public forum, actually give *more* credibility to Creationists, because the public's impression is then that evolution is *not* true and *cannot* stand up to scrutiny. Some evolutionists have recognized this and advised that Creationism should be met head-on, rather than being ignored. Eugenie Scott's book *Evolution vs. Creationism*, in which she strikes a dialogue with Creationists, is a case in point.[128] In our view, this is a sound practice. What is needed, as we have emphasized over and over, is a squaring off,

127. Alain Aspect performed a delicate experiment to differentiate between the two rival interpretations of quantum physics and, by implication, of nature itself: the deterministic view, espoused by Einstein and de Broglie, and the indeterminate view, espoused by the Copenhagen school. The experiment verified the predictions of the latter school.

128. Scott, *Evolution vs. Creationism*. Eugenie Scott is the executive director of the National Center for Science Education, an anti-Creationist organization. Unfortunately

in which *all* schools make risky predictions about several matters, and these predictions are then tested in a public way.

Summary

Creationists have taken two major approaches in the process of developing their theories. Typically they start with a core set of religious beliefs, erect scientific assumptions on those beliefs, and then develop their scientific theories to be in accordance with these assumptions. A minority has started from the other end. They have become unsatisfied with the empirical evidence for Neo-Darwinian theory and see in Creationists' ideas a new set of hypotheses they believe can better explain the data.

Creationists steadfastly maintain a history of 6,000 to 10,000 years (or so) for the earth. They do this by constructing theories to explain the apparent old age of the universe, utilizing concepts from general relativity that enable them to accept the old age of the universe while maintaining a young earth or by utilizing hypotheses about the speed of light, to shrink the age of the universe. They also advance theories to explain geological features of the earth that are in accordance with their belief that it is young. These theories are based on short-term, high-energy processes, associated with a global flood, which was responsible for observable geology as well as fossil deposits. They can point to evidence that thick sedimentary rock can both form quickly and be eroded quickly by high-energy processes similar to those envisioned to have occurred during the earth's history. They seek to discredit such absolute chronometers as radiometric dating, though evidence presented against such methods is mostly anecdotal. And they believe that the fossil record is actually a record of contemporaneous or nearly contemporaneous ecosystems.

The scientific work of the Creationists is far from complete, and many serious issues both theoretical and empirical remain unresolved. For example, the proposed flood should have left telltale signs on civilizations predating it, but it does not seem to have done so. While some type of flood is fairly well attested in world legend, explanations other than the worldwide flood envisioned by Creationists have been proposed.[129] Nonetheless, at least one branch of Creationism has made a series of scientifically testable predictions.

Scott's book makes only a superficial effort to engage Creationists and ignores most of their more sophisticated theories.

129. William Ryan and Walter Pitman, *Noah's Flood* (New York: Simon & Schuster, 1998).

Other than the argument against evolution based on thermodynamics, we have not discussed the arguments put forth by Creationists against Neo-Darwinian theory in this chapter. Those arguments, some of which have merit, are discussed in chapter 5. Some of those arguments are also addressed in chapters 7 and 8 in relation to the concerns of the Meta-Darwinian and the Intelligent Design schools regarding Neo-Darwinian theory. Finally, it is important to stress that the Creationist school differs from the Intelligent Design school in a number of key areas, despite sharing some common ground regarding the problems of Neo-Darwinism. The Intelligent Design school will be taken up in chapter 7.

Table 6.3 evaluates Creationism with respect to the ten criteria of a good scientific theory.

Table 6.3. Creationist school and the ten criteria of a genuine scientific theory

Criterion	Creationism
1. Compactness	The closest Creationists come is distinction between high-energy processes and slow uniformitarian processes. Otherwise Creationists seem to be moving along individual directions.
2. Simplicity	Many theories utilize ad hoc explanations, but some (e.g., Walt Brown's hydroplate theory) are more integrated.
3. Falsifiability	Can be falsified: experimental tests can be found for some of the hypotheses advanced.
4. Verifiability	Varies; difficult or impossible in some cases, especially regarding biblical Creation Week events and aspects of the flood. But can be done for some aspects of the theory.
5. Retrodiction	Most of the school's theories were devised to explain (retrodict) observations that seem to be at variance with biblical accounts. However, some (e.g., Walt Brown's hydroplate theory) do account for most observed geological and paleontological facts.
6. Prediction	Some falsifiable predictions have been made, especially by Brown; much more needs to be done, but serious Creationists are moving in this direction.
7. Exploration	Definitely suggests many avenues of exploration.
8. Repeatability	Many events proposed in the past cannot be repeated or simulated in any way, as they involve things such as one-time changes in physical constants. These events, however, may result in consequences observable today.
9. Clarity	Varies; some theories clear, others not.
10. Intuitiveness	Varies; many theories seem too contrived to be intuitive (but this depends on one's point of view vis-à-vis the Bible); others are less contrived.

7

THE INTELLIGENT DESIGN
SCHOOL

It is hard to resist the notion that the natural world has been designed. Even a cursory glance out the nearest window supplies ample evidence for those who wish to argue for design. The sparrow on the tree with its hollow bones and aerodynamic feathers seems to be perfectly designed to achieve the type of flight of which human engineers can only dream. In fact, design seems to be a common thread that runs through the whole of nature. Time and again, in cases that have been cataloged since the dawn of biology, nature reveals that (1) its inhabitants are remarkably suited to fit their environment and (2) the various parts and systems that constitute organisms are remarkably suited to work in concert with one another.

For centuries, the only explanation for such observations was a creator, an intelligent agent responsible for designing the complexities and peculiarities of the organisms that inhabit our world. Darwin's theory turned this inference on its head. While not denying that the world *appears* to be designed, Darwin claimed that the design inference is merely illusory. Chance mutations, driven by the engine of natural selection, have given rise to what Darwinists refer to as "designoids," objects that give the illusion of being designed.

As Darwin's theory has gained acceptance among the scientific community, the design inference in nature has been relegated to the status of mythology: only the untrained observer, based upon some fuzzy feelings of the supernatural, would still believe that the organisms we see around us have been specifically designed by a creator. At least that was the idea. Surprisingly, though, the design inference has failed to go gently into the night. Rather than be plowed under by Darwin's theory, the design inference has been repackaged by a small but vocal group, the school of Intelligent Design.

Unlike the natural theologians of Darwin's time, who advocated the design inference based upon quaint "Isn't nature just so grand and complex!" stories, the modern Intelligent Design school has focused on a detailed examination of the complexities of organisms. The aim of the Intelligent Design school is to demonstrate the improbability of these dynamic complex organisms arising through the chance mechanisms found at the heart of Darwin's theory. Their main conclusions are that (1) natural processes such as natural selection are unable to fully explain the origin and evolution of many organisms, and (2) the "illusion" of design is much more than that: it is the result of design being written into many natural systems. However, this position is problematic because (as we shall see) identifying design is an inherently tricky problem in biology. Despite this obstacle, the Intelligent Design school believes the evidence for a designer is there, if only scientists are willing to look for it.

Views of the Intelligent Design School

The Intelligent Design school is the youngest of the four major schools of thought. Its origin may be dated conveniently to the publication of Phillip Johnson's book *Darwin on Trial*, in 1991, though the school owes much to an earlier work by biochemist Michael Denton, *Evolution: A Theory in Crisis* (1985). Johnson criticized evolution on methodological and factual grounds and decried what he viewed as its suppression of contrary evidence, its blind faith in naturalism, and its aversion to alternative explanations. But criticism does not a theory make, so without a doubt the most important Intelligent Design publication was biochemist Michael Behe's book, *Darwin's Black Box* (1996), which set forth the challenge to Neo-Darwinism in strictly scientific terms: there are biological systems so complex that they could not have originated by chance. Behe's work endowed the Intelligent Design movement with positive content and focused the movement on key scientific questions that could be investigated. More recently, Behe's thesis was restated by mathematician William Dembski in several books, wherein he developed the notion of the design filter, which was particularly intended to ferret out cases of Intelligent Design in biology.[1]

1. P. Johnson, *Darwin on Trial* (Downers Grove: InterVarsity Press, 1991); M. Denton, *Evolution: A Theory in Crisis* (Bethesda, MD: Adler & Adler, 1986); M. Behe, *Darwin's Black Box: The Biochemical Challenge to Evolution* (New York: Free Press, 1996); W. Dembski, *No Free Lunch: Why Specified Complexity Cannot Be Purchased without Intel-*

Recent years have seen this school gain significant visibility in the public debate over evolution, particularly when it comes to teaching evolution in the classroom (see fig. 7.1). It has set up shop at its own think tanks and research organizations. The undisputed leader in the movement is the Discovery Institute's Center for Science and Culture,[2] a public policy think tank used by the Intelligent Design camp as their intellectual launching ground to produce books, talks, and videos to promote the teaching of Intelligent Design in schools. A research-oriented wing of the movement can be found at the International Society for Complexity Information and Design (ISCID),[3] a new professional organization that produces a journal, online discussions, and workshops to promote related research.

Figure 7.1. A popular press cartoon about Intelligent Design in the classroom. © Knight-Ridder/Tribune Media Information Services. All rights reserved. Used by permission.

ligence (Lanham, MD: Roman & Littlefield, 2002); and W. Dembski, *The Design Inference: Eliminating Chance through Small Probabilities* (Cambridge: Cambridge University Press, 1998).

2. The center's web site can be found at www.discovery.org/crsc.

3. The ISCID web site can be found at www.iscid.org. The ISCID journal, which is currently available only online, is titled *Progress in Complexity, Information and Design*.

To distance itself from the Creationist movement, the Intelligent Design school operates under a broad-tent philosophy and includes members with a wide range of religious and philosophical views. While there is some overlap in belief between the Creationist and Intelligent Design schools—specifically, the belief that naturalistic mechanisms cannot (fully) account for life as we know it—the Intelligent Design movement has gone to great lengths to distinguish itself from traditional Creationists. Listed below are some important differences.[4]

- The major focus of Intelligent Design attacks is not evolution as such, but the efficacy of the naturalistic mechanisms underpinning it, particularly random mutation and natural selection. Creationism criticizes nearly all aspects of evolutionary thinking.
- Anyone opposed to naturalism could potentially qualify as an ally of Intelligent Design. This includes believers in evolution from microbe to man, so long as this belief involves some intelligent, direct physical intervention sometime during the billions of years of evolution. Creationism requires a more narrowly circumscribed set of beliefs, with emphasis on natural history as interpreted literally from the Bible.
- The school generally believes in, or is publicly neutral on, the age of the earth (4.5 billion years) that evolutionists teach and accept. Young earth Creationists explicitly reject this position.
- The school *generally* avoids speculation on the nature of the designer. Rather, it focuses its efforts on detection of that designer's work.

There are other differences as well. Unlike Creationists, the majority of Intelligent Design school proponents accept the scientific work done to date in geology, biochemistry, paleontology, and genetics (however, their interpretation is often different from Neo-Darwinists'). In addition, many in this school do not dispute the majority of the fossil record, and some even go so far as to advocate some form of common descent, thereby avoiding many of the problems the Creationist school must address (see chap. 6).[5] For example, questions that plague Creationists, such as why certain pseudogenes are nearly identical between humans and apes, do not trouble the Intelligent Design camp. Many in this camp are content to attribute this to natural selection and common descent and move on

4. Based on material from the AiG web site, www.answersingenesis.org/docs2002/0830 _IDM.asp?srcFrom=aignews.

5. What this "common descent" actually entails is never explicitly stated, but it appears to mean that although organisms are related, they require the help of a designer to make it over the major hurdles of evolution.

to other matters, specifically the biochemistry of complex systems. The basic structure of the Intelligent Design position is shown in figure 7.2.

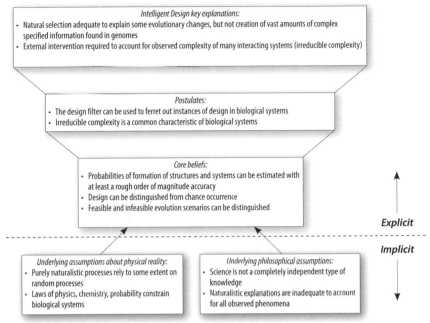

Figure 7.2. The logical structure of the Intelligent Design school.

Intelligent Design and Naturalism

For the most part, the Intelligent Design movement is willing to bend to the Neo-Darwinist position up until the point of breaking, conceding on everything from the age of the earth to, in some cases, common descent. Why then does the school pull back at the last instant and insist on planned intervention by a designer? One of the main reasons is that those within the Intelligent Design movement explicitly reject the *scientific naturalism* that is prevalent within the scientific community.[6] Scientific naturalism is the claim that natural causes are sufficient to explain *all* observable phenomena, which of course leads to the inference that no supernatural causes intervened in the creation or evolution of life. This is a philosophical position held by many leading proponents of Neo-Darwinian theory.[7]

6. Phillip Johnson, *Darwin on Trial* (Downers Grove, IL: InterVarsity, 1991), 124–34.

7. Richard Dawkins and Daniel Dennett are two popular authors who are committed to scientific naturalism, otherwise known as materialism. Some (the proponents of theistic evolution) argue that the distinction is misleading and short-sighted, because evolution can be viewed as God's creative action. That subject is beyond the scope of this book.

A more common and less divisive view is that of *methodological naturalism*, which claims that science can investigate *only* natural causes and must restrict itself to these. Although such a position does not remove God from the picture, it does remove God as a *primary causative* force in natural history whose effects can be detected in some measurable sense. Darwinist philosopher Robert Pennock, for example, defines science as the investigation of *natural* causes, thereby assuring that only natural mechanisms will be found during any *scientific* investigation of evolution. What the Intelligent Design school wishes to demonstrate is that, from a theoretical standpoint, such a scientific investigation of evolution, pursued on the basis of naturalistic assumptions only, will never succeed in explaining all biological phenomena. This is the crux of their position: some biological structures cannot be explained by natural causes alone. As a result, Intelligent Design researchers eschew both scientific and methodological naturalism when it comes to explaining the history of life.

This does not mean, though, that Intelligent Design researchers would expect to find supernatural causes for most phenomena, or that scientists in this camp believe it is appropriate to invoke design every time they encounter a difficult problem. Rather, the Intelligent Design school believes that natural processes *do* account for most of natural history; only certain evolutionary transitions cannot be made in this way. As a result, most biological research is quite unaffected by the assumption of Intelligent Design, because most research focuses on the structures, mechanisms, and operation of biological systems, not on their origin.

For many in this movement, the rejection of naturalism is the direct result of religious convictions. As a result, it has become fashionable to dismiss Intelligent Design as pseudoscience, a form of religion cleverly attempting to pass itself off as science. Pennock's book, for example, calls Intelligent Design the "New Creationists" or "Intelligent Design Creationists." Such labels make it easier for scientists to disregard their claims without further comment. Although many members of the Intelligent Design school despise such labeling, it may seem justified by remarks such as the following from Phillip Johnson:

> That God created us is part of God's general revelation to humanity, built into the fabric of creation. . . . That is why no natural mechanism has been discovered for the creation of new complex genetic information. No such mechanism exists. God created us.[8]

8. Phillip Johnson, *The Wedge of Truth: Splitting the Foundations of Naturalism* (Downers Grove, IL: InterVarsity, 2000), 152–53.

Such an assertion surely raises the question of whether the issue, for those like Johnson, has been resolved before the facts are even put on the table. Is there any evidence that would convince Johnson that God did not create us, or that purely natural mechanisms drove evolution? Of course, critics are quick to point out such comments: "This should come as no surprise to anyone for despite creationists' [IDers] protests to the contrary, it is patently obvious that their alternative hypothesis of intelligent design is not a scientific conclusion but a religious one."[9] While belief that the world has been designed has religious implications, critics would be wrong to dismiss Intelligent Design on this account alone, because scientific naturalism would have to be dismissed as well because it also has religious implications. What really matters is not the religious implications of the theory but whether one can conclusively demonstrate that naturalistic explanations fail to achieve their objective. If so, there is no other alternative but to conclude some form of intervention and design. If this turns out to be the case, it matters little whether this expanded type of explanation is called "science" or it is simply said that scientific (purely naturalistic) explanations have hard limits. Either way, design has entered the world of biological study.

This brings us to the other main factor that keeps the Intelligent Design school from fully embracing Darwin: a deep dissatisfaction with the empirical proof for Neo-Darwinian theory. This is an important point since it demonstrates that it is not merely a philosophical bias against naturalism that drives the school (although this seems to be an important factor). In fact, many Intelligent Design adherents have no inherent religious bias against naturalism and have been drawn to the school solely via the scientific evidence. Based upon their evaluation of the evidence, the main scientific claim that this school makes, and the one we will spend the remainder of this chapter investigating, is that although Neo-Darwinian theory can account for a limited amount of evolutionary change, neither Neo-Darwinian theory nor any other natural mechanism can explain the appearance of all biological structures. For example, the Intelligent Design school maintains that there are no natural mechanisms that can explain, down to the molecular details, how a flagellum evolved in an incremental fashion. In fact, they have many similar examples, such as how the first cell arose, how powered flight evolved, or how certain biochemical pathways evolved. This school maintains that Neo-Darwinian theory is not equipped to answer these questions. Since possible scenarios for these evolutionary events often lack hard experimental evidence, many

9. Robert Pennock, *Tower of Babel: The Evidence against the New Creationism* (Cambridge, MA: MIT Press, 1999), 213.

members of the Intelligent Design school are skeptical of or downright hostile toward the explanatory power of the Neo-Darwinist paradigm.[10] Given such frustrations, one goal of the Intelligent Design movement is to demonstrate the prudence of allowing Intelligent Design as a possible alternative explanation for the appearance of various biological systems. While this is obviously an uphill task, the school is attempting to do this by establishing a three-part argument:

1. There exist biological structures that, because they require the presence of multiple interacting parts to have any semblance of functionality, exhibit a property called "irreducible complexity" or "specified complexity."
2. Based upon our knowledge of naturalistic mechanisms, there is *no possibility* such systems could have formed by natural processes.
3. Therefore, the only way to explain the presence of such systems is to invoke an intelligence as their cause.

In determining whether the first two points are correct, two issues will continually resurface: (1) how does one *know* that an object is "irreducibly complex" and (2) how can one *prove* that such objects are beyond the scope of natural mechanisms? Answers to these questions are at the heart of the debate over Intelligent Design.

As we shall see, Intelligent Design research is mainly focused on demonstrating that certain complex structures are exceedingly improbable if left to natural mechanisms such as natural selection. This alone, however, is not sufficient to prove design. One must remember that natural selection is indeed the theory of the improbable. Richard Dawkins even titled a book about natural selection, *Climbing Mount Improbable*, to emphasize this very fact.[11] Therefore, the Intelligent Design camp cannot demonstrate just *improbability*, but rather they must demonstrate the *impossibility*, given the time and resources available for natural mechanisms to work. In fact, as we will soon see, proving design in nature can be done only if all naturalistic explanations have been ruled impotent.

10. Michael J. Behe noted in *Darwin's Black Box: The Biochemical Challenge to Evolution* (New York: Free Press, 1996), that he was unable to find any adequate peer-reviewed research papers examining the molecular details of how complex systems arose by a Darwinian mechanism. While some do exist, they are largely sequence comparisons demonstrating that *possible* genetic components were present in some evolutionary ancestor. How the structure in question was eventually constructed from these theoretical genetic precursors is left largely unaddressed. Some examples are discussed in more detail in chap. 5.

11. Richard Dawkins, *Climbing Mount Improbable* (New York: Norton, 1996).

Finding Design in Nature: Dembski's Design Filter

The scientific validity of Intelligent Design depends on its ability to detect design in nature. If no scientific method for detecting design in the natural world exists, then the biological theory of Intelligent Design can be discarded as theological speculation based on a biased worldview. At the forefront of the movement's effort to devise an acceptable empirical test for biological design has been and continues to be William Dembski, a mathematician by trade and a former faculty member of Baylor University's Polanyi Center. Dembski devoted an entire book to describing in detail a formula for detecting objects or events that have been designed.[12] In this book, *The Design Inference*, Dembski describes a procedure for determining whether an object is the result of design, a procedure Dembski has termed the "design filter." It is so named because it has been constructed to filter out alternative causal explanations where appropriate, such as chance or necessity, in order to conclude that an object was designed. It does this by identifying objects that exhibit what Dembski calls *specified complexity*.

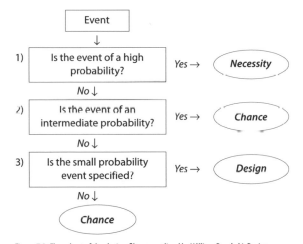

Figure 7.3. Flow chart of the design filter as outlined by William Dembski. Design occupies the last node of the filter (3) because it can be concluded only when all the alternatives have been eliminated. (William Dembski, *The Design Inference: Eliminating Chance through Small Probabilities* [Cambridge: Cambridge University Press, 1998], 36.)

To illustrate how the filter works to identify design, let us examine how it would deal with Mount Rushmore, an object that everyone agrees has been designed. Using the flow chart of the filter depicted in figure

12. William Dembski, *The Design Inference: Eliminating Chance through Small Probabilities* (Cambridge: Cambridge University Press, 1998).

245

7.3, answering the series of yes or no questions will ultimately place the event, in this case Mount Rushmore, into one of three causal explanations; chance, necessity (i.e., laws such as gravity have caused the event), or design. In other words, it answers the question of whether Mount Rushmore occurred via chance, via necessity, or via design.

The first question the filter asks is whether the event happened by necessity. This can be restated as: do the initial conditions dictate that an event must happen? If you hold a bowling ball two feet off the ground and release it, the initial condition (releasing the ball) coupled with the law of gravity dictate that it will fall to the ground. Although the decision to drop the bowling ball was made by an intelligent agent, once released, according to the law of gravity, the bowling ball must of necessity fall toward the earth. In contrast, there is no known law that would dictate that Mount Rushmore would have faces of four dead presidents on it. In this case, it is rather obvious that the unique formation of Mount Rushmore was not dictated by a particular set of initial conditions and laws. Such events pass through node 1 of the filter.

As both designed and chance events pass through node 1, they must be differentiated through the second and third nodes of the filter. At the second node, the filter asks what the probability of the event occurring is. If the probability of it occurring is intermediate to low, the event is attributed to chance. But if the event has an extremely low probability of occurring it passes through to the third node. In the case of Mount Rushmore, the probability of finding the complex likeness of four dead presidents on a mountainside is vanishingly small. Of all the mountain and rock formations in the world only one has such a formation. Since the event has such a low probability of occurring it passes through to the final node.

Events that reach the third node still can either be the result of chance or of Intelligent Design. To distinguish these, the third node asks if the event is *specified*, that is, does it conform to a nonarbitrary pattern that could be determined separately from the event? If the extremely low probability event is specified, it is attributed to design, while unspecified events are attributed to chance. What does this mean in the case of Mount Rushmore? It obviously conforms to a specified pattern, the facial pattern of four presidents. This nonarbitrary pattern is specified in that people can recognize it apart from seeing the Mount Rushmore carving. We recognize Washington's face, for example, in paintings and on dollar bills that have nothing to do with Mount Rushmore. Since Mount Rushmore is such a low probability event and it conforms to a preexisting pattern, we rightly conclude that it was designed. It exhibits what Dembski calls *specified complexity*.

Why do we recognize that the faces of Washington, Jefferson, Roosevelt and Lincoln were carved into the side of the mountain by skillful sculptors, but we do not think the rubble of rocks below them was intentionally assembled? Both are compatible with the laws of physics, but neither is determined by them. Both faces and the pile of rocks are complex. But only the faces tightly conform to a pattern that we recognize as meaningful, namely, the likenesses of four U.S. presidents.[13]

Another example will hopefully make this last step of the filter clear. Consider how the design filter would deal with the famous Balancing Rock in Digby, Nova Scotia, shown in figure 7.4. This rock forma-

Figure 7.4. A drawing of the Balancing Rock of Digby, Nova Scotia. While this is a highly improbable structure, it does not conform to a detachable pattern that is recognizable apart from the rock itself, therefore the design filter attributes its origins to chance.

13. Guillermo Gonzalez and Jay Richards, *The Privileged Planet: How Our Place in the Cosmos Is Designed for Discovery* (Washington, DC: Regnery, 2004), 217.

tion is also extremely improbable. However, the pattern that it forms is not specified; it is an arbitrary pattern unrecognizable to anyone apart from the rock itself. Since the highly improbable Digby rock formation does not conform to a specified pattern in the manner that Mount Rushmore does, the design filter would conclude that it is the result of chance.

It is important to note that the filter concludes chance in this case because of the simplicity of the explanation, even though the possibility exists that the event may have been designed. In fact, because of our limited knowledge, it is always possible that something that appears to us as having happened by chance actually was designed, and of course the filter is unable to help us in this regard. No one can completely rule out the possibility that thousands of years ago aliens constructed the Digby Balancing Rock. In fact, because of this limitation, the filter cannot definitively prove that anything actually happened by chance. Although such false negatives are certainly a limitation, they do not impede the ability of the filter to demonstrate that something was designed, which is its primary purpose:

> When for whatever reason an intelligent cause fails to make its actions evident, we may miss it. But when an intelligent cause succeeds in making its actions evident, we take notice. This is why false negatives do not invalidate the explanatory filter. The explanatory filter is fully capable of detecting intelligent causes intent on making their presence evident.[14]

Therefore, the intelligent cause must make its actions evident if we are to prove design. In addition, we need to have certain background knowledge that lets us recognize the specified pattern the intelligent agent has constructed. For example, a sequence of prime numbers may appear random if one has no idea what prime numbers are. A Chinese ideogram may appear as random squiggles if one does not know the Chinese language. As a consequence, there may be many intelligent actions that we may miss out of ignorance. These false negatives, again, do not invalidate the filter; they mean only that it cannot detect *all* cases of design.

What cases of design, then, is it good at detecting? In the case of most objects that have been designed by humans, such as the Mount Rushmore memorial, the filter appears to do an adequate job. The important question, though, is how does it fare with biological structures? It is to this question that we turn next.

14. William Dembski, "Redesigning Science," in *Mere Creation: Science, Faith and Intelligent Design*, ed. William Dembski (Downers Grove, IL: InterVarsity, 1998), 106.

Proving Design in Biology Using the Filter

When it comes to using Dembski's design filter to investigate biological phenomena, there are two generalizations on which most in the evolution debate would agree. The first generalization is that organisms such as you and I are not here out of necessity. Given the initial conditions of the universe, there is no law that dictates that the specific life-forms around us *must* exist or even that life *must* exist. There is no law that dictates certain bacteria should have a flagellum or even that the types of bacteria present in our world should exist. This is not to say that organisms do not follow and obey laws; they do. For example, Brian Goodwin has argued convincingly that physical laws can restrict or influence the forms that organisms may take.[15] However, while laws certainly can make some forms stable and other possible forms unstable, they do not strictly determine the specific life-forms that exist in the world around us. As a result, biological life is not attributed to necessity or law alone.[16]

The second generalization concerns specification, and most biologists would agree that organisms are specified entities. ("Specified," in this context, means conforming to a particular pattern, which can be identified or "specified.") Humans, for example, are made up of many interacting parts that conform to such a pattern. Only one fundamental arrangement makes up the proper functional human form, allowing, of course, for the infinite slight variations seen in the world. This pattern is not arbitrary, as in the case of the Digby rock, or ad hoc because it is absolutely required for life. This is a conclusion that nearly everyone, including most Darwinists, will grant. These two generalizations, which establish life-forms as contingent and specified, are not without their critics,[17] but in comparison to the last step they are widely acknowledged as legitimate conclusions.

There is one additional criteria necessary to identify design using the filter and it is the most controversial, and therefore it is where we

15. Brian Goodwin, *How the Leopard Changed Its Spots: The Evolution of Complexity* (New York: Charles Scribner's Sons, 1994).

16. The possibility exists, however, that this may not be correct. One position on the matter, "virtual creation," which goes back at least to Gregory of Nyssa (ca. 330–394), postulates that the initial conditions and the boundary conditions on the earth would always lead to the same or similar results. In modern scientific parlance this is referred to as convergent evolution. An excellent book outlining this position is Simon Conway Morris, *Life's Solution: Inevitable Humans in a Lonely Universe* (New York: Cambridge University Press, 2003). A more recent scientific article describing this possibility is: G. Vermeij, "Historical Contingency and the Purported Uniqueness of Evolutionary Innovations," *Proceedings of the National Academy of Sciences USA* 103 (2006): 1804–9.

17. See Howard Van Till, "E. coli at the No Free Lunchroom: Bacterial Flagella and Dembski's Case for Intelligent Design," at www.counterbalance.net/id-hvt/index-frame .html, for a critique of the specification argument.

will focus most of our attention. This criteria requires demonstrating that organisms and their component structures should be classified as extremely low probability entities. If it is granted that organisms are specified and it can also be demonstrated that they are extremely low probability events, the filter concludes that organisms must be the result of design. So the key step for design theorists is to demonstrate that an organism or a biological structure is an extremely unlikely event, that is, that it has an extremely low probability of occurring. If they are successful in demonstrating this point, they can conclude design.

How then is such a conclusion about probabilities drawn? It is at this point that design theorists encounter almost insurmountable difficulties. This is because assigning probabilities to the occurrence of biological structures is a nearly intractable problem, given the dearth of information regarding the emergence of those structures. Let us look at the bacterial flagellum, a favorite example of the Intelligent Design school (fig. 7.5), to illustrate the problem. Let us ask, What is the probability of a flagellum evolving? Readers who have difficulty getting started on this problem are not alone. Why? Because the answer is contingent on a staggering array of unknown variables: What proteins originally made up the first flagellum? In what organism did it originate? What proteins were available within that organism? And what, if any, evolutionary pathways were available? Because these answers are difficult, if not impossible, to ascertain, assigning a probability tends to generate more questions than answers.

Despite these daunting obstacles, Dembski has attempted to compute the probability of a flagellum emerging from scratch.[18] Since this is a structure that is composed of more than fifty different complex proteins, it is not surprising that the value he obtained for it emerging all at once is beyond human comprehension, 1 in 10^{2954}. This probability is so low that it is quite impossible for chance to bring it about, even were the earth a trillion times older than it is believed to be.

To understand why this is the case, note first that Dembski proposes what he calls the universal probability bound. What this bound represents is a probability below which any specified event must have been designed. For example, if the bound turns out to be 1 in 10^{50}, any specified event with a probability below this must have been designed.

In calculating this number, Dembski settles on 1 in 10^{150}. He arrives at this by starting with the estimated number of particles in the universe, approximately 10^{80}. Then he assumed that each could do something in the smallest time interval we know how to measure, Planck time, 10^{-43} second.

---

18. William Dembski, *No Free Lunch: Why Specified Complexity Cannot Be Purchased without Intelligence* (Lanham, MD: Rowman & Littlefield, 2002), 289–302.

Since the universe is believed to be 15 billion years old, or about 10^{18} seconds, there have been $10^{18}/10^{-43} \sim 10^{60}$ Planck time intervals in its history. Dembski multiplies 10^{80} by 10^{60} to get 10^{140} and then goes a bit higher with 10^{150} just to be safe. Basically he has multiplied the number of particles that exist by the possible number of interactions they could have.

Obviously, this number is a massive overestimate, since the larger things are, the longer it takes them to change, and even most elementary particles can't do anything in Planck time. So it does represent a solid upper bound on the number of "events" that could conceivably have happened in the history of the universe.

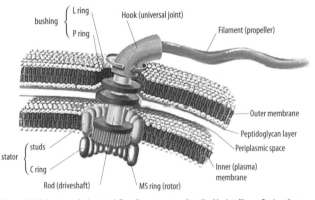

Figure 7.5. A drawing of a bacterial flagellum, a system described by Intelligent Design theorists as an example of specified complexity. The rotor is able to spin the filament in order to produce movement. The outer membrane, the peptidoglycan layer, and the inner membrane are not part of the flagellum but represent the membrane of the bacterium in which the flagellum is embedded. (From D. Voet and J. Voet, *Biochemistry*. 2nd ed. [New York: Wiley & Sons, 1995]. Copyright © 1995 by John Wiley & Sons, Inc. Used by permission.)

The probability that Dembski calculated for the flagellum is well beyond this bound. If his calculation for the flagellum is correct, there certainly was not enough time and matter (probabilistic resources) available in the universe to generate the flagellum. As Dembski explains:

> This is one instance where less is more, where having fewer probabilistic resources [the 1 in 10^{150} calculation] opens possibilities for knowledge and discovery that would otherwise be closed. Limited probabilistic resources enrich our knowledge of the world by enabling us to detect design where otherwise it would elude us.[19]

Dembski does not deny that if one could formulate and test all possible combinations of biological proteins, one would eventually hit on

19. Dembski, *No Free Lunch*, 100.

251

a combination that made a functioning flagellum. What he is arguing is that there is not enough time or resources in the universe for even a tiny fraction of these possibilities to have been assayed by evolution. As a result, he concludes that the flagellum could not have occurred by chance. He argues that the chance formation of the flagellum is therefore not just *improbable*, but that it is *impossible*. Hence, the flagellum must be the result of design, *if*—and this is a big *if*—the probability of the bacterial flagellum computed by Dembski is accurate. However, as we shall see, arriving at his number of 1 in 10^{2954} is not as straightforward as Dembski makes it out to be, and therein lies the challenge for the Intelligent Design movement. Before we look at his critics, we first need to discuss the other main component of the Intelligent Design argument, irreducible complexity.

Irreducible Complexity: The Biochemical Challenge to Macroevolution

While Dembski's design filter represents an ambitious attempt to detect design in nature, the resurgence of design-oriented thought in biology probably owes more to the work of Lehigh University biochemist Michael Behe than to any other single figure. Rather than employing a predominantly mathematical argument like Dembski, Behe's argument is based upon the wealth of new information pouring in from the fields of biochemistry and molecular biology. These fields have revealed that cells, rather than being simple balls of protoplasm as Darwin envisioned, are composed of interacting organelles, which in turn are composed of complex biochemical systems, which in turn are composed of specifically arranged macromolecules. Armed with this new arsenal of information, Behe argues that nothing but design could account for the presence of the Lilliputian biochemical world found in all cells.

> The result of these cumulative efforts to investigate the cell—to investigate life at the molecular level—is a loud, clear, piercing cry of "design!" The result is so unambiguous and so significant that it must be ranked as one of the greatest achievements in the history of science.[20]

Behe's case for this position is made in probably the best-known and most widely read treatise on Intelligent Design, *Darwin's Black Box*. As he rummages through various biochemical systems in his book, Behe encounters what he terms *irreducible complexity*, a characteristic of biological systems that he believes indicates design. For a definition of

20. Behe, *Darwin's Black Box*, 232.

an irreducibly complex system, let us return to William Dembski, who has improved on Behe's original concept:

> A system performing a given basic function is *irreducibly complex* if it includes a set of well-matched, mutually interacting, nonarbitrarily individuated parts such that each part in the set is indispensable to maintaining the system's basic, and therefore original function. The set of these indispensable parts is known as the *irreducible core* of the system.[21]

The example of an antique pocket watch is useful for illustrating such a system. If the minimum function of the watch (the core function) is to tell time, then many of the pieces of the antique pocket watch are not essential for the core function, for example, the chain, the glass covering, and the gold casing. Of course, these things are nice additions that will help preserve the watch and possibly make it more fashionable, but they are not involved in keeping time. There is, however, a set of interacting gears and dials that are arranged in a specific manner that is absolutely essential for the watch to be functional. If any of these gears is removed, *the watch can no longer function at all*. This is an important point: the watch doesn't retain *some* of its ability to keep time, it has *no* time-keeping ability at all. Not only is the watch unable to keep time once a gear is removed, *the new arrangement serves no new function*. Removing a gear from a watch does not turn it into a remote control or a calculator. Removing a part of a truly irreducible complex system removes *any function that depends on the remaining interacting parts*.

Behe's controversial claim is twofold. First, he argues that many biological systems exhibit irreducible complexity, and second, he argues that such systems could not have resulted from the gradual step-by-step assembly that is required by Neo-Darwinian theory. The reason that irreducibly complex systems cannot be built by natural selection rests on the fact that for natural selection to maintain a system, that system *must* have the ability to perform a beneficial function. If a system has no beneficial function, natural selection will not maintain it just in case it might develop such a function in the future. Thus, having half of an irreducibly complex system does the cell no good; neither does having 95% of an irreducibly complex system because, according to Behe, you must have the whole system or you have no function. To demonstrate his point, Behe uses a number of biochemical examples, one of which is the eukaryotic cilium.

As in all his examples, Behe starts by asking the question, What proteins are *absolutely necessary* for such a system to have any minimum

21. Dembski, *No Free Lunch*, 285.

functionality? This allows him to identify the irreducible core of the structure.

> Ciliary motion certainly requires microtubules; otherwise there would be no strands to slide. Additionally it requires a motor, or else the microtubules of the cilium would lie stiff and motionless. Furthermore, it requires linkers to tug on neighboring strands, converting the sliding motion into a bending motion, and preventing the structure from falling apart. All of these parts are required to perform one function: ciliary motion. . . . Ciliary motion simply does not exist in the absence of microtubules, connectors, and motors. Therefore we can conclude that the cilium is irreducibly complex—an enormous monkey wrench thrown into its presumed gradual, Darwinian evolution.[22]

Why? Because, as Behe claims, all three components must be present to have any minimal function, in this case ciliary movement. In fact, all known cilia have *at least* these three components. As a result, if an organism had only two of these proteins, they would be of no benefit to the organism in forming cilia; it would be just as immobile as an organism that lacked these two proteins. Without a function, in this case a minimal ability to have controlled movement, there is no reason the two proteins would be maintained such that a third, and then possibly a fourth or fifth protein could be added to the system in a gradual Darwinian fashion. Such incomplete systems that lack a function will not survive the selection process. They are dead ends, or "road kill," as Behe describes them.

As a result, Behe argues that the only manner by which the origin of such a system can be fitted into a naturalistic framework is by proposing the appearance of all three specifically interacting proteins simultaneously during the course of evolution. This is highly unlikely, in fact so unlikely that the Intelligent Design school would argue that it falls under the realm of the chance events that will not happen in the lifetime of our planet, given the mode by which natural selection operates.[23] Even Darwin knew this: "If it could be demonstrated that any complex organ existed which could not possibly have formed by numerous, successive, slight modifications, my theory would absolutely break down."[24]

In the cilium, Behe believes he has one such "organ." Furthermore, he claims that many systems—such as the blood-clotting pathway, certain

22. Behe, *Darwin's Black Box*, 65.
23. This is identical to the last step of the design filter, in which a specified event, a cilium, of extremely low probability does not happen by chance. According to the Intelligent Design school, the best inference is then design.
24. This refers to Darwin's famous quote on gradualism in *The Origin of Species*, 175.

metabolic pathways, and the cell's protein assembly machinery—also fit the bill of irreducible complexity. Michael Denton goes a step further, arguing that even proteins themselves are irreducibly complex because they are made up of "a complex web of electronic or electrostatic interactions . . . which ultimately involve virtually every other section of the amino acid chain in the molecule."[25] Because of this, Denton believes that the gradual transformation of an existing protein into one with a novel function is virtually impossible for many of the same reasons Behe has described above for the case of the cilia:

> The impossibility of gradual functional transformation is virtually self-evident in the case of proteins. . . . To change, for example, the shape and function of the active site . . . in isolation would be bound to disrupt all the complex intramolecular bonds throughout the molecule, destabilizing the whole system and rendering it useless. Recent experimental studies of enzyme evolution largely support this view. . . . The general consensus of opinion in this field is that significant functional modification of a protein would require several simultaneous amino acid replacements of a relatively improbable nature.[26]

Denton's argument is an overstatement, given that many amino acids in a protein sequence are not critical to the protein's function and can be readily mutated without adversely affecting protein function. Despite this caveat, it is true that certain amino acids are absolutely critical for any specific protein to function. If, as Denton argues, all of these critical amino acids must be present in a specific order to have any minimal functionality, then proteins, one of the essential biological building blocks, would appear to be irreducibly complex and beyond the reach of cumulative selection.

What Does It Prove?

If one grants Behe (and Denton as well) his point that some biological systems are irreducibly complex, does this prove definitively that biological organisms are the product of design? If natural explanations and chance, as Behe claims, cannot readily explain irreducibly complex systems, must they therefore be designed? Even Behe pulls back from this conclusion: "There is no magic point of irreducible complexity at which Darwinism is logically impossible. But the hurdles for gradualism become higher and higher as structures are more complex, more inter-

---

25. Michael Denton, *Nature's Destiny* (New York: Free Press, 1998), 340.
26. Michael Denton, *Evolution: A Theory in Crisis* (London: Burnett Books, 1985), chap. 13.

dependent."[27] In fact, Behe admits the possibility that other naturalistic explanations could account for those systems that pose difficulties for evolution, although he concludes that this is unlikely:

> Might there be an as-yet-undiscovered natural process that would explain biochemical complexity? No one would be foolish enough to categorically deny the possibility. Nonetheless, we can say that if there is such a process no one has a clue how it would work. Further, it would go against all human experience, like postulating that a natural process might explain computers.[28]

Notice that Behe merely claims, without going into the probabilities, that natural mechanisms are extremely unlikely to have produced the biological structures he has deemed to be irreducibly complex. It may be unlikely (a fact most Darwinists admit), but it is not necessarily impossible. That requires further proof.

Developmental and Functional Integration

While irreducible complexity and the design filter remain at the forefront of the Intelligent Design movement, they are not the only biological avenues the school has pursued. Another less appreciated area Intelligent Design advocates have pursued deals with development. To see its relevance to the Intelligent Design issue, let us examine how one might construct a working flagellum (see fig. 7.5). First of all, it is *possible* that a flagellum could be built in a test tube if a human intelligence added the correct proteins in the correct ratio to the tube, granted that a cell membrane upon which the flagellum could assemble was provided as well. Furthermore, by providing energy to the tube as well as the proper ions, the long whiplike region could possibly rotate in a fashion similar to that seen in an outboard motor. In this case, construction of the flagellum would be identical to the reverse engineering techniques used to construct many man-made machines.

But unlike this theoretical test tube flagellum, the flagellum in the cell must be constructed without any assistance. In order to do this, there are a few issues that the cell must somehow address. Because too little and in some cases too much of any of the needed proteins could bring the whole process down, the cell must regulate the amounts of each component. The cell must even acquire the material to build these components on its own. In addition, the cell-driven assembly of the flagellum cannot occur at random times and in random locations in the

27. Behe, *Darwin's Black Box*, 214.
28. Ibid.

cell; rather, it must occur in a coordinated fashion during development and then be targeted to the membrane, the site at which the flagellum functions. Regulating this entire process requires a set of coordinated genes that must operate independently of the flagellum's structural genes, but without which these structural genes would be worthless.

This raises a significant evolutionary problem that Intelligent Design theorists are quick to point out. How does one developmentally regulate a new system without knowing the beneficial function the system will provide? Unfortunately, one cannot know the beneficial function of a new system unless it is regulated properly during development. The problem becomes more and more complex in higher organisms such as vertebrates, as Jonathan Wells argues, because they have much more elaborate developmental patterns:

> [Developmental] gene sequences confer selective advantages only if they program the development of useful adaptations. If a primitive animal possessed [developmental] genes but lacked all of the adaptations now associated with them, then those genes must have originated prior to these adaptations. How then did [these] genes evolve?[29]

If a process for developmental regulation were in place before the biological structure to be regulated appears, the developmental genes would have to have had a function separate from the process they currently regulate. What this would be is anyone's guess because nothing is regulated in the same fashion as a flagellum. While it is true that similar genes are used to regulate different developmental processes in different organisms (see chap. 3), each developmental process is unique, despite the fact that some of the regulatory proteins may be similar. Evolutionarily, this is a critical problem that all too often goes unaddressed.

In addition to the difficulties of explaining developmental regulation, another problem Intelligent Design researchers have seized upon is the necessity of integrating any novel feature into the context of a functioning organism. For example, as described in chapter 5, many theorists posit that eyes began as a simple light-sensitive spot. If this is true, there is no reason to assume that the appearance of the ability to sense light would be of any use to an organism unless the organism had some way of integrating this new information into its current way of making a living. It may be intuitive to us humans that the ability to sense light would be useful to an organism; we find it quite helpful because we are wired to use this information. We know what to do with it because we have entire regions of our brain dedicated to processing just this

29. Jonathan Wells, "Unseating Naturalism: Recent Insights from Developmental Biology," in *Mere Creation*, 578.

information. However, if an organism has never before sensed light, the organism would have no clue as to how to use this information correctly. Should it swim toward light? Should it swim away from it? Should it do a little dance? Even once it "chooses" one of these options, carrying it out isn't as simple as it seems. To respond to light in any one of the manners described above, the organism must already have an integrated pathway in which the exposure to light triggers locomotion in a certain direction. How this would come into existence before the light sensitive spot existed is again anyone's guess. But one can be sure that the spot is relatively useless unless this pathway or a similar one is already in place. Just as with developmental regulation, it is not clear how you can get one (the light sensitive spot) without the other (some type of functional integration). Natural selection would seem to need both to occur simultaneously in order to succeed, and the probability of this occurring is extremely low.

Problems for the Intelligent Design School

While the Intelligent Design school has continued to pursue its work on different fronts, it has encountered stiff resistance from the scientific community, in part because of Intelligent Design's inherent bias toward design. While bias on both sides is an important issue, there are also scientific reasons the Intelligent Design school has met with resistance, mainly because many researchers are critical of Dembski's and Behe's arguments. In fact, these criticisms are now found in college textbooks on evolution, which is proof of both the inroads the Intelligent Design movement has made with the general public as well as the seriousness with which the biological community has taken the Intelligent Design threat.

Does the Design Filter Really Work?

When we examined Dembski's contribution to the Intelligent Design movement, the design filter, we found that it appears to function adequately in detecting objects that have been designed by humans, such as the Mount Rushmore monument. However, it seems to encounter difficulty when it enters the realm of biological structures. Why the difference between man-made and biological structures? The problem is that biological structures differ radically from the man-made objects that fall under the purview of such disciplines as archaeology and cryptography. When an archaeologist uncovers a detailed wood carving or a painted piece of pottery, he or she concludes design because archaeologists have direct knowledge of how these types of objects are normally created. We all know from experience that humans make detailed wood carvings and decorative pottery, and all such objects we have encountered have been made by intelligent agents or by machines that were designed by intelligent agents. Even in the search for extraterrestrial intelligence (SETI), we look for patterns that we know from experience to be produced by intelligent agents, namely humans. If we discovered a radio signal transmitting the sequence of all prime numbers, we know that generating this sequence requires intelligence. Why? Because in our experience it has always been associated with human intelligence.

In contrast to the foregoing examples, we have never seen a biological structure being designed from scratch by any intelligent agent.[30] We know from experience how detailed wood carvings originated; we do not know from experience how biological entities originated. As Robert Pennock explains, this lack of experience limits us:

> The world is full of complexities of all sorts. To identify something as an intelligent signal from among these will require not only that it stand out in sharp contrast to ordinary complexities but that it is recognizable by reference to a pattern that we know to be characteristic of our own intelligent signaling.[31]

Biological entities, then, may be specified (conform to a nonarbitrary complex pattern), but we cannot claim to have familiarity with the origin of the pattern. For example, we do not know precisely how the first squirrel came into existence in the same manner in which we know how detailed wood carvings come into existence. Given our experience with detailed wood carvings, we know how unlikely it is that such a thing could be created by chance. The same cannot be said of a squirrel.

To illustrate this difficulty, let us reexamine the probability that the bacterial flagellum came into existence by chance. Dembski's calculation adds up the proteins necessary, about fifty, and looks at the odds of their appearing in one fell swoop from scratch. He also looks at the difficulty of localizing the proteins in the right place and then configuring them in the right orientation to make a functioning flagellum. This is unquestionably a worst-case scenario. What Dembski's calculation ignores is that *every* biological structure originates within a certain context, that is, within an organism at a specific time and environment in natural history. This

---

30. Craig Venter and colleagues are attempting to create a new life-form from an existing bacterium. This, however, would not satisfy the criterion of creating a new life-form from scratch because the bacterium from which they are starting supplies most of the design work.

31. Pennock, *Tower of Babel*, 254.

context may already provide some of the material for the flagellum, such that it does not have to go from nothing to everything all at once. (Of course, these proteins must be performing another useful function for the cell in order to be maintained by selection.) The context may even provide for some localization of some of the parts to the membrane.

In addition, certain proteins, such as those that make the rings found in the bacterial flagellum, may spontaneously form a stable configuration based upon local chemical interactions if enough subunits are present. This is analogous to the spontaneous formation of viral capsids from a mass of viral coat proteins. In simple viruses, if enough viral coat proteins are present, a capsid is constructed spontaneously through electrostatic interactions between the surfaces of the individual nonsymmetrical subunits (fig. 7.6). Larger, more complex capsids are composed in a similar manner but require scaffolding proteins, proteins that help order and stabilize the viral capsid during assembly. In both cases though, the structure appears to result spontaneously from local interactions when enough structural and scaffolding subunits are available.[32] Thus, once the needed proteins are present, the configuration problem may not be as drastic as Dembski assumes.

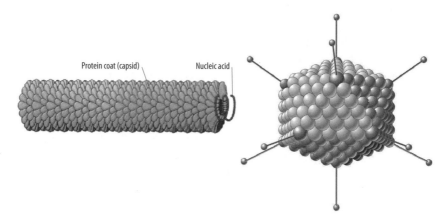

Protein coat (capsid)   Nucleic acid

Figure 7.6. On the right is an example of an icosahedral virus, adenovirus, while on the left is an example of a cylindrical virus, tobacco mosaic virus. The 3D capsid is made of repeated individual subunits that interact spontaneously to create various structures. (From G. Karp, *Cell and Molecular Biology: Concepts and Experiments* [Hoboken, NJ: Wiley, 2005], figs. 1.20 and 1.21. Copyright © 2005 by John Wiley & Sons, Inc. Used by permission.)

There is another reason Dembski's estimated probability may be an extreme overestimate, and it has to do with the manner in which he came up with the probability for each of the fifty or so proteins that

32. For a tutorial on viral assembly, see www.tulane.edu/~dmsander/WWW/224/Replication224.html.

make up the flagellum. In his book *No Free Lunch*, Dembski discusses a 113-amino-acid protein called URF13 that appears to have emerged spontaneously from a region of unused DNA in a species of maize. At first glance this appears to be a case in which Dembski's hallmark of design, specified complexity, emerged by purely natural mechanisms. Dembski dismisses this idea, though, arguing that the probability of the URF13 protein forming from scratch isn't that low. His reasoning is telling:

> Any time one strings together a sequence of amino acids, one is likely to obtain some three-dimensional structure that includes alpha helices [these are very common structures found in proteins and they are found in URF13] since these are easy to form.[33]

This position undercuts his main point, though, namely that it is impossible to get biological proteins to evolve from scratch. Dembski is right that getting a stretch of amino acids to form alpha helices is not very improbable. In fact all proteins are composed of alpha helices and/or beta-pleated sheets, because random stretches of amino acids tend to form these structures in solution. Since these essential structures are quite common, getting functional proteins from scratch may not be as hard as Dembski admits. In the case of the URF13 protein, the initial presence of alpha helices could occur by chance. The alpha helical structure in these proteins may have some minimal function that then is gradually improved on such that it reaches an optimal fitness level. The URF13 data should give Dembski reason to pause, but apparently it does not. In fact, he tries to have it both ways. In the flagellum calculation he makes no reference to the ease of forming alpha helices when calculating the improbability of forming the proteins in question, while with the URF13 protein he is quick to point out that alpha helices are easy to form. If it is true for URF13, the same must then be true of the flagellum proteins. Surely then this information would affect the probability he has calculated for the flagellum proteins.[34]

Unfortunately, when Dembski calculates his probability, he ignores all of these caveats. They certainly would affect the probability he has calculated, but no one knows to what degree because (1) we do not know how probable it is that a sequence of amino acids would spontaneously form an alpha helical or beta-pleated sheet structure, (2) we do not know how much spontaneous self-assembly occurs during the building of the flagellum, and (3) the historical context in which the

33. Dembski, *No Free Lunch*, 218.
34. It is important to note, however, that to make the formation of the flagellum reasonably plausible, one would have to propose mechanisms that reduce Dembski's calculation by nearly 2,900 orders of magnitude—a nontrivial task.

flagellum originated is probably forever beyond the ability of science to reconstruct. This last point is the most important because it is the one problem that may be intractable. Consider what information would be needed to reconstruct the historical context for even a single protein: What was the original structure of the protein? How much could it vary its sequence or structure without losing its original function? Could it have performed two functions? How did the original version arise? Was it the result of a gene duplication? How essential was the original function of the protein to the organism during the period of conversion to a flagellum protein? Were there other partially redundant proteins that could take the place of this protein? In addition, one needs to answer other global questions such as: Which proteins were actually needed to form a working flagellum at the time it originated? Are these the same proteins that currently compose the flagellum? How were the genes that encode these proteins regulated? What selective pressures were present at the time of formation? And the list goes on and on. As a result, proving design or proving Darwinian selection via the design filter is no easy task because coming up with an accurate probability estimate is nearly impossible.

Despite this serious problem, discussion of Dembski's filter has helped in pinpointing the issues that must be resolved if we are to ever understand biological origins. These issues cannot be ignored or pushed to the side by any partisan in the debate. The Intelligent Design school can rejoice in the massive probabilities they conjure up for various biological structures, but if these probabilities ignore the contingencies described above, they will fail to be persuasive. To be sure, the goal of a quick and easy way empirically to detect design in biological systems is attractive, but in reality this is a much more difficult and tedious problem. The history and function of any biological system must be peeled apart in sufficient grueling detail if one is to have any hope of concluding design. It is now time to examine some of these details.

Are Things Really Irreducibly Complex?

While Dembski has focused on the mathematics, Behe has focused on the biological details, more specifically the molecular details. Given the biological nature of Behe's arguments, Neo-Darwinists are much more comfortable engaging him than they are Dembski, so critiques of the Intelligent Design movement in college textbooks focus almost entirely on Behe's notion of irreducible complexity. While some critics have attempted to dismiss the notion of irreducible complexity entirely, many other critics have tackled the issue head-on. For those who have done so, the argument for co-option is the most common explanation for how irreducibly complex systems could have evolved.

To understand this argument, let us revisit the cilia example. As Behe pointed out, all cilia have three types of proteins that are required for the cilium to have any ability to function, namely, microtubules (tubulin), connectors (nexin), and motors (dynein). While it is true that these three types of proteins are needed to form any minimally functioning cilia, they need not be considered useless to the cell apart from their potential role in forming cilia. In fact, these proteins are used in other cellular processes. Microtubules, for example, serve many functions in the cell, from forming structures needed for cell division to supporting and maintaining the shape of the cell. Motor proteins are also needed for many functions, including transport of material through the cell and the separation of chromosomes during cell division. In fact, to perform these functions these motor proteins actually bind to microtubules (fig. 7.7). Thus, these proteins could have evolved individually, all the while being retained by natural selection because they performed a useful function apart from forming a cilium. In this scenario, all three would not have to appear simultaneously. These proteins would have first arisen to perform a function unrelated to the cilium and, over time, they could have gradually been co-opted for use in a minimally functioning cilium.

Figure 7.7. Motor proteins are shown moving down a microtubule (the long rod-like structure) carrying vesicles. In the cell, transport vesicles and other molecules are shuttled along microtubules via these motor proteins. (From G. Karp, *Cell and Molecular Biology*, fig. 9.6. Copyright © 2005 by John Wiley & Sons, Inc. Reprinted with permission.)

While co-option is a possibility, Behe is quick to point out the difficulties of co-opting proteins from other distinct systems. In Behe's opinion, one cannot readily co-opt a motorlike protein already being used by the cell in a separate process, a connector being used in another, and a microtubule arrangement from still a third because they will not fit together with any specificity. As a biochemist, he knows that proteins have very specific binding and regulatory domains such that proteins being used in one system are not likely to fit lock-and-key with a protein from an unrelated system, much less with an array of proteins necessary for the functioning of an irreducibly complex system. Behe uses the example of a mousetrap to illustrate his point:

> Suppose you wanted to make a mousetrap. In your garage you might have a piece of wood from an old popsicle stick (for the platform), a spring from an old wind-up clock, a piece of metal (for the hammer) in the form of a crowbar, a darning needle for the holding bar, and a bottle cap that you fancy to use as a catch. But these pieces couldn't form a functioning mousetrap without extensive modification, and while the modification was going on, they would be unable to work as a mousetrap. Their previous functions make them ill-suited for virtually any new role as part of a complex system.[35]

Behe's point must be taken seriously. Having the required *type* of proteins present in the cell does not mean they will fit together with the other proteins necessary to build the system with any specificity. If the three proteins are doing other tasks, there is no reason to assume they will be able to interact properly with one another to form a cilium. If the motor protein is initially involved in transporting cellular vesicles down microtubules, there is no assurance it will be able to slide two microtubules in relation to each other, a feature necessary to form a functioning cilium. In addition, any mutation that makes the motor protein able to bind multiple microtubules could significantly alter its original ability, the binding of a cellular organelle or vesicle. This could lead to a selective *disadvantage*. Furthermore, until the cilium is functioning, mutations that facilitate some of the various parts fitting together with limited precision will not be favored by natural selection.

One way in which this difficulty may be circumvented is by taking advantage of the molecular redundancy within cells. Many molecules in the cell, when deleted, do not lead to significant adverse effects because another molecule is similar enough in structure and function to replace it (at least partially). As discussed in chapter 5, there are mechanisms by which certain genes can at random be duplicated in the genome such that there are two copies of a specific gene.[36] Sections of chromosomes and even entire chromosomes can be duplicated.

---

35. Behe, *Darwin's Black Box*, 66.

36. Gene duplications can be mediated by defects that occur during meiosis as well as by transposons, which appear to be remnants of retroviruses. They can "jump" them-

Regardless of how it is supplied, the duplicated gene(s) then has some freedom to mutate without significantly altering the old function that the original gene would still be performing. These mutations could theoretically lead to the acquisition of an altered or novel function in the duplicated gene. In regard to cilia, biology professor Kenneth Miller argues as follows:

> Behe scoffs at the notion that a biochemical system adapted for one purpose might be adapted by evolution for a totally different function. . . . He dismisses, for example, the notion that the parts of a cilium, including proteins like dynein [motor protein] and tubulin [microtubule protein], could have evolved by gene duplication even though similar forms of dynein and tubulin are used for other purposes in the cell. Most cell biologists will be unconvinced by his explanations of why the cilium could not have been assembled from proteins originally used for other purposes.[37]

The fact that cells have multiple types of dynein motor proteins and tubulin microtubule proteins supports the notion that gene duplications can give rise to new versions of these proteins. There are other examples, such as the crystalline proteins that make up a large portion of the eye lens. These appear to be degenerative remnants of a diverse group of duplicated genes.[38] While such data support the argument that redundant proteins can assume novel functions, it doesn't *prove* that they can assemble into irreducibly complex structures. It is important to remember that in irreducibly complex structures, the redundant proteins not only have to be present but also must be able to interact with one another and assemble into the proper structure. To do this, they must be expressed at the same time and sequestered in the appropriate area of the cell. This organizational work is paramount to the performance of the system. But what does the organizational work of building these complex systems? How do all the components of, say, the cilium, even if they exist in the cell, come together during evolution to form a new complex structure, that is, a functioning cilium?

Intelligent design critics claim that the components don't have to come together all at once; rather, they can come together piece by piece *despite* their apparent irreducible complexity. How? To understand this argument one must first recognize that the *current* organization of a structure such as a cilium may not adequately represent how such a structure was

---

selves in and out of the DNA molecule, copying themselves and often neighboring genes, thereby leading to gene duplications.

37. Kenneth Miller, "Review of Darwin's Black Box," *Creation/Evolution* 16 (1996): 36–40.

38. S. Freeman and J. C. Herron, *Evolutionary Analysis*, 3rd ed. (Upper Saddle River, NJ: Pearson-Prentice Hall, 2003), 100–101.

originally formed. Objects that are now irreducibly complex, such as the cilium, may not have been so when they originated:

> An irreducibly complex system can be built gradually by adding parts that, while initially just advantageous, become—because of later changes—essential. The logic is very simple. Some part (A) initially does some job (and not very well, perhaps). Another part (B) later gets added because it helps A. This new part isn't essential, it merely improves things. But later on, A (or something else) may change in such a way that B now becomes indispensable. This process continues as further parts get folded into the system. And at the end of the day, many parts may all be required.[39]

Although such a possibility exists and could explain the presence of seemingly irreducibly complex systems, the Intelligent Design school would question how one protein could ever perform the function of a cilium.[40] Cilia are large structures that require the presence of a number of molecules arranged in an orderly fashion. The same holds true with the blood-clotting pathway and the flagellum and many other biological entities that Behe believes are irreducibly complex. In all these cases, a minimum number of proteins would need to be present for the system to have *any minimal function*, no matter how small. It is highly unlikely that a single protein would be unable to perform the function in question by itself.

To address this issue, critics turn once again to the possibility of co-option, only this time they argue that cells can co-opt groups of proteins rather than just co-opting one protein at a time. Suppose that A and B work together to perform a function; call it "function 1." If protein C is added to the mix, then A, B, and C could work together to perform a new function; call it "function 2." In this case, A and B are co-opted together to help perform function 2 once protein C is added to the mix. Thus, A, B, and C would be irreducibly complex in regard to function 2, but if C is removed, A and B can still perform function 1 or, it is argued, they could have done so in the evolutionary past. Let's return to the cilia example to illustrate this point. It is possible that the motor and

39. H. A. Orr, "Darwin v. Intelligent Design (Again)," *Boston Review* (December–January 1996): 7, 29.

40. Critics of Intelligent Design attempt to use this line of reasoning to demonstrate that a mousetrap composed of only one piece could be functional, thereby undermining the idea that this structure, used by Behe as an example of an irreducibly complex system, is irreducibly complex. While it is possible to imagine such a mousetrap, the single piece that performs the function of the mousetrap would have multiple domains that do different functions. One holds the bait, another is the trap, and so on. The molecule would be irreducibly complex; it would require multiple interacting domains to have any function at all. The problem of irreducible complexity still remains, although now at the molecular level.

microtubule proteins of the cilia initially were involved in a transport system, a system that would provide a beneficial function to the cell. If this transport system became partially redundant and a third protein, the linker, was added, it is possible that a cilia-like structure could form. While this idea is intriguing, even some Intelligent Design critics argue against this idea as a general principle:

> We might think that some of the parts of an irreducibly complex system evolved step by step for some other purpose and were then recruited whole-sale to a new function. But this is unlikely. You may as well hope that half of your car's transmission will suddenly help out in the airbag department. Such things might happen very, very rarely, but they surely do not offer a general solution to irreducible complexity.[41]

Others, though, are more confident that such wholesale co-option could have occurred routinely during evolution. In fact, this is what many mean when they argue that natural selection is a grand tinkerer, cobbling together new systems from the systems that happen to be working presently in the cell.

> Feathers are adapted for flight, but that is not their only function. By trapping a layer of air next to the body surface they also assist in thermo-regulation. Feathers may have evolved first to help small warm-bodied proto-birds retain their body temperature. Only later did they assist in flight. Their function then changed, or at least became more complex.[42]

While this is a possibility, evidence is needed to back up this claim regarding co-option. Remember that *positing* a possible scenario is different from *proving* a scenario is feasible in the real world, where the consequence of failure is death. Likewise, *disproving* a possible scenario is different from demonstrating that a specific scenario is merely unlikely.

Could Co-option Work? We Don't Know

At this point two things should be evident: (1) the design inference suffers from an absence of empirical evidence, and (2) because of this the arguments and counterarguments can (and probably will) go on ad infinitum, with little or no likelihood of resolution. What is needed is some hard data, which unfortunately is hard to come by.

For example, despite the tremendous amount of active research into the structure and function of the flagellum, biochemists are still trying

41. Orr, "Darwin v. Intelligent Design (Again)," 29.
42. Mark Ridley, *The Problems of Evolution* (New York: Oxford University Press, 1985), 38.

to understand just how it works. As a result there is not enough information available to postulate exactly how the structure evolved gradually or even if it could have evolved gradually. The same holds true of the classic evolutionary example, the human eye, which is discussed in chapter 5. Despite this lack of data, both proponents and critics of Intelligent Design usually display little caution in their statements. Proponents are quick to assume design, while critics are quick to assume the validity of co-option scenarios, when the actual data are far from clear.

One major problem is that it is difficult to verify what, if any, intermediary structures any given irreducibly complex system could have been co-opted from. In the case of the bacterial flagellum, for example, the main proposed intermediary step is the type III secretory system, which is found in many bacteria and is used to deliver effector proteins into host cells. These effector proteins modulate the function of the host cell to the advantage of the bacterium.[43] This system is simpler in form and uses many proteins that are homologous to those used to assemble the bacterial flagellum (fig. 7.8). Unfortunately for Intelligent Design critics, it appears that, if anything, the type III secretory system evolved from the flagellar system[44] as "flagella are very ancient organelles, predating by far the targets for bacterial pathogenesis [what the type III system is used for]—plants, mammals, etc."[45] A further problem is that the type III system explains only one step in what was surely a long convoluted process. To get something as complex as the flagella there must be countless more intermediate forms.

Despite the current difficulties in demonstrating an indirect evolutionary route to the flagellum via wholesale co-option, it is clear that all the evidence is not in. In fact, some may be forever lost, shrouded behind the veil of natural history. As a result Intelligent Design critics like Ken Miller caution that we cannot make judgments out of ignorance:

> Before [Darwinian] evolution is excoriated for failing to explain the evolution of the flagellum, I'd request that the scientific community at least be allowed to figure out how its various parts work.[46]

Miller is correct on this point. It would be a bit rash to argue that it must have been designed if we don't even know how it works. In fact, our

43. G. V. Plano, J. Day, and F. Ferracci, "Type III Export: New Uses for an Old Pathway," *Molecular Microbiology* 40 (2001): 284–93.

44. L. Nguyen et al., "Phylogenetic Analyses of the Constituents of Type III Protein Secretion Systems," *Journal of Molecular Microbiology and Biotechnology* 2 (2000): 125–44.

45. Robert Macnab, "The Bacterial Flagellum: Reversible Rotary Propeller and Type III Export Apparatus," *Journal of Bacteriology* 181 (1999): 7149–53.

46. Kenneth Miller, *Finding Darwin's God: A Scientist's Search for Common Ground between God and Evolution* (New York: HarperCollins, 1999), 148.

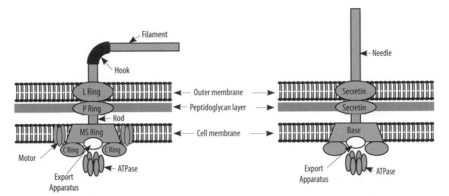

Figure 7.8. A side-by-side comparison of models of the type III secretory system on the left and the bacteria flagellum on the right. While the visual similarities are apparent, especially in the membrane-spanning region, not all of the proteins that compose these structures are conserved. (Plano, Day, and Ferracci, "Type III Export," 284–93.)

ignorance of many aspects of cell biology and cellular behavior should give all partisans in the debate reason for pause. Why? Because, it is possible that new research will reveal information that helps explain how the flagellum evolved (a situation supporting Darwinists), but it is also possible that it will reveal additional difficulties for evolutionary processes (a situation supporting Intelligent Design). As the physicist Stephen Barr has said, "we simply don't know all of nature's tricks, a lesson that both Neo-Darwinians and Intelligent Design enthusiasts should take to heart."[47]

While at present the wholesale co-option argument has difficulties explaining certain structures such as the flagellum, it does fare better with other irreducibly complex systems such as the cilium (described above) and the blood-clotting pathway. Probably the best compilation of these arguments can be found in Kenneth Miller's book, *Finding Darwin's God*.

The blood-clotting pathway is worth investigating because it is an example of an irreducibly complex structure that Behe argues is beyond the reach of natural selection. The proteins involved in blood clotting float around in the blood in an inactive form. To become active and begin to form a clot, they must be cleaved, at the appropriate time, by activator proteins. The proteins in the clotting pathway must then be shut off at the appropriate time to avoid clotting the entire circulatory system. While the blood-clotting pathway involves many proteins, Miller correctly asks what is the simplest set of proteins necessary to get any minimal clotting ability. Basically he settles upon two proteins: a protein that can actually form the clot and a protease protein that can cleave and thereby activate

47. Stephen Barr, "The Miracle of Evolution," *First Things* 160 (2006): 30–33.

the clotting protein. To build his co-option case for indirect evolution, Miller, citing the work of Russell Doolittle,[48] points out that the main protease protein in the human clotting pathway is similar to a digestive protein involved in breaking down food in the gastrointestinal tract. Both proteins are in the protein family known as serine proteases and, based upon sequence analysis, the digestive protein appears to be more ancient. As a result, one component of the clotting system, the activator protein, was available via co-option. In addition, there are many soluble proteins in the blood for reasons that have nothing to do with clotting[49] that could have possibly been substrates for the co-opted protease.

So the raw material was available, but how could it clot blood? Suppose a functioning serine protease was inside healthy cells and the clotting protein existed in an inactive form in the fluid outside of the cell.[50] Once the cell was damaged by a cut, the protease could have spilled out and activated the "clotting" protein by cutting it in two. Cutting it could have exposed sites on the protein that caused it to stick to other cut fragments, thereby forming a primitive clot. This relatively simple mechanism is basically how a clot forms in the human circulatory system: an inactive protein called fibrinogen is cleaved to form fibrin, which is the sticky protein that forms the clot.

At this point Behe might argue that you would also need a protein to shut this process off or the clot would spread through the body. However, this may not be necessary given the fact that the two main mechanisms for stopping the major protease in human blood clotting are remarkably simple. First, much of the protease gets trapped inside the clot, and second, much of the remaining protease is washed away and diluted by the flow of blood through the site of the wound.

The simple system that Miller describes here is similar to the clotting mechanism found in some invertebrates. While this seems to be a plausible scenario, it is not definitive proof, and Miller is upfront about this:

> Now, it would not be fair, just because we have presented a realistic evolutionary scheme supported by gene sequences from modern organisms, to suggest that we know exactly how the clotting system has evolved. That would be making far too much of our limited ability to reconstruct the details of the past.[51]

48. R. F. Doolittle, "The Evolution of Vertebrate Blood Coagulation: A Case of Ying and Yang," *Thrombosis and Hemostasis* 70 (1993): 24–28.

49. Many of these proteins are there to maintain osmotic balance between the blood and the other body fluids. Other proteins are there to assist in transporting molecules through the blood.

50. Serine proteases are found in cells and have a number of functions including breaking down old proteins.

51. Miller, *Finding Darwin's God*, 158.

Even so, Miller is confident that we are on the right track and certainly this type of research goes a long way toward demonstrating that biologically feasible indirect routes for the origin of *certain* irreducibly complex systems actually exist. Does this mean that indirect routes will be found for every irreducibly complex system? Of course not, but it is possibile that indirect routes will be found for some (or even all) irreducibly complex systems. The ultimate answer awaits further research.

This situation makes it extremely difficult to definitively conclude that specific biological systems were designed. There is always the possibility that future knowledge of biochemical structures or pathways or even new theories along the lines of Stuart Kauffman's complexity and self-organization theories (see chap. 8) will give natural selection an added boost.[52] But there is also the possibility that they may not. Increased knowledge could strengthen the Intelligent Design position or weaken it. Unfortunately, this increase in knowledge may not be as inevitable as we would like. If, as is probably the case in many instances, an understanding of the evolutionary process is closed to us due to an irremediable lack of knowledge—for example, missing fossils or lack of biochemical data from fossils—then the possible details of the evolution of an irreducibly complex system such as the flagellum may never be fully known. If this is the case, we can never know for sure if co-optable intermediaries ever existed or if they could have been feasible stepping-stones on the way to a functioning flagellum. If these gaps persist, should we abandon the search for natural causes? Certainly not, but should we expect natural causes to be able to explain *everything*? The answer to this question is beyond science and is rooted in one's philosophy. This, of course, keeps the debate going.

Assessment of the Intelligent Design School

The main focus of the Intelligent Design movement has been to establish that natural mechanisms are impotent in explaining the origin of structures that exhibit specified complexity. If this is done conclusively, the Intelligent Design school can claim that an intelligence directly intervened during the course of natural history. Given this strategy, the Intelligent Design position is dependent primarily on *negative* evidence to make its case, that is, demonstrating that natural mechanisms cannot build certain structures.

52. Howard Van Till, "E. coli at the No Free Lunchroom: Bacterial Flagella and Dembski's Case for Intelligent Design," www.counterbalance.net/id-hvt/index-frame.html.

This is not to say that positive evidence gathered in everyday experience cannot support the Intelligent Design position, but it should be apparent that this is not nearly enough. The *positive* evidence on which the Intelligent Design school draws is that all objects exhibiting specified complexity (or irreducible complexity) for which we know the full causal history (e.g., books, houses, songs) are the result of design. From this fact, the school then makes the inference that specified complexity equals design. Therefore, if organisms display specified complexity, they must have been designed. Using this positive evidence, the school claims that it is not out of ignorance—that is, a God-of-the-gaps approach—that they infer design. Rather, they infer design because things like DNA fit a certain pattern of specified complexity shared by other designed systems such as computer codes and encyclopedias.

Although the reasoning has some persuasive force, it is not a deductive argument but rather is (according to the Intelligent Design school) an inference to the best explanation. Certainly it is evident from everyday experience that specified complexity can be achieved via design; this we know. However, how do we know that natural processes cannot achieve it as well? To demonstrate that natural processes cannot produce specified complexity, one needs *negative* evidence demonstrating the impotence of natural mechanisms on this account. Not surprisingly therefore, Behe and Dembski go to great lengths to demonstrate that natural mechanisms *cannot* easily account for such systems. As Dembski explains:

> Design theorists are *not* saying that for a given natural object exhibiting specified complexity, all the natural causal mechanisms so far considered have failed to account for it and therefore it had to be designed. [This is a God-of-the-gaps type reasoning.] Rather they are saying that the specified complexity exhibited by a natural object can be such that there are compelling reasons to think that *no natural mechanism is capable* of accounting for it.[53]

This is precisely what Dembski set out to do in his book *No Free Lunch*—namely, to prove that some biological systems cannot be obtained by natural processes given the limited resources of the universe. He sought to come up with what he calls "in-principle theoretical objections for why the specified complexity in biological systems cannot be accounted for in terms of purely natural causes."[54] The problem with this approach is that it relies on probability calculations that in turn rest upon assumptions that are difficult to justify. To do these calculations, one must be able to establish the approximate probabilities of these systems forming

53. Dembski, *No Free Lunch*, 331; italics added.
54. Ibid., 332.

272

through natural means, taking into account the more imaginative routes posed by critics. Furthermore, if certain possible routes are not understood because of our ignorance of biology or because of other unknown contingencies, calculating accurate probabilities may not be possible.

Despite these problems, Intelligent Design theory has been able to create quite a stir among the biology community, who have realized to their collective dismay that the notion of design holds considerable sway over the average person on the street. Behe's work in particular has gathered a large following among the general public as well as among college students. While some have tried to dismiss Behe's work and Intelligent Design in general as pseudoscientific, this is disingenuous. The claims that Behe makes can be investigated scientifically even if the conclusions he derives from them go beyond science.[55] We have done so in this chapter, and many biologists and biology textbooks have done so in their critiques of the Intelligent Design school. These critics have posited scenarios, some of which are plausible, for how some of these systems could have arisen by natural selection. In certain cases they have demonstrated from sequence analysis that the necessary genes were available and could have been co-opted. While such evidence does not prove Intelligent Design theorists wrong, it does support the notion that natural mechanisms may in theory be able to handle the problem. The critical step is demonstrating the biological viability of these scenarios, and in most cases this has not been done.

In practice, disproving any specific Intelligent Design claim—for example, demonstrating that the cilium or the blood-clotting pathway can originate via purely natural mechanisms—is not fatal to the theory because Intelligent Design theory does not rest on any one particular example. Rather, it rests upon numerous systems that the school claims to be intelligently designed. Critics argue that this arrangement insulates Intelligent Design from falsification. If natural mechanisms are found to explain one supposedly irreducibly complex system such as the cilium, the school can always fall back on other systems. In practice, however, if Intelligent Design claims of irreducible complexity are continually falsified or shown lacking, the theory will collapse under its own weight.

A cursory reading of the Intelligent Design literature makes it clear that many in the school believe that design in nature has been proven

55. Intelligent Design theory makes claims that can be investigated scientifically. In this sense it has a scientific component. The conclusions it makes, though, that a higher intelligence intervened in natural history, step outside the bounds of science given that science is usually defined as the study of natural causes. This conclusion, however, does not invalidate the scientific question Intelligent Design theory is asking: Can natural mechanisms account for all biological structures? This question must be taken seriously and it should be investigated scientifically.

or is at the very least the best possible explanation for certain biological structures. Based upon what we have discussed here, these individuals overstate their case. While the Intelligent Design school has not been able to prove that certain structures have been designed, its adherents have been able to garner the attention of the scientific community. They have done this largely by exposing the weak points of evolutionary theory to a relatively receptive public, forcing the scientific community to respond to their efforts.

The receptiveness of the public to the Intelligent Design movement should not be surprising. By questioning the efficacy of natural mechanisms, the school gives God a more active role in creation and evolution. In a country that is dominated by practicing Christians, it is little wonder that such a theory would gain a foothold in the public square. Many Christians who are uneasy with the notion of a seven-day creation event find in the Intelligent Design school a happy medium. It is a theory that gives God an active role in creation but still gives credence to most scientific discoveries about the age of the earth and the fossil record. This middle-of-the-road position gives it a unique appeal and has helped fuel its popularity. But science is not a popularity contest. In fact, common wisdom is often wrong when it comes to the truth of scientific theories. What really matters is whether Neo-Darwinian theory can successfully answer the questions posed by the Intelligent Design school and whether Intelligent Design theorists can actually demonstrate conclusively the impotence of natural mechanisms.

Table 7.1. Comparison of Intelligent Design and Neo-Darwinian explanations of observed evidence related to evolution

Fact	Explanation		Crucial test or experiment	Relative importance in the debate
	Neo-Darwinian theory	Intelligent Design		
Fossil progression [historical evolution]	Successive life-forms existed over the estimated 3.8 billion years from the Precambrian period	Successive life-forms existed over the estimated 3.8 billion years from the Precambrian period	No dispute	Little
Characteristics of fossil record: Cambrian explosion, absence or paucity of transitional forms	Belief in the imcomplete nature of the fossil record and speculation that extremely rapid evolution possible under certain conditions	Can be explained by invoking design	Recovery of additional fossil forms. Specification and test of conditions under which rapid evolution can occur	Important
DNA coding method nearly identical in organisms	Common descent	Common descent with input from designer	N/A: Rational but not scientific explanation proposed	Little
Similarities in DNA sequences	Common descent with modifications; tree or bush model	Common descent and similarities in function require similarities in DNA	N/A: Rational but not scientific explanation proposed	Little
Biological structural/ physiological similarities in organisms	Common descent with modifications; tree or bush model	Common descent and similarities in function require similarities in structure/physiology	N/A: Rational but not scientific explanation proposed	Little
Specified complexity of biological systems such as bacteria flagellum	Evolved via co-option or other natural mechanisms	Evidence for the need of invoking design	Identify a viable biological pathway to show that evolution of such characteristics/ systems is possible	Extremely important
Emergence of higher taxa (phyla, classes, etc.)	Can obtain new genetic information through random mutation	Limited ability to incorporate new information; emergence of complex specified information cannot be accounted for by natural mechanisms	Determine the level of genetic information that can be added to organisms through natural means	Extremely important

Table 7.2. Intelligent Design school and the ten criteria of a genuine scientific theory

Criterion	Intelligent Design theory
1. Compactness	Not very compact, incorporates many diverse mechanisms both natural and supernatural
2. Simplicity	Sacrifices simplicity for flexibility; can account for natural mechanisms as well as design
3. Falsifiability	Can be falsified if plausible natural mechanisms can be devised to explain all candidates for irreducible complexity, but this is difficult because list of candidates is open-ended
4. Verifiability	Only in negative sense—that is, all efforts to explain irreducibly complex systems by other means fail
5. Retrodiction	At present no way to reliably distinguish those structures that were result of intervention from those that arose by natural means; relies on normal biology to explain most of the biological world
6. Prediction	Does not yet commit to many directly falsifiable predictions; usually speaks only in general terms
7. Exploration	Definitely suggests exploration of organisms to determine if they can be product of natural processes
8. Repeatability	Not a problem because usual scientific method applies
9. Clarity	Design filter, irreducible complexity is fairly clear, though application is not always so
10. Intuitiveness	Notion of design is intuitive, although it is difficult to identify in biology

8

THE META-DARWINIAN SCHOOL

The education of most biologists tends to cultivate a belief in the adequacy of the Neo-Darwinian paradigm to account for the history of life on earth. While most biologists are aware that there are unresolved issues within the study of evolutionary biology, these details do not directly affect their work.[1] As a result, most biologists assume that Darwin's theory will remain the preeminent explanation for the history and evolution of life, and they are content to leave any messy details to someone else, namely experts within the field of evolutionary biology. Nonetheless, unresolved problems with the orthodox Neo-Darwinian paradigm, discussed in chapters 5–7, have begun to be acknowledged by researchers within the field of evolutionary biology. The questions that have been raised challenge some of the Neo-Darwinian theory's fundamental assumptions:

- Is most evolutionary change gradual?
- Is most evolutionary change adaptive or is it neutral?
- Does selection play a large role in generating novelty or does novelty originate via other processes?

1. Many research programs assume evolutionary linkages between organisms in order to compare gene sequences and find homologous genes in different species. Despite this, most researchers are not directly involved in investigating these evolutionary linkages that are for the most part taken for granted. For example, a researcher looking for a specific enzyme in humans may use data from other organisms in order to identify target sequences for searching genomic databases, but the researcher does not investigate directly the proposed linkages between those organisms and humans.

- If novelty does originate predominantly via other mechanisms, what precisely is the role of natural selection?

Biologists who have asked these questions and found Neo-Darwinian theory lacking are left with two options. The first is to discard the theory outright, which is not a particularly attractive option given (1) the intellectual investment scientists have made in the theory, and (2) the risk that rejecting Neo-Darwinian Theory would bolster the credibility of Creationist critics. Not surprisingly, most scientists who are dissatisfied with the strict Neo-Darwinian account of evolution have not taken this route.

The second and more attractive option is to supplement Neo-Darwinian theory with additional naturalistic explanations. This allows scientists to remain committed to naturalistic explanations without restricting them to a single one-size-fits-all paradigm. Those who take this route are the focus of this chapter. Because they do not reject Darwinian mechanisms as a source of evolutionary change, but instead wish to supplement them with additional naturalistic mechanisms, we refer to this disparate group of evolutionists as "Meta-Darwinists." We have coined this term as an umbrella designation to cover all naturalistic theories of evolution that advocate mechanisms over and above natural selection working at a gradual rate on random mutations. What we will see, though, is that the Meta-Darwinists have widely varying views on the efficacy of natural selection, making them difficult to characterize using generalities. Some believe that natural selection is quite efficacious at building novel structures and needs only minor assistance, while others in this grouping believe natural selection accounts only for minor change, and that the heavy lifting is performed by other mechanisms. Despite this heterogeneity, all members that fall under this rapidly expanding grouping believe that evolution can be explained by purely natural mechanisms.

Logical Structure of the Meta-Darwinian Theories

The theories of the Meta-Darwinian school are the most diverse of any of the four major schools (table 8.1). Given this diversity, it is not surprising that the different Meta-Darwinian theories are often in disagreement as to the dominant mechanism that is actually driving evolution. Some Meta-Darwinian theories seek to work hand in hand with natural selection, while others relegate natural selection to a minor role. Despite this, most Meta-Darwinian theories share the belief that new species or adaptations could have arisen suddenly,[2] in line with the fossil evidence but in stark contrast to Darwin's ideal of gradualism. Variations on this theme are now widespread and have been advanced by many reputable scientists. The Meta-Darwinian school dates from the 1970s, when its first members began publishing works advocating the infusion of new ideas into evolution theory. Probably the most popular of these new ideas, and the one that has found the most widespread support, is the theory of punctuated equilibrium. Others are less well known but nevertheless seem to be promising avenues of research.

Table 8.1. The major ideas grouped under the umbrella heading of Meta-Darwinism

Meta-Darwinian theories

Theories	Supplemental mechanism	Well-known advocates
1. Punctuated equilibrium	No new mechanism; just a novel way of looking at the tempo of evolution.	Steven J. Gould, Niles Eldredge
2. Hierarchical selection	Groups of organisms, species, and groups of species can all be the subject of natural selection. Organisms and genes are no longer the sole objects of selection.	Steven J. Gould, Niles Eldredge
3. Exaptation	Many structures emerge as nonadaptive by-products of developmental or evolutionary processes. These are subsequently used for some benefit to the organism.	Steven J. Gould, Niles Eldredge
4. Neutral theory	Most variations at the molecular level are neither advantageous nor disadvantageous. These neutral variations are the major driver of evolutionary change.	Motoo Kimura
5. Developmental mutations: evo-devo	Minor mutations in developmental genes can cause large affects on the organism.	Sean Carroll, Jeffrey Schwartz
6. Morphogenic fields	The morphogenic field (interactions between cells and their environment) determines available developmental pathways and helps direct evolutionary change.	Brian Goodwin
7. Self-organization/ complexity theory	Molecules can spontaneously organize into autocatalytic networks that have order (cells).	Stuart Kauffman
8. Endosymbiosis	Organisms can acquire large amounts of foreign DNA from other species or fuse with other organisms to create new species.	Lynn Margulis

The logical structure of Meta-Darwinian theories is shown in figure 8.1. This is a composite; individual members of the school may reject some of the assumptions or explanations, but in general the diagram reflects the beliefs of the school. The Meta-Darwinists hope that shoring up Darwin's theory with additional braces will allow us to maintain the basic framework of naturalistic evolution and simultaneously tackle

2. This term can mean a variety of things. For some theories "suddenly" means a few hundred thousand years, while for others it might mean a few generations.

the more vexing problems that some believe to be beyond the reach of Neo-Darwinism as currently understood. If this is done successfully, the problems of Neo-Darwinism could be resolved in a manner that does not resort to any form of supernatural intervention.

For the sake of clarity, we will focus on the leading ideas put forth by Meta-Darwinists and will evaluate each in turn. As each idea is presented, we invite readers to think about some important questions:

- Does the proposed idea actually solve the problems facing Neo-Darwinism as described in chapter 5?
- Does the proposed idea leave a major role for natural selection?
- Is there an experimental test that can differentiate the proposed idea from others and from Neo-Darwinism?
- Does the idea represent what is really happening in nature?

We now turn to the eight theories grouped under the Meta-Darwinian umbrella, beginning with the best-known theory, punctuated equilibrium, or "punk-ek," as it is commonly known.

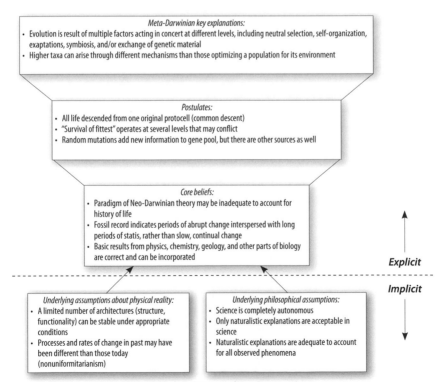

Figure 8.1. The logical structure of Meta-Darwinian theories.

Punctuated Equilibrium

Fossils occupy a place of privilege in the study of evolution, in large part because they hold a major key to the story of evolution. Unfortunately, it is not an easy task to decipher the story the fossils are telling. The fossil story reads much like a Faulkner novel; it is a story told by many voices, some more reliable than others, some missing altogether. Piecing together these disparate stories is critical to the understanding of evolution, but the difficulty of this task makes comprehending a Faulkner novel look like child's play in comparison.

The interpretation or reading of the fossil record has much to do with one's notion of how evolution proceeded. Darwin's view was that species accumulate small successive changes over long periods of time, a view known as *phyletic gradualism*. Darwin's commitment to an extremely slow rate stemmed in part from the fact that any mention in his day of rapid evolution smacked of design intervention. To unambiguously distance himself from this implication, he made a point of advocating only gradual and continuous change involving "numerous transition stages." Such a view made certain predictions of the fossil record, mainly that paleontologists should be able to identify sequential transitional forms within many lineages, for example, fish gradually changing into amphibians, dinosaurs into birds. Unfortunately, in Darwin's time neat sequential progressions delineating the orderly gradual transition of one species into another species were exceedingly rare. Darwin himself acknowledged this problem:

> The abrupt manner in which whole groups of species suddenly appear in certain formations, has been urged by several paleontologists—for instance, by Agassiz, Pictet, and Sedgwick—as a fatal objection to the belief in the [gradual] transmutation of species. . . . If numerous species, belonging to the same genera or families, have really started into life at once, the fact would be fatal to the theory of evolution through natural selection.[3]

Darwin refused to be deterred, though, claiming that the lack of transitional forms was the by-product of an imperfect fossil record.[4] Darwin's response challenged paleontologists to find the type of gradual change that his theory predicted. Unfortunately, by the time the modern synthesis had been codified in the middle of the twentieth century, examples of gradual changes in the fossil record remained few and far between. As Stephen J. Gould explained, paleontologists were in a peculiar situation:

> They could be good Darwinians and still acknowledge the primary fact of their profession [the absence of gradations in the fossil record]—but only at

3. Charles Darwin, *The Origin of Species* (New York: Mentor Books, 1958), 311.
4. Ibid., 441.

the price of sheepishness or embarrassment. No one can take great comfort when the primary observation of their discipline becomes an artifact of limited evidence rather than an expression of nature's ways.[5]

Eager to be good Darwinists, paleontologists began to suspect that the subjective methods by which fossils had been classified up until the middle of the twentieth century were obscuring the underlying gradual evolution that Neo-Darwinian theory demanded. If the gradations in the fossil record were so fine as to be "imperceptible," a subjective evaluation of the fossils would tend to miss subtle changes. To overcome this limitation, new objective statistical methods that could simultaneously compare changes in a variety of characteristics were developed. Armed with these new methods, paleontologists of the late twentieth century set out to reinvigorate their discipline by reconciling the fossil record with Neo-Darwinian gradualism. What their work revealed in overwhelming abundance, even after rigorous objective analysis, was stasis. In other words, failure:

> Generations of paleontologists learned to equate the potential documentation of evolution with the discovery of insensible intermediacy in a sequence of fossils. In this context, stasis can only record sorrow and disappointment. Paleontologists therefore came to view stasis as just another failure to document evolution. Stasis existed in overwhelming abundance, as every paleontologist always knew. But this primary signal of the fossil record, defined as an absence of data for evolution, only highlighted our frustration—and certainly did not represent anything worth publishing.[6]

If stasis was disappointing for these paleontologists, the relative rarity of intermediaries or transitional forms was even more so.[7] If the new paleontological techniques were meant to vindicate Darwinian gradualism, they had failed in their mission. How then was this new data to be reconciled with Darwinian theory?

A New Paradigm

Enter Gould and Eldredge and their famous 1972 paper, in which they introduced the idea of punctuated equilibrium (PE) as an alternative

5. Stephen J. Gould, *The Structure of Evolutionary Theory* (Cambridge, MA: Belknap, 2002), 750.

6. Ibid., 759–60.

7. To be clear, intermediate or transition fossils were found but the record was not teeming with them as Darwin's theory would predict. A discussion of a number of transition forms is found in chap. 5.

to phyletic gradualism. What they argued was that evolution works in rapid bursts or spurts and that for the majority of evolutionary history species are rather stable, that is, they exist in equilibrium. Every so often though, this equilibrium is disrupted by a brief punctuated period of rapid change, and it is during these punctuated periods that most new species form.

Because it is more consistent with the fossil record, Gould and Eldredge's idea has achieved widespread acceptance within the field of paleontology during the past thirty years. As a result, no longer are stasis and the sudden appearance of forms a sign of inadequate fossils or methods; rather, in the context of punctuated equilibrium, both take their place as legitimate data. But while paleontologists rejoiced, reconciliation was not quite complete. If Gould and Eldredge are correctly interpreting the fossil record, then the Darwinian model of speciation, in which species climb the fitness scale via insensible intermediates, is inadequate because it does not explain the geologically rapid speciation events observed in the fossil record. Moreover, Darwin's model, while not inconsistent with stasis, has to be reconciled with the preponderance of stasis.

How Does Punctuated Equilibrium Resolve These Problems?

To resolve the problem associated with rapid speciation, Gould and Eldredge turned to an idea championed by the renowned evolutionary biologist Ernst Mayr. Mayr argued that most species form via allopatric speciation, a process that occurs through the physical isolation of a portion of a larger population.[8] If a small group were to be cut off from the parental population, theoretically it would allow for speciation to occur relatively quickly in this side group because new mutations would have an easier time spreading through that smaller population. As a result, evolutionary change could be concentrated in rapid bursts occurring in small isolated populations. Once these isolated populations had changed significantly, they could then spread out into a wider range as a distinct novel species. The expected evolutionary branching of punctuated equilibrium (dominated by allopatric speciation) versus phyletic gradualism (the Neo-Darwinian notion) can be shown schematically (fig. 8.2).

The notion of allopatric speciation offers a potential explanation for the punctuation, but what about the second problem, that of stasis? If, as Neo-Darwinism maintains, natural selection is continually scrutinizing organisms, keeping the good and removing the bad, why then do some organisms remain unchanged for millions of years? To explain

8. Other forms of effective isolation exist, such as *sympatric speciation*, in which a subpopulation shares space with a larger population but achieves reproductive isolation.

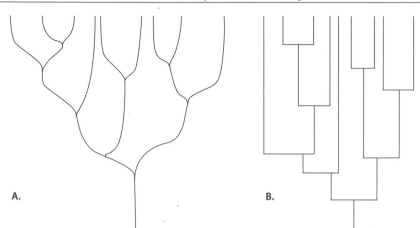

A.

B.

Figure 8.2. The difference in speciation branching of phyletic gradualism (A) and the rapid type of allopatric speciation (B) that is the hallmark of punctuated equilibrium. In both A and B the branching starts with one species at the bottom and ends with nine at the top. Horizontal movement indicates the amount of change in the species. In figure A, the sloped and curving lines indicate the continual gradual nature of speciation events. In figure B, the abrupt splitting of a lineage by a horizontal line marks the rapid nature of speciation events.

this, Gould and Eldredge advocate a new view of species, one radically different from the Neo-Darwinian view. According to this view, species are stable entities resistant to change rather than infinitely fluid entities as implied by Neo-Darwinian theory:

> If we view a species as a set of subpopulations, all ready and able to differentiate but held in check only by the rein of gene flow, then the stability of species is a tenuous thing indeed. But if that stability is an inherent property both of individual development and the genetic structure of populations [populations have an inherent stability], then its power is immeasurably enhanced, for the basic property of homeostatic systems, or steady states, is that they resist change by self-regulation. That local populations do not differentiate into new species, even though no external bar prevents it, stands as strong testimony to the inherent stability of species.[9]

While there are certainly counterexamples in which local populations do differentiate into new species (the multitude of *Drosophila* species that inhabit the islands of Hawaii), such counterexamples tend to involve widely disparate ecosystems such as the ecological niches on the Hawaiian Islands.

9. Niles Eldredge and Stephen J. Gould, "Punctuated Equilibria: An Alternative to Phyletic Gradualism," in *Models of Paleobiology*, ed. T. J. M. Schopf (San Francisco: Freeman, Cooper & Co., 1972), 114–15.

The idea that species have internal stabilizing forces runs counter to the thinking of many Neo-Darwinists.[10] Nonetheless, there is evidence for it from breeding and selection experiments.[11] This concept of species, while fundamentally different from the Neo-Darwinian view, fits within the context of punctuated equilibrium and has, as we will see, important implications regarding evolutionary theory.

What Is Really New about Punctuated Equilibrium?

The theory of punctuated equilibrium has changed the way many view the temporal process of evolution, from one of continuous gradual change to one in which species arise in geological moments (tens of thousands of years) and persist in stasis for millions of years. It is important to note, though, that what is offered by the theory is nothing revolutionary; rather, it is merely *a new way of viewing the temporal process of evolution* within a thoroughly Darwinian framework. As Gould states: "Punctuated equilibrium emerges as the expected scaling of ordinary allopatric speciation into geological time, and does not suggest or imply radically different evolutionary mechanisms at the level of origin of species."[12] In other words, while punctuated equilibrium may rely on a slightly different type of speciation, it relies on the same familiar Darwinian *mechanism*, that is, random mutation operated on by natural selection.

This point is often missed, given the term "punctuation," which makes it seem as if speciation with punctuated equilibrium occurred instantaneously or magically—something that sounds eerily similar to Creationism. The punctuation in punctuated equilibrium, however, refers to what is seen in the fossil record and can correspond to tens of thousands of years, an immense expanse of time when compared to the human life span. For example, if a paleontologist is digging for fossils, he or she may find a fossil (*A*) in a lower strata and a related but anatomically different one (*B*) in the strata immediately superimposed on it. The paleontologist, intrigued by this find, may search for transitional forms that lead from *A* to *B*. Unfortunately, in most cases the paleontologist comes up empty-handed. Does this necessarily mean *A* spontaneously changed into *B*, or that *B* leaped into existence? Of course not. One must remember that even though the *geological* distance between *A* and *B* is only one strata layer, the *temporal* distance may be 100,000 years (or even more). Geologically, 100,000 years is but a blink of an eye, and this is the main reason punctuated equilibrium theorists believe transitional forms are

10. Richard Dawkins, *The Blind Watchmaker* (New York: W. W. Norton, 1996), 247.

11. For more on how species demonstrate this inherent stability, see the section "Limits of Breeding Experiments: Genetic Homeostasis" in chap. 3.

12. Gould, *Structure of Evolutionary Theory*, 76.

absent from the fossil record: the change is happening too quickly for the lens of geology to capture.

This brings us to an important point. If the pattern of punctuated equilibrium in the fossil record accurately represents the tempo of evolution, then speciation must be explained within a compressed time frame. Instead of having millions of years to spawn new species by the phyletic gradualism of Neo-Darwinian theory, Darwinian mechanisms now have only a few hundred thousand years. If one believes that the Darwinian mechanisms have difficulty explaining the emergence of novelty under a strict Neo-Darwinian framework, then the theory of punctuated equilibrium compounds the problem by forcing the Darwinian mechanism to operate within a shorter time frame.

Given that it does not propose any new *mechanisms* for speciation, merely changes in the *type* of speciation, punctuated equilibrium is little more than a different way to describe the pace of evolutionary change. Unfortunately, it is not clear that punctuated equilibrium accurately describes the pace at which evolution proceeds. To understand why, let us look at Gould's hypothesis regarding speciation. Gould believed that most new species originate in small subpopulations that have become physically isolated from the range of the parental population. As a result, he argued that such transitional species would not be found because they are "beyond the borders of the parental range that provides the exclusive source for standard paleontological collections."[13] As a result, they are "seen" in the fossil record only when they return to the original parental range, and they return fully formed as distinct species. Thus, what punctuated equilibrium offers is not evidence of the existence of missing transitional forms, but rather a new reason why they cannot be found.

The problem here is that arguing that speciation events take place in the periphery, away from the prying eyes of paleontologists, undercuts Gould's main point about punctuation. If you can't observe the speciation because it is always happening elsewhere in the fossil record, it is impossible to infer the rate at which it is occurring. While a new form may appear suddenly in the parental range, it could have been evolving quite slowly over millions of years in the periphery. There is no way of knowing this. All one sees is the end product once it has migrated back into the parental range. Therefore, punctuation in the fossil record does not necessarily demonstrate that speciation normally occurs rapidly in geological time, as the theory of punctuated equilibrium so boldly asserts.[14] This does not disprove the theory,

13. Ibid., 780.
14. This argument is made in Daniel Dennett, *Darwin's Dangerous Idea: Evolution and the Meanings of Life* (New York: Simon & Schuster, 1995).

but it does bring into question whether it adequately describes the temporal sequence of evolution. It also makes it difficult to prove or disprove Gould's thesis. If punctuated equilibrium advocates argue for rapid speciation in the periphery and Neo-Darwinists argue for gradual changes, there is no way to resolve this issue until fossils are actually found for the transition in question. If there is no way to find such species, as Gould seems to imply, then it insulates Gould's theory from falsification.

Hierarchical Levels of Selection

While arguments over the tempo of evolution continue, the theory of punctuated equilibrium opens up another avenue by which selection may act: selection upon a species as a whole. According to Neo-Darwinian theory, species are not stable entities because they are constantly changing and adapting over time (although at variable speeds). Hence, any description of a species is purely arbitrary, since it is describing only a temporal snapshot of an extremely fluid group of organisms that are constantly moving through design space. What today looks like a finch could in 10,000 years cross the boundary imposed by systemists and become an entirely different bird.

The theory of punctuated equilibrium offers a different notion of species, one that opens up interesting possibilities for evolutionary theory. In the punctuated equilibrium model, species have relatively quick births (punctuation events) and thereafter spend the majority of their existence in stasis (equilibrium). During this time in stasis the species may "produce offspring" by giving rise to daughter species. As such, according to Gould, species themselves can be recognized as Darwinian individuals, entities on which selection can act:[15]

> If new species usually arose by the smooth transformation of an entire ancestral species, and then changed continuously toward a descendant form, they would lack the stability and coherence required for defining evolutionary individuals. The theory of punctuated equilibrium allows us to individuate species in both time and space; this property (rather than the debate about evolutionary tempo) may emerge as its primary contribution to evolutionary theory.[16]

15. For a good overview of the problem of defining species in biology, see essays under the heading "Problems of Classification" in Michael Ruse, *Philosophy of Biology* (New York: Macmillan, 1989).

16. Stephen J. Gould, "Darwin and the Expansion of Evolutionary Theory," *Science* 216 (1982): 380–87.

What Gould is claiming is that because species are stable entities, selection can theoretically act directly on the species as a whole rather than on just the individual members of the species. Let us explore this a bit further. Darwin claimed that selection works on individuals and the emergence of new species was the result of the accumulation of selective effects working on individuals. Others, the ultra-Darwinists such as Richard Dawkins, have reduced this to the level of genes, with all of evolution explained by selection operating on genes. The adequacy of such single-level explanations is what this new view of species challenges. This view opens up the possibility that selection operating at other levels, particularly at the level of the species, can drive evolutionary change.

Not surprisingly, this understanding of selection, known as hierarchical selection, is highly controversial because it disrupts the strict extrapolation from micro- to macroevolution that is central to Neo-Darwinian theory and because it is difficult to demonstrate experimentally. To demonstrate that selection can in principle act on higher-level entities, the higher-level entities—the species, the clade, the multispecies community, or even a higher taxonomic unit such as genus or family—must meet two criteria:

1. They must have a distinct birthpoint, a distinct deathpoint, and sufficient stability in between. (This classifies them as individuals. Most species meet these criteria under PE theory.)
2. They must generate variable numbers of similar offspring. (This classifies them as Darwinian individuals, giving them potential status as entities subject to selection.)[17]

Since species, as understood under punctuated equilibrium, usually meet these criteria, there is at least the possibility that selection could act on them. In addition, it seems that demes (subpopulations within species), clades, and higher taxa could in certain cases be considered stable enough entities to be subject to selection. David Sloan Wilson and Elliot Sober go so far as to claim that even multispecies communities can be subject to selection.[18] In this framework there exists a hierarchy of levels of selection (fig. 8.3), all of which can in principle be levels at which selection operates if they meet the conditions outlined above.

17. E. Lloyd and S. J. Gould, "Individuality and Adaptation across Levels of Selection: How Shall We Name and Generalize the Unit of Darwinism?" *Proceedings of the National Academy of Sciences USA* 96 (1999): 11904–9.

18. D. S. Wilson and E. Sober, "Reviving the Superorganism," *Journal of Theoretical Biology* 136 (1989): 337–56.

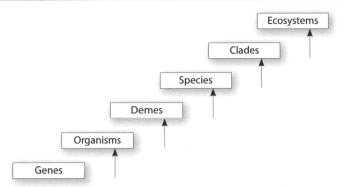

Figure 8.3. The different levels on which natural selection could conceivably act are shown in a hierarchy, with genes at the bottom and whole ecosystems at the top.

While it may be possible in theory, application of the notion of higher-level selection is not straightforward. One can often reduce the selective pressure exerted on the *species* to the mere sum of the effects of selection on the *individuals within the species*. For example, the reason that one species goes extinct may not be attributed to any property of the species as a whole, but rather may be the result of each individual in the species having a "bad" gene. Therefore, to determine that selection is acting at a specific level one must satisfy the following strict criteria:

1. There must be a property that emerges at that level that cannot be explained merely in terms of the levels below it.
2. There must be an effect of the property on the fitness of the species, deme, clade, or higher-level unit.

To illustrate exactly what this means, it is useful to borrow a hypothetical example from Wilson and Sober.[19] Imagine a species of insect that lives in groups in small ponds. These insect groups spend time at a specific pond, and when it dries up they disperse randomly and interbreed with neighboring groups. Imagine that there are two types of insects that make up the population: type A, which can detoxify the water in the pond but therefore have fewer offspring: and type *a* insects, which cannot detoxify the pond and therefore have more energy to produce more offspring. In this case:

The trait [A] is disfavored by natural selection operating within groups [A's produce less offspring than *a*'s], at the same time it is favored by natural selection operating between groups [each group needs some A's to detoxify

19. Ibid.

289

the water]. We therefore have a conflict between levels of selection, and the outcome depends upon their relative strengths.[20]

Depending on the strengths of the respective levels of selection, the *A* trait may die out (if within-group selection wins out) or it may be maintained at a high level in the population (if between-group selection dominates). *A*'s will do worse in generating offspring than *a*'s, but the *a*'s need some *A*'s around to detoxify their water. As a result, determining the selective advantage of the gene requires an understanding of properties that emerge only at the level of the group or deme. In addition to this hypothetical example, Wilson and Sober also describe how multilevel selection could work in cellular slime molds and social insects.[21]

The main problem with most examples of hierarchical selection is that they are purely hypothetical, as in the case of the pond insects described above. The other problem is that often higher-level selection can be explained by invoking selection only at the level of the individual. This makes verifying cases of higher-level selection difficult and causes most biologists to continue to question its relative importance.[22] To illustrate this, let us examine two possible explanations for the differential survival of mollusk species during the Cretaceous period, one that invokes species selection and one that invokes selection only at the level of the individual.[23] Researchers have found that, based upon the fossil record, mollusk species that produced larvae that floated in the seas for longer periods of time had wider geological ranges and gave rise to more daughter species relative to species that had larvae that floated for short periods. One possible explanation for this phenomenon is that the wider geological range of the floating species, a species-level trait, allowed for more genetic variability since these species occupy more diverse niches. This increase in genetic diversity as compared to the mollusks with short floating larva may have allowed the species to resist extinction and reap a greater degree of evolutionary success.

That, however, is not the only possible explanation. It is also possible that the advantage the long-floating larval species had was merely because the trait was advantageous to the individual organism. It is quite plausible to assume that the longer a mollusk floats the more food it will find and the more likely it will be to find a nice spot to take up residence. If this is true, then species-selection is not needed,

20. Ibid., 343.
21. Ibid., 337–56.
22. Dawkins, *Blind Watchmaker*, 267.
23. D. Jablonski, "Heritability at the Species Level: Analysis of the Geological Ranges of Cretaceous Mollusks," *Science* 238 (1987): 360–63.

and the evolutionary success can be attributed to individual selection. Unfortunately, there appears to be no way of distinguishing between these two possibilities in extinct organisms, and therefore the simplest explanation, selection operating at the level of the individual, tends to be preferred.

The ultimate impact hierarchical selection theory has on shaping evolutionary theory remains to be seen, but it appears to offer little in terms of new mechanisms to drive evolution. This is because selection at any level of the hierarchy still proceeds via changes in gene frequency. The novel idea here is that selection can operate on each level independently, but this actually poses problems for evolution because different levels of selection can either be in conflict or they can be uncorrelated. For example, selection between different individuals of a species may have no effect on higher-level trends, such as species survival. Thus, "any trait built at another level must be viewed as effectively random with respect to its operation at a new focal level."[24] For example, selection among individuals of a species for members that have dark brown fur versus light brown fur may have no effect on whether the species survives. It is possible that the species could in fact survive equally well with both colors of fur, yet it just so happens that the females tend to prefer males with dark brown fur. In this case the fur color may change based upon sexual selection at the level of the individual, but it would have no effect on species survival. Indeed, if, as punctuated equilibrium theory states, a species remains stable throughout its life span, competition and selection at the level of the individual may not be causing any directional change during most of the life span of the species.[25] This claim, of course, is suspect as described in the above section on punctuated equilibrium. Regardless of whether this claim is correct, the question of how complex traits are built is not addressed by models involving hierarchical selection. In fact, no one even suggests that species selection or any other higher level of selection can build complex structures such as the eye. As Dawkins points out:

> The most it can do is to choose between various alternative complex machineries, given that those complex machineries have already been put together by true Darwinian selection.[26]

Theoretically, it can select one complex species or group over another, but it fails to explain how these complex species or groups originated in the first place. When cornered to explain this, Gould, a champion of

24. Lloyd and Gould, "Individuality and Adaptation," 11907.
25. Gould, "Darwin and the Expansion of Evolutionary Theory," 385.
26. Dawkins, *Blind Watchmaker*, 269.

hierarchical selection, almost always reverts to the step-by-step mechanisms of Neo-Darwinian theory: "the model of PE . . . does not deny that allopatric speciation occurs gradually."[27] As a result, one is still left with the basic Darwinian mechanism even if hierarchical selection is taken into account.

Exaptation versus Adaptation

At the start of their seminal 1979 paper, Stephen J. Gould and Richard Lewontin lamented that "an adaptationist programme has dominated evolutionary thought in England and the United States during the past 40 years."[28] What does this mean and why did it represent a problem for Gould and Lewontin? First, let us describe the adaptationist program to which they are referring. This is the tendency to explain the emergence of any advantageous trait during the course of evolution as having been a direct adaptation initially beneficial to the organism. For example, the emergence of the eye would have begun with the emergence of a functioning light-sensitive spot and over time, through the result of gradual improvements, a full-fledged eye emerged. The important point to note is that at each step along the way the structure was a successful adaptation for some specific function; in this case the initial structure and all of the intermediates were adaptations for "seeing."

Gould and Lewontin put forth an alternative possibility, namely, that many traits originate in a *non*adaptive state, meaning they initially had *no* selective advantage. In this way, certain traits were not built gradually by natural selection acting on minute random variations, but rather appeared initially as chance by-products and were then honed by natural selection into something useful to the organism. An analogy would be the remains of a sheet of plywood after you have cut a circle out of it. The circle is what you selected but the by-product, the sheet with the hole in it, may be co-opted for some other purpose. For example, your child could prop it up and use it as a hoop to jump through or with a few modifications it could be used as a brace to prevent your garbage can from falling over.

To describe such traits, Gould and Elisabeth Vrba coined the term "exaptation," which is defined as "characters, evolved for other usages [to make a circle out of plywood] (or no function at all), and later 'coopted'

27. Stephen J. Gould and Niles Eldredge, "Punctuated Equilibria: The Tempo and Mode of Evolution Reconsidered," *Paleobiology* 3 (1977): 115–51.

28. Stephen J. Gould and R. C. Lewontin, "The Spandrels of San Marco and the Panglossian Paradigm: A Critique of the Adaptationist Programme," *Proceedings of the Royal Society of London B* 205 (1979): 581–98.

for their current role [a hoop to jump through]."[29] Thus there is a fundamental difference between an *adaptation*, a feature that has arisen and been refined for a particular role by natural selection, and an *exaptation*, a feature that has arisen with no connection to its present use and was subsequently refined for a new role by natural selection. Exaptations therefore represent another pool of variations besides random genetic variations that natural selection has at its disposal for creating new features.

One of the factors that inspired Gould and Lewontin to advocate exaptations was the endemic use of just-so stories to postulate adaptive significance to many human behaviors such as blushing, rape, and infanticide. To their dismay, many of these stories stretched the bounds of credibility. In addition to being proffered with little or no proof, such just-so stories seemed to be force-fitted into the adaptationist program at the same time that other, more rational, possibilities were restricted. For example, it can be argued that blushing was selected for in humans because it is beneficial for a social animal like humans to be able to detect when someone else in the social group is lying. While this is possible, a more likely explanation views blushing as an exaptation, a by-product of the fact that blood is red and that the stereotypical response of the nervous system to embarrassment or stress is increasing blood flow to certain regions of the body. The increased blood flow in response to stress is a selective advantage to the individual under stress; any other social advantage blushing offers is likely secondary and has nothing to do with the "evolution" of blushing behavior.

Although Gould's theory of exaptations does indeed challenge the many just-so stories within evolutionary biology, Neo-Darwinian theory is not seriously threatened by it. Neo-Darwinian theory can easily incorporate the addition of exaptations. The problem that Gould and Lewontin saw was that most present-day Neo-Darwinists do not take these exaptive alternatives seriously. In fact, most Neo-Darwinists believe that natural selection operating on random mutation is of sufficient power and under so few constraints that it holds nearly exclusive sway in the design of organisms. To be sure, other factors may play a role, but that role is so minor as to be insignificant. Gould and Lewontin explain the manner in which Neo-Darwinists deal with alternative ideas:

> You acknowledge the rival, but circumscribe its domain of action so narrowly that it cannot have any importance in the affairs of nature. Then, you often congratulate yourself for being such an undogmatic and ecumenical

29. Stephen J. Gould and Elisabeth Vrba, "Exaptation—A Missing Term in the Science of Form," *Paleobiology* 8 (1982): 4–15.

293

chap. We maintain that alternatives to selection for best overall design have generally been relegated to unimportance by this mode of argument.[30]

Gould is attempting to make evolutionary theory more expansive, or in his words, more "ecumenical." He sees an overemphasis on natural selection acting solely on small mutations within organisms as problematic:

> Darwinian evolutionists have known this all along in their heart of hearts, and have tended to escape the resulting paradox by a leap of faith into the enabling power of geological time to accomplish anything by accumulation of small inputs. Most events of microevolutionary adaptation—that is, of ordinary Darwinian natural selection in the organismal mode—work against evolvability [macroevolution] by locking organisms into transient specializations.[31]

Something must be able to overcome this tendency toward local optimization and its negative effect on the organism's ability to evolve. Gould argues that it is the exaptive pool. If an organism has large numbers of currently nonadaptive features at its disposal, these represent potential avenues for selection to embark on that could move the organism away from one specific optimization. In essence, altering exaptive features for novel purposes represents a method for creating novelty that is not completely restricted by local selection pressures. If the exaptive features arise randomly, as the definition suggests, organisms with large numbers of these have much more flexibility and can "exapt the rich potentials supplied by . . . [these features] emplaced into the exaptive pool."[32] This seems to be the most promising aspect of the idea of exaptations.

Do exaptations really augment the ability of natural selection to create novel structures and fuel macroevolution, as Gould argues? To investigate this, we must look at the two classes into which exaptations fall to see how they can help evolution.

The first class of exaptations consists of structures that arise as random by-products of evolution. Gould and Lewontin choose an architectural example, the spandrels[33] of San Marco (Venice) to illustrate this class of exaptations; however, we will focus solely on biological examples. The first example deals with snails and concerns the use of a space called the umbilicus for brooding their eggs. Snails that grow by means of a tube coiled around an axis generate a cylindrical space along that axis,

30. Gould and Lewontin, "Spandrels of San Marco," 585.
31. Gould, *Structure of Evolutionary Theory*, 1294–95.
32. Ibid., 1295.
33. A spandrel is the approximately triangular space between two adjacent arches, which may be at any angle with respect to each other.

the umbilicus.[34] This space appears to have arisen as a nonadaptive by-product of winding a tube around an axis; that is, if you wind a tube around an axis, you get an umbilicus (fig. 8.4). Some species of snails take advantage of the umbilicus and use it as a brooding chamber to protect their eggs. It appears that the initially nonadaptive umbilicus was eventually co-opted to serve a specific purpose. How can one be sure this is what has happened? Could not the snails have evolved the coiled tube in order to protect their young in the umbilicus it would create? This second explanation seems unlikely:

> The cladogram of gastropods [snails] includes thousands of species, all with umbilical spaces . . . but only a very few with umbilical brooding. Moreover, the umbilical brooders occupy only a few tips on distinct and late-arising twigs of the cladogram, not a central position near the root of the tree. We must therefore conclude . . . that the umbilical space arose as a spandrel [exaptation] and then became coopted for later utility in a few lines of brooders.[35]

We will turn to other cases in a moment, but first it is worth pointing out that calling this an exaptation does nothing to help explain the emergence of the complex structure of the snail shell (a structure on which the umbilicus depends for its very existence), nor does it explain how the brooding behavior emerges.

The second class of exaptations consists of structures that were used initially for one function and then were gradually changed to be able to perform another unrelated function. This process is called co-option.[36] An example of this type of exaptation is the acquisition of human language and symbolic thought. According to most researchers, the human brain reached its present size hundreds of thousands of years ago in the skulls of hunter-gatherer hominids in the African savannahs. Despite the fact that selective pressures were much different for these early members of our species than they are for us today, the human brain seems well suited for modern life. Since the brain could not have preadapted itself in anticipation of the need to develop written language, understand quantum physics, or argue about evolutionary theory, many claim that our ability to perform these tasks is a side consequence of increased brain size. In other words it is an exaptation, or "the myriad and inescapable consequences of building any computing device as complex as the human

34. Stephen J. Gould, "Exaptive Excellence of Spandrels as a Term and Prototype," *Proceedings of the National Academy of Sciences USA* 94 (1997): 10750–55.

35. Ibid., 10753.

36. This notion is discussed at length in chap. 7, as it is the main defense of Neo-Darwinists when challenged by an apparently irreducibly complex system.

Figure 8.4. The umbilicus can be seen in the snail shell on the right. The umbilicus runs through the center of the coil and is used by some species of snails as a brooding chamber for the young.

brain."[37] Basically, the argument goes that if you build a brain big enough to perform the tasks primitive hunter-gatherers performed, you will have the raw materials present for the acquisition of language. Even Noam Chomsky, the founder of modern linguistics, sees the origin of language as an exaptation, an accidental gift of increased brain size:

> The brain did not get big so that we could read or write. . . . The universals of language are so different from anything else in nature, and so quirky in their structure, that origin as a side consequence of the brain's enhanced capacity, rather than as a simple advance in continuity from ancestral grunts and gestures, seems indicated.[38]

Whether human language represents an evolutionary exaptation is debatable, but irrespective of the outcome of the debate, regarding language acquisition as a fortuitous accident does little to explain the origin of the complexity involved. Getting the brain to the size necessary to support language may have been an exaptation, a by-product of increasing the brain size to aid in hunting or memory, but our understanding of how language evolved is not enhanced by calling language an exaptation. Language is more than mere brain size; it is equally dependent on the proper organization and structure of different regions of the brain. How this organization either formed gradually or sprang fortuitously into existence must be addressed.

The upshot is that exaptations seem to be useful explanatory tools up to a point. As the examples illustrate, exaptations can indeed provide *more raw material* (such as increased brain size or an umbilicus) for natural selection to work with. However, calling something an exapta-

37. Gould and Vrba, "Exaptation," 14.
38. Stephen J. Gould, "Tires to Sandals," *Natural History* (April 1989): 14.

tion does not explain how the *organizational work* that is needed to drive macroevolutionary events (such as the development of language or even coiled snail shells) is done. This requires a more detailed examination of the structure or behavior in question.

Neutral Theory of Evolution

The neutral theory of molecular evolution has been hailed by some as the most important innovation within evolutionary theory since Darwin, while others have dismissed it as mere evolutionary noise. This theory, originally put forth in the 1970s by Motoo Kimura, focuses on the importance that neutral changes at the molecular level have in driving evolution. The debate surrounding Kimura's theory has created two opposing camps: the selectionists, who subscribe to Neo-Darwinian theory, and the neutralists, who believe neutral genetic changes are the main force that drives evolution.

To understand the crux of this debate, one must first understand the different types of molecular changes that can occur in a gene following a mutation. Most protein encoding genes contain regions that are extremely important for the function of the protein, for example, the binding site region of a DNA-binding protein or the active site for an enzyme that breaks down starch. If these regions are mutated, the protein is unable to perform its function. Such *negative* mutations are disadvantageous to the individual and would be selected against and removed from the population. If, however, a chance mutation occurred within these regions that actually *helped* the protein bind more specifically to the DNA, or degrade starch more effectively, it would be favorable to the individual. Such a *positive* mutation would be selected for and could possibly spread throughout the population.

In addition to positive and negative mutations, a third type, *neutral* mutations, can occur. These mutations have no effect on the protein's ability to perform its function and would be expected to occur predominantly in areas that are not in contact with the binding site or the active site. Such a mutation would be considered neutral and would have no selective advantage or disadvantage. Such neutral mutations, because they offer no selective advantage, rarely spread throughout the population.[39] Since it is clear that all three types of changes occur in nature, the debate is mainly one of relative frequencies as well as the ability of selectionist versus neutralist theories to explain the data.

39. Neutral mutations get fixed within a population with a probability of $1/2N$ with N being the population size. The bigger a population, the less likely a neutral mutation will spread.

297

Kimura's Theory

Kimura's neutral theory does not deny the role of natural selection in evolution but it does postulate neutral mutations as the driving force in evolution:

> At the molecular level most evolutionary change and most of the variability within species are not caused by Darwinian selection but by random drift of mutant alleles that are selectively neutral or nearly neutral. The essential part of the neutral theory is not so much that molecular mutants are selectively neutral in the strict sense as that their fate is largely determined by random genetic drift. In other words, the selection intensity involved in the process is so weak that mutation pressure and random drift prevail in molecular evolution.[40]

Kimura does not hold the position that all molecular changes are neutral; he allows for "a great deal of negative selection and some positive selection."[41] What he does claim is that most mutations do not have enough of an effect on the phenotype to be either selected for or against; thus they are seen as neutral by natural selection.

The impetus for this theory emerged primarily from two lines of observation. When examining the amount of sequence variation across mammalian species, Kimura found that the amount of variation appeared to be more than could be accounted for by natural selection. The reasons for this have to do with the substitutional load associated with natural selection, originally proposed by the evolutionary biologist J. B. S. Haldane in the early twentieth century. In essence, the load here is the cost the species has to pay in order to get rid of all the disadvantageous mutants in the population. If there is only one bad gene, then only those having it need to be removed. If there are two bad genes, then all those with either of these genes must be removed via selection, and so on. If there are too many bad genes being selected at any one time, the load becomes too high to tolerate and the species will degenerate, as it will be unable to rid itself of all the bad genes. In fact, Haldane calculated that only one gene substitution could be tolerated for every three hundred generations without adversely affecting the species. But the amount of variation that Kimura calculated from direct observations of DNA sequence data was orders of magnitude greater than this, making the *calculated* substitutional load too high for any of the mammalian species he was studying to be able to tolerate. His response was to propose that the vast majority of the observed sequence

40. Motoo Kimura, *The Neutral Theory of Molecular Evolution* (Cambridge: Cambridge University Press, 1983), 34.
41. Ibid., 54.

changes were selectively neutral and thus would have no effect on the substitutional load.

The second line of evidence that supported Kimura's premise was the high numbers of polymorphisms (genes that exist in more than one form) within a single species. Polymorphism in most species is thought to be between 10% and 50%. According to Neo-Darwinian theory, to maintain such a high amount of polymorphism within a species, there must be a selective advantage that heterozygotes have over homozygotes, a condition known as *overdominance*. To maintain this at even 10% of the genes within a species is difficult because of the high segregation load this imposes on the species. This load refers to the work required to remove the inferior homozygotes that will be produced in each generation. According to Kimura,

> If 2000 overdominant loci are segregating, each with 1% heterozygote advantage, and if the selection is carried out by premature death of less fit homozygotes, each individual must produce on the average roughly 22,000 young in order to maintain the population number constant from generation to generation. It is evident that no mammalian species can afford such reproductive waste.[42]

If a more realistic number of young were produced, the amount of polymorphisms in the population would dwindle as one version of the gene would become fixed. Given the fact that polymorphisms exist in high amount, to circumvent the problem of the segregation load, Kimura proposed that most polymorphisms are selectively neutral. Neutral mutations would have no effect on the segregation load because they would be immune to selection, and therefore could be maintained in high amounts.

Criticism of the Neutral Theory

Kimura's belief that neutral mutations dominate evolution stands in stark contrast to the claims of Neo-Darwinian theory, namely, that large numbers of molecular changes have an effect on the phenotype and are therefore subject to selective pressure. Since Kimura put forth these ideas, selectionists have responded with various criticisms of his mathematical models. Maynard Smith devised a model in which natural selection could fix one allele much faster than one per three hundred generations, while others have argued that other mechanisms could account for the high amount of polymorphisms.[43] Other research has supported Kimura,

42. Ibid., 28.
43. See Mark Ridley, *The Problems of Evolution* (New York: Oxford University Press, 1985).

such as simulation experiments that have demonstrated the robustness of biochemical networks, that is, that networks are not sensitive to small changes such as new polymorphisms.[44]

Most work on this issue is theoretical or mathematical, and it is not always easy to distinguish which model is more accurate when it comes to live organisms. As a result, it is very difficult to verify whether Kimura's theory or natural selection is the major evolutionary force driving change in functional genes:

> It has long been known, for instance, that rates of evolution in enzymatic proteins are slower in the "active site," the part of the protein where its function is carried out, than in the outlying parts. Kimura explains this by supposing that the exact amino acids are crucial in the active site, but not so crucial elsewhere in the molecule. Changes are more likely to be neutral away from the active site, where evolution accordingly is faster. Evolutionary changes, when they occur, are still neutral, but neutral changes are not equally likely in all parts of the molecule.[45]

Darwinian selectionists, on the contrary, can argue that the changes in the periphery are advantageous and are therefore driven by positive selection. In addition, because the active site has more stringent functional constraints, changes to the active site are less likely to be advantageous. As a result there is less change in this region of the protein and more change in the periphery. One can see that both theories have the ability to explain the data. While both sides agree that negative selection maintains the integrity of the active site, Darwinists argue that positive selection is driving change outside the active site while neutralists argue that these changes are neutral. Unfortunately, this type of explanatory flexibility makes it difficult to distinguish between the two theories, as Neo-Darwinian Mark Ridley notes:

> Now that we have incorporated the idea of negative selection into neutralism [Kimura's theory], it has become as vague and versatile as selectionism. It can explain any relative frequencies of genes, or base triplets, by declaring that negative selection is at work.[46]

Kimura's neutral theory suffers from the same vagueness that Neo-Darwinism can suffer from. Rather than attempt to resolve the dispute here, we shall examine whether neutral selection helps to supplement natural selection.

44. N. Barkai and S. Leibler, "Robustness in Simple Biochemical Networks," *Nature* 387 (1997): 913–17.
45. Ridley, *Problems of Evolution*, 70.
46. Ibid., 71.

Efficacy of the Neutral Theory

Under Kimura's neutral theory, the predominant mechanism by which genes spread is genetic drift. Genetic drift describes the random change in the genetic makeup of a population that occurs by chance as new individuals are born and older ones die. For example, there may be more blue-eyed genes in your generation than in your parent's generation, not due to any type of selection, but simply because the blue-eyed alleles had the luck of the draw. Such drift is more likely to change the genetic composition within a small population because such populations are statistically much more susceptible to random fluctuations. While neutral changes can alter the genetic composition of the organisms in these populations, since these changes by definition do not affect the fitness of phenotype, they should have no role in driving the evolution of new traits. Ernst Mayr likens these ongoing neutral changes to a backdrop of evolutionary noise that plays no role in building new structures.

> When a genotype, favored by selection, carries along as hitchhikers a few newly arisen and strictly neutral alleles, it has no influence on evolution. This may be called evolutionary "noise," but it is not evolution. However, Kimura is correct in pointing out that much of the molecular variation of the genotype is due to neutral mutations. Having no effect on the phenotype, they are immune to selection.[47]

Here Mayr agrees with Kimura that most of the changes at the molecular level are neutral (thereby helping to explain high rates of molecular change and polymorphisms), but at the same time he discounts the ability of neutral mutations to do anything useful. Since they are silent, it is hard to see how these mutations will help explain the emergence of new phenotypes, which, according to Dawkins, is what a theory of evolution should do: "What I mainly want a theory of evolution to do is explain complex, well-designed mechanisms like hearts, hands, eyes and echolocation."[48] Kimura's theory cannot do this. It can explain why there is so much variation at the molecular level, but it cannot explain how changes at the *phenotypic* level occur, since neutral mutations do not affect the phenotypic level.

Realizing this point, Kimura proposes a way in which neutral mutations can build novelty. He argues that neutral mutations at the molecular level act in a similar fashion to exaptive structures at the phenotypic level. Just as exaptations, which were discussed in the previous section, originate in a neutral state, Kimura proposes that most mutations do

47. Ernst Mayr, *What Evolution Is* (New York: Basic Books, 2001), 199.
48. Dawkins, *Blind Watchmaker*, 265.

so as well, but they can become advantageous later due to fluctuating circumstances:

> We should not overlook the possibility that some of the "neutral" alleles may become advantageous under an appropriate environmental condition or a different genetic background; thus, neutral mutants have a latent potential for selection. . . . [They] can be the raw material for future adaptive evolution.[49]

Kimura thus admits that although neutralism dominates at the molecular level, at the phenotypic level structures are built by selection. In this sense, the neutral theory provides nothing new in terms of building adaptations. It may be able to help by providing the first few steps, but without selection it can do nothing:

> [Although] some early mutational steps in the evolution of a new function may be selectively neutral, sooner or later an advantageous mutation must occur, otherwise some already established steps may become lost by mutation and random drift and the process would be disrupted.[50]

Thus, at the end of the day, Kimura's neutral theory does not offer any new mechanism by which evolution can proceed. This is because, like Darwinists, neutralists "assume that natural selection shaped the phenotypes, and that random genetic change provided the raw material of evolution."[51] The only difference is that the neutral theory presents a possible way to get more of this raw material.

Development: Evo-Devo

In order to reconcile the punctuated fossil record with the gradual accumulation of small beneficial genetic mutations, it would be convenient if one could identify a manner by which small genetic changes could have large-scale effects on an organism. The small genetic change would be consistent with Darwinian theory, and the large-scale phenotypic change would be consistent with the fossil record. Mutations that affect early development seem to fit the bill perfectly. One would expect such mutations to have larger phenotypic consequences than mutations that affect only the adult. Certainly disrupting the formation of an organ

---

49. Kimura, *Neutral Theory of Molecular Evolution*, xiii.
50. Wen-Hsiung Li, *Molecular Evolution* (Sunderland, MA: Sinauer, 1997), 429.
51. Phillip Johnson, *Darwin on Trial* (Downers Grove, IL: InterVarsity, 1991), 98.

would have more drastic effects on the phenotype than tweaking the organ once it has formed.

This potential of developmental mutations was first championed by Richard Goldschmidt, the infamous architect of the widely ridiculed "hopeful monster" hypothesis (see chap. 2). While his ideas regarding drastic genetic rearrangements, that is, the hopeful monster, have been rejected, Goldschmidt correctly recognized that minor genetic mutations that affect early development could have large effects on the structural organization of the organism. However, one must keep in mind—as Goldschmidt did—that early development is tightly integrated, so there is a delicate balance these mutations must strike:

> The individual developmental processes are so carefully interwoven and arranged in time and space that the typical result is only possible if the whole process of development is in any single case set in motion and carried out upon the same material basis. Changes in this developmental system leading to new stable forms are only possible as far as they do not destroy or interfere with the orderly progress of developmental processes.[52]

Even though the developmental process is highly structured and constrained such that it can only be altered in defined directions, there should be ways of bumping it into these available trajectories:

> But how about the possibility of occasional successful mutational changes acting upon earlier developmental processes? Would such a change, if possible at all without breaking up the whole system of the orderly sequence of development . . . have the consequence of changing the whole organization and bridging with one step the gap between taxonomically widely different forms?[53]

If this possibility is correct, then we have a theory in which a single or a few small mutations affecting the developmental program of an organism could lead to a punctuated change in the phenotype of the organism. Here, perhaps, Darwinian micromutation meets the punctuated story of the fossil record.

Goldschmidt's ideas were put forth long before the genes that control early development were understood. Since then, researchers have uncovered a wealth of information concerning the genes that turn on and off various developmental processes, much of which supports Goldschmidt's basic tenet, that is, that developmental mutations lead to large-

52. R. Goldschmidt, "Some Aspects of Evolution," *Science* 78 (1933): 539–47.
53. Ibid.

scale changes in the organism. The most studied of these developmental (homeotic) genes is a group of genes called the Hox genes,[54] which were originally characterized by Ed Lewis in his studies of *Drosophila*.[55] These genes are important in controlling the segmentation of flies, so that antennae end up on the head and wings on the thorax (fig. 8.5). Mutations in these genes can lead to some very surprising phenotypes, such as an extra pair of wings or an extra pair of legs. For example, defects in the *antennapedia* gene, which is normally turned on in the thorax (and to some extent in the abdomen), can lead to an extra set of legs growing from the head of the fly (fig. 2.5). The transformed legs found on the *antennapedia* mutant are structurally similar to the normal legs found on the thorax; only the position in which they appear is abnormal. The reason for this is that homeotic genes "regulate other genes that actually determine segmental structure and function."[56] In other words, it acts as a master control switch, turning on the genes for making legs wherever *antennapedia* happens to be found in the developing fly. If the gene is turned on in the head, it will signal legs to be built there.

Another surprising fact about the homeotic genes is their near ubiquity within the animal kingdom. While flies have one Hox gene complex that regulates the identity of body segments, mammals have four complexes and certain ray-finned fishes have seven.[57] They have relatively conserved sequences, and they are used to specify identities along the anterior-posterior axis of the organism. For example, while the cluster in flies specifies the anterior-posterior axis of the fly body, a Hox cluster in the mouse specifies the anterior-posterior axis of the brain stem. Such functional and structural similarity is remarkable, as is the fact that the order of the Hox genes on the respective chromosomes is quite similar from species to species (see chap. 3).

The conserved function of these genes, their ubiquity in bilateral animals, and their ability to induce large-scale transformations make them likely candidates for the genes Goldschmidt required. Changing the number of homeotic genes, the location in which a homeotic gene is expressed, or the time during development at which it is expressed could contribute to alterations in the body plan during evolution. This can lead to many changes such as alterations in vertebrae number in vertebrates or changes in the number of appendages in arthropods. As Sean Carroll, an evo-devo advocate, has correctly pointed out, "body parts that

54. See chap. 2 for more details regarding Hox genes.
55. E. B. Lewis, "A Gene Complex Controlling Segmentation in *Drosophila*," *Nature* 276 (1978): 565–70.
56. Lewis, "Gene Complex Controlling Segmentation in *Drosophila*," 569.
57. V. Prince, "The Hox Paradox: More Complex(es) than Imagined," *Developmental Biology* 249 (2002): 1–15.

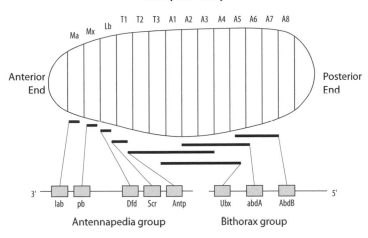

Figure 8.5. The expression pattern of the various Hox homeotic genes in the fly embryo. Each gene is expressed in only a subset of the embryo. These genes specify the anterior/posterior axis of the fly.

are often modular and constructed of similar units often differ between species largely in number and kind."[58] These body parts are controlled by homeotic genes, and so altering their expression could cause changes in the number and type of these modular parts during evolution. This could explain how snakes have evolved hundreds of vertebrae (changes in number) or how arthropods have developed extra walking legs or swimmerets in different body segments (changes in types).

While the modularity of body structures as well as the conservation of homeotic genes is highly suggestive of their role in producing variable forms during evolution, it would be more conclusive to identify a homeotic mutation creating novel successful variants in the wild.[59] Unfortunately, finding examples of evolution caught in the act is never an easy task. Despite the difficulty, a successful example has been found in a natural variant of the plant *Clarkia concinna*, which instead of having four sepals and four petals has eight sepals and no petals. This particular homeotic mutant, unlike most others, has no developmental abnormalities and no reduction in fertility, and it has organs similar to the wild-type strain. The identifiers of this mutant correctly claim that it "demonstrates that a large morphological difference governed by a simple genetic change can become established in a natural population."[60]

58. S. Carroll, *Endless Forms Most Beautiful* (New York: W. W. Norton, 2005), 27.
59. This is particularly important given the fact that homeotic mutants devised in the lab are usually less fit than their wild-type counterparts.
60. V. S. Ford and L. D. Gottlieb, "*Bicalyx* Is a Natural Homeotic Floral Variant," *Nature* 358 (1992): 671–73.

How Do They Spread?

There is no question that modern genetic research has demonstrated that changes in developmental genes can cause dramatic large-scale changes in organisms. The main problem with such large changes, in terms of their usefulness for evolution, is that they are usually detrimental to the organism. In fact, none of the known Hox *Drosophila* mutants is considered to be a successful improvement over the wild-type form. This may be because of the limits of laboratory research, or it may be because large-scale changes that cause organisms to move away from their local optimal region of "design space" normally are not advantageous.

> Large leaps sideways in a fitness landscape will almost never be to your benefit; wherever you currently find yourself, you are where you are because this has been a good region of Design Space for your ancestors—you are near the top of some peak or other in the space—so, the bigger the step you take (jumping randomly, of course), the more likely you are to jump off a cliff.[61]

Despite this, given the millions of years evolution has had at its disposal, one would expect that not all homeotic mutants would have jumped "off a cliff." It is possible that in some cases such mutants would have landed quite nicely on a nearby peak. Once this occurs, though, there is still one more hurdle to clear. The novel homeotic mutant is an individual drastically different from its parents. In fact, such an unlikely individual would be so different from its nonmutated siblings that finding a potential mate to propagate this new lineage could be problematic.

Paleoanthropologist Jeffrey Schwartz has sought to address this problem by examining the manner in which genetic change arises within a population. Schwartz argues that most mutations arise in the recessive state. In this way developmental mutations can initially spread through the population without having any affect on the phenotype, as many individuals will have one recessive mutant allele and one dominant wild-type allele.

> As this process proceeds for a number of generations, the mutated recessive allele spreads throughout the population but is unexpressed because it is in the heterozygous state. At some point, the population becomes so saturated with heterozygotes that they begin to produce homozygotes for the mutated recessive allele, . . . [these mutants] will mate with each other and produce more of their kind.[62]

61. Dennett, *Darwin's Dangerous Idea*, 288.
62. J. H. Schwartz, *Sudden Origins: Fossils, Genes, and the Emergence of Species* (New York: Wiley, 1999), 365.

If the mutation did not adversely affect survival, such a scenario, which is consistent with Mendelian genetics and natural selection, would lead to numerous potential mates arriving on the scene contemporaneously. Schwartz speculates how the eye could have evolved in one fell swoop, eliminating the need to invoke transition forms:

> Since it is likely that the mutation that produced the original allele for having eyes arose in the recessive state, the genetics and the morphology produce the following evolutionary picture: with two dominant alleles, no eyes; with one dominant and one recessive allele, no eyes; but with two recessive alleles, fully functional eyes housed in bony eye sockets. Although it may have taken a number of generations for the *Rx* gene to spread throughout the population in which it appeared, once homozygotes for it were produced, they would have had completely useful and fully formed eyes—not shallow depressions in the front of the head, or even half eyes, but actual eyes.[63]

A subpopulation with eyes could then be established within only a few generations. Given such a scenario, transition forms become unnecessary and their absence from the fossil record can be explained by the simple assertion that they never existed.

Limits of Developmental Mutations

The genetics of Schwartz's argument are sound, and mutations could certainly arise and spread through a population in the manner he has described. The major problem with his version of eye evolution is a misunderstanding of what these developmental genes actually do. These developmental genes do not *create* structures—they merely *regulate* them: the creation of the eye depends on the structural proteins that make up the many components of the eye. These proteins are what actually become the eye, and if they are absent, a gene such as *Rx* that regulates the development of the eye is worthless in terms of making an eye. You can express the *Rx* gene all you want in a plant, and it will never develop vertebrate eyes. You can put it into a bacterium or a sponge, and neither will develop vertebrate eyes. The regulatory (*Rx*) gene does not make the eye; structural and regulatory genes downstream of the *Rx* gene do. The *Rx* gene merely puts this process in motion, but it can do so only if the proper arrangement of structural genes is already present.

Let us be clear: mutations in developmental genes alter structures that are *already present*; they do not create *new* structures on their

---

63. Ibid., 369.

own. Take for example the evolution of the feather. The genes involved in regulating the major steps in feather formation, BMP4 and *noggin*, require that the proteins needed to construct the barbs and the rachises (shafts) of feathers, the feather-specific keratins, and so forth, as well as downstream regulatory genes, be already in place. Mutations in BMP4 and *noggin* can greatly disrupt the orderly pattern by, for example, converting barbs into rachises, but they do not build new structures.

Another example is useful to drive this point home. Sean Carroll has looked at the evolution of insect wings from crustacean gill branches. The notion that wings evolved from gill branches is supported by the fact that both of these structures are regulated by the same homeotic genes named *nubbin* and *apterous*.[64] Additional support is found in the fact that the gills in certain crustaceans have become modified to the point that they are very similar in pattern and structure to insect wings. While it may be that this transition occurred, one must realize the homeotic genes *nubbin* and *apterous* had only a partial role in driving the change, because whether one has wings or gill branches depends on what genes and proteins *nubbin* and *apterous* activate. It is these factors that actually build the structures in question, and it is these factors that must be altered to transition from gill branches to wings. Exactly how this is done is not addressed by the evo-devo scenario.

Based upon these examples, it is easy to see what homeotic genes actually are and what they are not. They are not eye-building genes or leg-building genes. Rather, they are highly versatile molecular switches that can turn on certain developmental pathways. These switches have nothing to do intrinsically with the structure they turn on, any more than a power switch has anything to do with the light it turns on. As a result, the same developmental gene can be used in one organism to turn on the formation of an eye and in another the formation of a portion of the hindbrain. Developmental biologist Jonathan Wells stresses this point: "If the same gene can "determine" structures as radically different as a fruit fly's leg and a mouse's brain or an insect's eyes and the eyes of humans and squids, then that gene is not determining much of anything."[65] The genes that actually do the heavy lifting are downstream of genes like Rx. They are the genes that encode for ion channels of the photoreceptors and crystallins of the lens. It is the functional organization of these gene products that must be explained.

64. Carroll, *Endless Forms Most Beautiful*, 175–77.
65. Jonathan Wells, "Unseating Naturalism: Recent Insights from Developmental Biology," in *Mere Creation: Science, Faith and Intelligent Design*, ed. William Dembski (Downers Grove, IL: InterVarsity, 1998), 56.

Morphogenetic Complexity

The structure of an organism depends critically on the developmental process by which it is built. As a result, knowing how developmental processes were built during evolution as well as why certain ones prevailed could go a long way toward understanding the mechanisms driving evolution. While some biologists argue that the developmental processes we can study today in living organisms are merely historical accidents (they just happened to have survived by chance because the organisms they built survived), developmental biologist Brian Goodwin prefers a more robust interpretation of why certain developmental pathways have survived. Goodwin believes such pathways are favored from the start because of the intrinsic properties by which organisms are built. It is not just selection and chance that carve out certain developmental pathways; there are laws of organization that make certain pathways more robust and stable.

To understand what these laws may be, we first must understand Goodwin's view of development. Goodwin sees development as the process of breaking symmetries. During development, one starts with a mass of nearly symmetrical cells that then break their initial symmetry and become specialized along the anterior-posterior axis (often under the direction of Hox genes). Once this occurs, limb buds on each side begin to differentiate from the tissue around them in order to generate a specific limb structure, further breaking the symmetry of the organism. Goodwin argues that the areas where symmetries get broken, the morphogenetic fields, are the drivers of development.

Most biologists view a morphogenetic field as a section of the developing organism that has been carved out by regulatory genes such that a specialized structure (e.g., a leg) can develop. Goodwin's view is quite different. Rather than focus on the regulatory genes involved in forming these fields, Goodwin is interested in all of the components of the field and how they interact as a complex whole.

> Biologists often invoke genes as the repository of all developmental information, the ultimate source of all the instructions for the development of the embryo. But genes can only respond to their immediate biochemical environment in a cell. They do not know where they are in a tissue unless something tells them. So we come back to the morphogenetic field as the source of spatial information.[66]

Goodwin is correct that genes alone do not determine development, and there is ample evidence for the influence of other factors such as

---

66. Brian Goodwin, *How the Leopard Changed Its Spots: The Evolution of Complexity* (New York: Charles Scribner's Sons, 1994), 150.

temperature, ionic currents, electric fields, and the extracellular matrix. Goodwin's favorite example is the developing limb bud:

> In the limb bud the whole system of cells embedded in the extracellular matrix, plus the epithelium (the surface layer of cells), makes up the morphogenetic field. This is also an excitable medium that can spontaneously change its state and generate spatially nonuniform patterns [break the symmetry], because each of the cells has an excitable cytoplasm, and they communicate with one another mechanistically and chemically through the extracellular matrix. So a pattern is bound to arise.[67]

Since the developing limb bud consists of many excitable components interacting in a dynamic fashion, one of two things will happen: either a stable pattern will emerge from these interactions or there will be chaos. Chaos would be discarded by natural selection because it would lead to an unstable structure; as a result, the various morphogenetic fields found in organisms tend to display consistent patterns. In the case of the limb bud, this pattern appears to be a robust one. In all mammalian limbs, the initial portion of the limb consists of a single bone at the center: the humerus in the arm and the femur in the leg. Goodwin argues that the dynamics of the developing limb bud are restricted by physical laws such that only specific patterns can emerge from the roughly cylindrical initial outgrowth of the limb bud. The physical laws serve to attract the developing limb bud along a specific trajectory, much as a marble is attracted to the bottom of the bowl once it is released from the edge. Goodwin is unsure of what these physical laws are, but the key point is the robustness they confer:

> Genes would have to search long and hard to find a region of parameter space where something other than a single element arose first in such limbs. So they don't bother—they sensibly go with the robust, generic form and modify it secondarily to suit various purposes.[68]

Goodwin goes on to argue, based mainly upon the equations and simulations of G. F. Oster, J. D. Murray, and P. K. Maini, that the components of the developing limb can break symmetry in three manners: condensation, bifurcation, and splitting (fig. 8.6).[69] He argues that they do this not because of the specific genes involved, but because the dynamic environment of the morphogenetic field is such that only these possi-

67. Ibid.
68. Ibid., 151.
69. G. F. Oster, J. D. Murray, and P. K. Maini, "A Model for Chondrogenic Condensations in the Developing Limb: The Role of the Extracellular Matrix and Cell Tractions," *Journal of Embryology and Experimental Morphology* 89 (1985): 93–112.

bilities emerge as stable outcomes given the physical constraints. The complex interactions among the various components can develop into these three patterns spontaneously. Other possibilities, such as dividing the bud into four branches, do not form as spontaneously.

Another example of a robust morphogenetic field that establishes a complex structure is the eye field. The genes of vertebrates, insects, and crustaceans vary considerably, but they all have evolved eyes of various capacities. Goodwin and others believe that this fact demonstrates that eyes are not improbable, but rather are the almost expected result of animal development. However, our understanding of all the factors involved in generating this pattern is anything but clear.

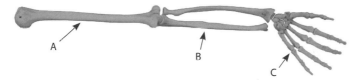

Figure 8.6. The three developmental possibilities for developing limb buds. Bones can condense at the middle of a limb bud (A) as in the humerus, they can bifurcate into two branches (B) as in the radius and ulna, or they can split into sections (C) as in the phalanges of the fingers.

What is known is that the folding and migration of cell layers is critical to the development of an eye. Furthermore, it is probable that the inverted disc pattern that is characteristic of the developing eye may emerge within the morphogenetic field as a result of the dynamic interactions between different tissues and the environment within the developing eye field. What this morphogenetic field determines, however, is the gross structure, the inverted-disc folding pattern that allows the eye to have its shape (fig. 8.7). What it does *not* determine is how the cells are organized to function as an eye, nor does it explain how all the various protein components of the eye originated. Goodwin has nothing to say regarding how photoreceptors with rhodopsin, transducin, and gated-ion channels are produced within the field. This is an important point. While the morphogenetic field may drive the formation of the gross structure such as an inverted disc, it does not address how the organism builds the complex detailed dynamical structures, such as photoreceptors and muscles, that exist within the field.[70]

Nonetheless Goodwin's notion of the morphogenetic field can explain why certain structures are favored in evolution. For example, the limb patterns described above may act as attractors to pull the developing limb bud in a specific direction. Getting away from these attractors may be energetically difficult, so selection simply will favor the robust forms.

70. Goodwin, *How the Leopard Changed Its Spots*, 167.

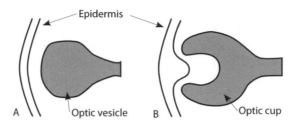

Figure 8.7. The morphogenetic field of the developing vertebrate eye. The eye starts as a protrusion known as the optic lobe. Once it enlarges, it comes in contact with the overlying epidermis, and deforms inward to form an inverted disc as pictured.

So once all the proteins needed for the formation of bone and muscles in the limb bud are available (again, a sizeable problem not addressed by the theory), new breaks in symmetry could bump the limb bud from one attractor to the next (from condensation to a bifurcation). In this manner, patterns, once established, could be bumped very quickly during morphogenesis to favor a novel symmetry-breaking form, as long as this new form is robust. Theoretically, this would allow for the rapid evolution of different patterns within the limb bud, a process that may speed evolution.

Complexity Theory and Self-Organizing Systems

Cells are dynamically organized structures that contain much complex information. They are composed of many integrated parts organized in a specific fashion to allow for survival in a dynamic world. The organizational work that goes into making a cell function is staggering: tens of thousands of genes, proteins, and metabolites interact extensively with one another in order to create the dynamics of a living cell. The field of proteomics, which has as its task the categorization of all the protein interactions within the cell, is merely beginning its daunting quest to gain insight into *how* cells are organized. To make matters more complex, many of the cellular components such as mitochondria, ribosomes, and flagella are tightly organized and regulated structures that could be studied by proteomics in their own right. Explaining the origin of the organizational dynamics at each level and how they interact between levels, often in a nonlinear manner, is a daunting task. Many researchers feel that Darwinian mechanisms alone seem inadequate to explain such complexity, so they have proposed an alternative, namely that this spectacular organization was not built gradually but rather emerged spontaneously in one fell swoop:

312

A particularly striking property of these complex systems is that even chaotic behavior at one level of activity—molecules or cells or organisms—can give rise to distinctive order at the next level—morphology and behavior. This has resulted in one of the primary refrains of complex studies: order emerges out of chaos. The source of large-scale order in biology may therefore be located in a distinctive type of complexity of the living state.[71]

In other words, the organizational work, the order of life, is an emergent property of the underlying complexity of living entities. According to most complexity theorists, large-scale evolutionary novelties like eyes and cells could not have originated by the gradual accumulation of mutations, but rather they are the result of emergent order—novelty that springs up in an instant. This idea is at the heart of what has come to be known as *complexity theory*.

Order in Dynamic Systems

Explaining the emergence of order as a spontaneous property of matter may seem a bit far-fetched, but such ordered phenomena turn out to be more common than one would expect.

The story begins in the 1950s, when Russian chemists Boris Belousov and A. M. Zhabotinsky discovered a chemical reaction, the Belousov-Zhabotinsky (BZ) reaction, which generated spontaneous ordered patterns that change over time. As the major reaction proceeds,[72] one of the reactants, bromous acid, is generated as one of the products. This positive feedback is coupled with a separate inhibitory reaction and the oscillation between these two generates an ordered pattern in which concentric waves move out from different regions of the dish, annihilating one another on contact. The oscillation keeps the reaction from reaching equilibrium, and as a result the reaction is able to generate a dynamically ordered pattern spontaneously. This pattern appears to arise for "free" since it is not specified by the molecules within the reaction.

Such patterns also emerge in living organisms. A good example is the pattern formed by slime molds when the individual amoeboid cells that make up a slime mold aggregate to form a multicellular fruiting body. When the individual amoeboid cells sense a lack of nutrients in the environment, they begin to secrete a signaling molecule, cAMP, to notify the neighboring cells that it is time to form an aggregate. The

71. Ibid., ix–x.
72. The major reaction is the following: $HBrO_2 + BrO_3^- + 3H^+ + 2Ce^{3+} \rightarrow 2\ HBrO_2 + 2Ce^{4+} + H_2O$.

Figure 8.8. A cartoon of the spontaneous pattern generated by slime mold aggregation as individual amoeboid cells begin to migrate in response to extracellular signals to form a fruiting body.

cAMP molecule is propagated from cell to cell in the form of a wave as each cell responds to the cAMP signal by making more of it (a step analogous to the positive feedback of the BZ reaction). To prevent the environment from being swamped by cAMP though, the cells that are activated quickly follow the initial cAMP production with the secretion of an enzyme that breaks down cAMP (a step analogous to the inhibitory reaction in the BZ reaction). This pattern of interactions allows cAMP to move in a wave from the point the signal originated. The interaction between these different cellular responses oscillates, much like the reaction in the BZ reaction, causing the cells to migrate in distinct patterns as they converge to form a slime mold aggregate (fig. 8.8). The pattern is not determined by some master blueprint or some pattern-forming gene. Rather, it arises out of the complex interactions between the component parts, that is, the cAMP, the cAMP receptors, and the enzyme used to break down cAMP. The spatial and temporal interactions between these parts define a field from which order emerges spontaneously:

> There is no plan, no blueprint, no instructions about the pattern that emerges. What exists in the field is a set of relationships among the components of the system such that the dynamically stable state into which it goes naturally—what mathematicians call the generic (typical) state of the field—has spatial and temporal pattern.[73]

73. Goodwin, *How the Leopard Changed Its Spots*, 51–52.

The dynamically stable state into which the field enters (in the case of the slime mold it is the spiral wave structure of slime mold aggregation) is often called an *attractor* because of its robust nature. Minor perturbations and sometimes major alterations within the field are not usually sufficient to disrupt formation of the characteristic pattern.[74] For example, the density of cells in the slime mold populations or the cAMP concentration in the propagating waves can vary significantly without altering the underlying pattern. The idea that order emerges in this fashion, robust and stable within systems displaced from equilibrium, means that natural selection is left with the task of fine-tuning what is built by an entirely different process. In this sense, complexity theory describes the formation of structures while natural selection maintains them. This is certainly a different view than the Neo-Darwinian perspective, but does it really have the ability to generate the order and complexity found in living things?

Problems with Complexity Theory

At first glance, slime mold aggregation seems to be an example of order for free; no blueprint necessary. Just supply the cells with energy to keep them from reaching equilibrium and a pattern spontaneously emerges. There is a cost, though, and it has to do with the fact that the slime mold cells must have all of the interacting components present in the correct spatial and temporal order to generate the pattern. The exact amounts may be irrelevant given the robust nature of these systems, but certainly in the absence of cAMP, the slime mold cells will not form the patterns associated with aggregation. They also need a mechanism for secreting cAMP, receptors for cAMP, a mechanism for secreting phosphodiesterase, and a mechanism to desensitize themselves to cAMP levels in the fluid around them immediately after exposure. How all of these field components came to be aligned to interact properly is not clear. The basic problem is that without the interacting parts there is no order, yet without order there is no need for the interacting parts.

The gradual accumulation of the components necessary to build this dynamic field appears to require some type of selection. But it is hard to see how receptors for cAMP would be useful if there were no mechanism in place for secreting it during times of starvation and vice versa. To build this system gradually, some co-option scenario for the gradual generation and preservation of all of these components would be required (see chap. 7 for a discussion of co-option). This is

74. Barkai and Leibler, "Robustness in Simple Biochemical Networks," 913–17.

an important point to realize: while complexity theory may explain how order can emerge spontaneously once certain components are in place, it doesn't explain how all the components got there in the first place.

Stuart Kauffman and Complexity Theory

Spurred by the order that emerges in slime mold patterns and the BZ reaction, many have turned to complexity theory as a way to explain other more complex events during the course of evolution. Stuart Kauffman has been at the forefront of this effort, and his ideas on how complexity theory can explain the origin of life, though outside the topic of evolution, provides a good example of what complexity theorists advocate.

Rather than positing that life originated as a simple self-replicating molecule, Kauffman believes that life originated wholesale, as an ordered dynamic network of molecules. According to Kauffman:

> There are compelling reasons to believe that whenever a collection of chemicals contains enough different kinds of molecules, a metabolism will crystallize from the broth. If this argument is correct, metabolic networks need not be built one component at a time; they can spring full-grown from a primordial soup.[75]

Kauffman's idea that a dynamic interacting network is responsible for the emergence of life finds its origin in the primitive hypercycles first proposed by Manfred Eigen and Peter Schuster (see fig. 8.9).[76] The hypercycle depicted in figure 8.9 links the four words together such that the replication of each word depends on the presence of another word in the cycle. When this occurs, the set is considered *autocatalytic*.

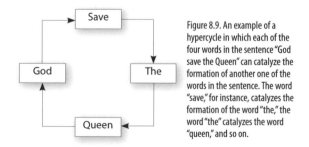

Figure 8.9. An example of a hypercycle in which each of the four words in the sentence "God save the Queen" can catalyze the formation of another one of the words in the sentence. The word "save," for instance, catalyzes the formation of the word "the," the word "the" catalyzes the word "queen," and so on.

75. Stuart Kauffman, *At Home in the Universe: The Search for Laws of Self-Organization and Complexity* (Oxford: Oxford University Press, 1995), 45.

76. J. M. Smith, "Hypercycles and the Origin of Life," *Nature* 280 (1979): 445–46.

Theoretically, such autocatalytic sets are possible, as they have been developed in the lab.[77] Developing such sets in the lab, however, doesn't answer the origin-of-life question: What is the likelihood of a group of complex autocatalytic molecules forming spontaneously from the chemicals that existed on the early earth, the aptly named primordial soup? While this may seem improbable at first glance, Kauffman believes it is not so difficult to achieve the catalytic closure seen in autocatalytic sets. In fact, he argues that autocatalytic sets will emerge with near certainty if enough chemicals are mixed together.

The gist of his argument is as follows. Every chemical added to a potential autocatalytic set may possibly be formed by a few different reactions, depending on the other molecules present. For example, if only the molecules *AA* and *B* are in the mix, *AAB* can be formed only by adding a *B* to the *AA* molecule. If, however, the molecule *AABB* is also present, *AAB* can also be formed by cleaving off a *B* from the *AABB* molecule. Thus, the more molecules that are added to the mix, the more possible reactions exist that could lead to the formation of *AAB*. Eventually enough molecules will be present such that at least one of the reactions for forming *AAB* will be catalyzed by another molecule in the soup. As Kauffman claims:

> When the number of catalyzed reactions is about equal to the number of [chemicals], a giant catalyzed reaction web forms, and a collectively autocatalytic system snaps into existence. A living metabolism crystallizes. Life emerges as a phase transition.[78]

An example of a simple autocatalytic set is shown in figure 8.10. All the molecules present in this set are catalyzed by other molecules that are present. This represents catalytic closure, and Kauffman argues it occurs with near certainty when roughly 10,000 molecules are present, so the figure is not representative of the molecular diversity needed to achieve such a set. The obvious question to ask is, why 10,000 molecules? What is so special about this level of diversity? Let's see how Kauffman gets that number. He starts by figuring the probability that a random molecule could catalyze the formation of any other random molecule. He argues that this is a one in a million possibility. Now, if the primordial

77. A simple autocatalytic system in which *A* catalyzes the formation of *B* and *B* catalyzes the formation of *A* from simple precursors has been done by Gunter von Kiedrowski. While an autocatalytic set was derived, the set was not confused with a living cell. As Stuart Kauffman admits, "autocatalysis and molecular reproduction are necessary for life, but not yet sufficient. Life possesses deeper, still more mysterious properties" (*Investigations* [Oxford: Oxford University Press, 2000], 47).

78. Kauffman, *At Home in the Universe*, 62.

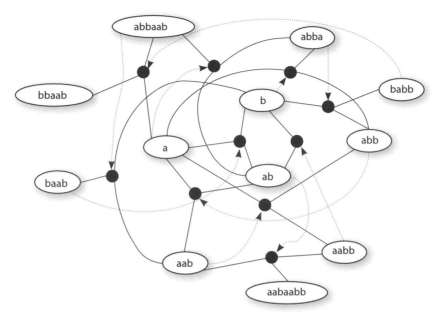

Figure 8.10. An autocatalytic set. Molecules are indicated by ellipses. Solid lines connect reactant molecules with product. A dashed line indicates the reaction a particular molecule can catalyze. In this small set, all the reactions can be catalyzed by another member of the autocatalytic set.

soup is made of only molecules X and Y, there is only a one in a million probability that X would catalyze the formation of Y. Obviously this is a problem in generating catalytic closure. Small networks will likely never exhibit catalytic closure.

> A critical diversity of molecules must be reached for the system to catch fire, for catalytic closure to be attained. A simple system with 10 polymers in it and a chance of catalysis of one in a million is just a set of dead molecules. Almost certainly, none of the 10 molecules catalyzes any of the possible reactions among the 10 molecules. Nothing happens in the inert soup.[79]

To get around this problem, Kauffman postulates a huge population of molecules as a starting point. If this occurs, then the number of possible reactions that can occur increases immensely as described above. If you have 1,000 molecules present, they can obviously react in many more ways than if you have 10 molecules. Kauffman argues that when you get roughly 10,000 different complex molecules, there will be so many different possible reactions (trillions of them, in fact), that each molecule in the set

79. Ibid., 64.

will be able to catalyze one of these.[80] Once this critical threshold is met, that is, a sufficient diversity of molecules is present, an autocatalytic net is bound to emerge "as an expected property of the physical world."[81] As a result, Kauffman is optimistic that such autocatalytic networks can explain the origin of life. In his view, life is the expected outcome of chemistry: just dump enough chemicals together in a localized environment and life crystallizes. There are, however, a few problems with this idea.

Paying for It

The main problem that Kauffman ignores in his scenarios is how exactly this critical threshold of molecule density is reached. How do the thousands or tens of thousands of molecules necessary to reach catalytic closure form? There is nothing directing their formation, so apparently this is left to chance; however, the chance formation of macromolecules is fraught with problems.

For example, forming large proteins, which are chains of amino acids, is an uphill battle that requires energy. That is because polymerization of amino acids in the primordial soup must compete with the reverse reaction of breaking the polymers apart back into the individual amino acids. Concentrations of amino acids (or nucleotides in the case of DNA and RNA) would have to be absurdly high to favor the chance formation of any large polymer, not to mention the problem of avoiding the myriad interfering cross-reactions that would tend to lead to insoluble macromolecules. Another serious problem is that just stringing together a random array of amino acids does not result in a functioning macromolecule. One needs an ordered array, a specific sequence of the building blocks in order to have any meaningful function. If only alanines or glycines are strung together, a functional protein does not result. Moreover, simple amino acids such as glycine and alanine would dominate in the soup, making the chance construction of a diverse molecule exceedingly difficult.

Kauffman acknowledges these problems, but addresses only some of them. His main solution is to propose that dehydration would favor the polymerization of large molecules. Removal of water molecules does indeed decrease the likelihood that the reverse reaction, the cleavage of proteins into amino acids, would occur. However, the reduction of water reduces the mobility of the molecules that do form, and thus

80. Remember the one in a million probability that any molecule could catalyze any specific reaction. If there are trillions of reactions, then it is highly probable that each molecule will be able to perform one of these possible reactions. Using these numbers, it would be highly improbable that any specific molecule *didn't* catalyze a reaction.

81. Kauffman, *At Home in the Universe*, 64.

finding their catalytic partner in the soup becomes a very difficult task.[82]

There is yet another problem with Kauffman's model, and it stems from the fact that autocatalytic sets differ in some very fundamental ways from all known forms of life. All "living" systems must be able to self-replicate, while at the same time they must be able to store and transmit some type of genetic information. In terms of these autocatalytic sets, there is no mechanism for storing information. If, during bacterial cell division, a certain protein is not passed on to a daughter cell, this is not necessarily a problem because the cell has the information in the DNA and a central processing pathway (translation and transcription) to produce that protein. Because autocatalytic sets lack such a mechanism, the autocatalytic set is much more fragile and is easily subject to degeneration. Imagine an autocatalytic set with a thousand molecules. If by chance one of these molecules, molecule X, failed to make it into the daughter protocell (this is quite likely given the random nature of the division of such sets), then any molecule, say molecule Y, whose formation is catalyzed by X will not be able to be maintained. Subsequently any molecule dependent on Y will disappear and so on until the entire autocatalytic set dissipates. A lack of information storage capacity dooms these autocatalytic sets because without it, no evolution could occur. As a result, until Kauffman's model and others like it address these deficiencies, they should be looked at with some skepticism when it comes to explaining the origin and evolution of life.

Endosymbiosis: A New Way of Looking at Evolution

Biologist Lynn Margulis has become world renowned for her endosymbiotic theory of evolution. This theory turns Darwin's theory on its head by focusing on associations and cooperation between organisms, rather than on competition between them, as the driving force behind evolution. The theory is the result of Margulis's analysis of symbiotic relationships between organisms, particularly symbiotic relationships that are to the mutual benefit of both organisms involved. For example, the bacteria in our gut are involved in a mutually beneficial relationship with us: we supply them food and they supply certain vitamins and metabolites. Margulis argues that such symbiotic relationships actually can drive evolution when one member of such a symbiotic relationship acquires either part or all of the other symbiont. Such

---

82. W. Bradley, "Designed or Designoid," in *Mere Creation*, 33–49.

incorporation is referred to as endosymbiosis or the large-scale acquisition of genomes.[83]

The classic and best-established example of endosymbiosis is the origin of mitochondrion, an organelle needed to maintain energy supplies for eukaryotic cells. Mitochondria are responsible for metabolizing carbohydrates and fats to carbon dioxide and water in order to produce energy for the cell. Margulis proposed that these organelles were once free-living prokaryotic cells that happened to be internalized by another cell. Following internalization, a symbiotic relationship developed that made the internalized cell specialized and indispensable to its host. What evidence does she cite? Like prokaryotic cells, mitochondria have their own circular DNA that is synthesized throughout the life cycle. The mitochondrial DNA also is very promiscuous, much like prokaryotic DNA.

> Genetic recombination in these organelles is far more reminiscent of phage and bacterial sexuality than it is of eukaryotic nuclear sexual behavior. Pairing and recombination of mitochondrial DNAs seem to occur often and at random.[84]

Mitochondria also have their own ribosomes that, based upon drug sensitivity, resemble prokaryotic ribosomes.[85] Finally, John and Whatley have made the case that mitochondria resemble a specific free-living aerobic bacteria, *Paracocus denitrificans*.[86] In total, the evidence strongly points to the conclusion that mitochondria were *not* built up gradually in a Darwinian fashion but were acquired as an entire functioning prokaryotic entity. While most researchers agree with this assertion, Margulis goes a step further, claiming that nearly *all* of speciation, even at the level of plants and animals, is the result of the symbiotic acquisition of genes or genomes.

> The creative force of symbiosis produced eukaryotic cells from bacteria. Hence all larger organisms—protocists, fungi, animals, and plants—originated

83. Margulis was not the first to propose the importance of symbiosis in evolutionary history, but she has assembled more empirical support for symbiogenesis than any of her predecessors. Paul Portier's 1919 book *Les Symbiotes* proposed that all organisms other than bacteria were formed by the association of previously independent organisms, which is in essence Margulis's position. Similar ideas were advanced by American biologist Ivan Wallin in the 1920s. Although Portier met with much skepticism because of a lack of experimental evidence, Margulis has made her reputation by positing concrete examples of endosymbiosis.

84. Lynn Margulis, *Symbiosis in Cell Evolution* (New York: W. H. Freeman, 1981), 219–21.

85. Ibid., 206–7.

86. P. John and F. R. Whatley, "*Paracoccus denitrificans* and the Evolution of Mitochondrion," *Nature* 254 (1975): 995–98.

symbiogenetically. But creation of novelty by symbiosis did not end with the evolution of the earliest nucleated cells. Symbiosis still is everywhere.[87]

She goes on to posit a few examples: "The surfaces of virtually all algae and leaves, the bark of plants, and the skin of animals have their particular microbiota. Every kind of community contains symbiotic associations; partners come from all of the higher taxa."[88] There is no doubt that symbiotic relationships abound in nature, but there is scant evidence that these symbiotic relationships lead to the creation of novelty as Margulis argues. The notion that endosymbiosis and the acquisition of genomes lies at the heart of plant and animal evolution must be supported by more than examples of symbiosis. Rather than merely showing that two organisms have become mutually dependent on each other one must show that genomes have fused and created a new organism.

In fact, the main evidence she posits to support her theory that the fusion of different genomes can account for major steps in the evolution of the higher taxa does not come from symbiosis but rather from the cross-fertilization of nonsymbionts. The data utilized by Margulis come from the work of marine biologist Donald Williamson, who believes that sexual encounters between individuals from different species can lead to novel forms.

> Williamson posits that sexual encounters . . . sometimes occurred between individuals of very different classes and phyla. Occasionally they were spectacularly successful. Such successful matings between very distantly related animals occurred infrequently, some thirty to fifty times in 541 million years. This means a fertile, successful outcome happens roughly once in 10 million years.[89]

Williamson has no direct evidence that this has occurred or that it has been evolutionarily important. He does, however, posit that the lack of correlation between the larval and adult forms of many related species is evidence of such cross-fertilization. For example, some hydrozoans alternate between a polyp stage, in which they resemble a vase with tentacles sticking out the top, and a medusa stage, characterized by the upside-down-bowl appearance with tentacles sticking down. Williamson argues that the hydrozoans with both stages have acquired the medusa stage via genome acquisition through cross-fertilization, although he cannot directly support this claim.

87. Lynn Margulis and D. Sagan, *Acquiring Genomes: A Theory of the Origins of Species* (New York: Basic Books, 2002), 55–56.
88. Margulis, *Symbiosis in Cell Evolution*, 164.
89. Margulis and Sagan, *Acquiring Genomes*, 166.

To investigate the ability of cross-fertilization to drive evolution, Williamson has done preliminary work in mating two distantly related aquatic animals, the sea squirt *Ascidia mentula* (a chordate) and the sea urchin (an echinoderm). Many of the offspring from this cross developed like normal sea urchin larva but with the addition of adhesive discs that are characteristic of *A. mentula*. This novel combination is not stable, however, and the majority of these cross-phylum hybrids die within ninety days. The hybrids that do survive have characteristics of just the sea urchin, casting uncertainty on this as a major mechanism in evolution.[90]

In the realm of plants, the notion that interspecies hybridization played a major role in evolution doesn't fare much better. While the literature is full of cases of fertile offspring resulting from crosses of plant species that share the same genus, merely being of the same genus does not guarantee success.[91] In fact, infertile offspring result in a similar amount of cases of interspecies hybridization. In addition, the morphology of the successful hybrid tends to be either intermediate (a blend of the two parental species) or it closely resembles one of the parental species. One does not normally cross plant species and get fertile plants that have novel plants or leaves. In this manner, plant interspecies hybridization can result in variations on a theme, more species within a certain genus, but it appears on its own to be ill-suited to build novel structures.

The notion that mitochondria were once free-living prokaryotes is nearly universally accepted, and this does help explain a major step in evolution: the creation of eukaryotic cells. It may also explain the origin of the nucleus and plastids, which may have emerged in a similar fashion. What it does *not* explain is how the machinery in the plastid or the machinery in the mitochondrion was first assembled. Endosymbiosis in all of these cases involves the *loss* of informational content. A mitochondrion contains less information than a free-living prokaryote; in fact, mitochondria could be considered degenerate prokaryotes. For example, mitochondrial ribosomes include proteins that are encoded within the nucleus of the host cell, while prokaryotic ribosomes are self-sufficient. Because of this, if one wants an explanation of how the biochemical process of the TCA cycle (see chap. 5) and the electron transport cycle emerged—both integral parts of the mitochondria—endosymbiosis does not offer any new information. It starts with a cell, a prokaryote that has these pathways already intact, and incorporates them within a new

90. Ibid., 165–84.
91. See Darwin, *The Origin of Species*, chap. 9; and Peter H. Raven, Ray F. Evert, and Susan E. Eichhorn, *Biology of Plants*, 6th ed. (New York: W. H. Freeman, 1999), chap. 12.

relationship. Endosymbiosis can put together novel combinations of complexity, but generating the underlying complex structures is another matter. Although it represents an intriguing evolutionary mechanism and there is no doubt that it has had a role in evolution, endosymbiosis appears to have its limits.

Conclusion

If there is one notion permeating all the Meta-Darwinian theories described in this chapter, it is the belief that the biological world is not merely the result of small random mutations coupled with selection blindly surveying the entire realm of structural possibilities. If one accepts some or all of these additions (endosymbiosis, exaptations, developmental mutations, multilevel selection, complexity theory, etc.), evolution is no longer reducible to natural selection working on small mutations. Rather, evolution involves a complex set of interactions. However, to date no one has actually synthesized a new theory of evolution that takes all of these (or most of them) into account. Such a task would be formidable indeed, and, needless to say, goes beyond the scope of this book.

Nonetheless, it is possible to give some idea of what such a theory would be. Referring to table 8.2, which summarizes the theories discussed in this chapter, we see that the new, more general theory would rely on multiple sources of proximate order and information, many of which are capable, in theory, of larger-scale changes than Neo-Darwinism predicts. Self-organizing networks, endosymbiosis, and developmental mutations are all naturalistic mechanisms that stand poised to tackle the difficulties of building novel forms within a naturalistic framework.

While all of the proposed supplementary theories have difficulties, and some have yet to gain widespread acceptance, it is important to remember that they are still relatively recent advances and are theories in the process of evolving. The ultimate fate of these ideas surely will vary. In concert, they may be able to overcome some or all of the difficulties facing natural selection discussed in chapter 5, but this is by no means certain.

Regardless of which, if any, of these ideas ultimately survive the test of time, they have cast a doubt on the primacy of natural selection. As Gould states:

> What if adaptation does not always record the primacy of natural selection, but often arises as secondary fine tuning of structures arising in

324

Table 8.2. Summary of Meta-Darwinian theories

Theory	Type of order/ new information	Source of order/new information		Role of natural selection	Change capacity
		Original	Proximate		
Endosymbiosis	Construction of complex organisms by integration of existing cells and structures	Unclear; possibly fusion of small genetic elements	Other cells absorbed and integrated	Fine tunes systems (FT)	Large-scale
Self-organization/ complexity theory	Order arises spontaneously as property of systems and networks	Unclear; still need to assemble components of the system	Attractor for nonlinear system	FT	Large-scale
Morphogenetic fields	Only certain developmental pathways are stable, based upon dynamic interactions within morphogenetic fields	Random mutation	Robust patterns imposed by physical laws or attractors	FT to major role	Large-scale
Developmental mutation: evo-devo	New organism arrangements arising from changes in master control genes	Random mutation	Small changes in control genes	FT to major role	Large-scale
Neutral theory	New raw material arising and propagating through population by chance, which can be used by natural selection	Random mutation	Neutral mutation, propagated through population	FT to major role	Small increments
Exaptation	New raw material resulting as by-product of other processes, or from slight modification of existing material	Random mutation	By-product of existing processes	FT to major role	Large-scale
Hierarchical selection	Dynamics of higher-level entities sets up selection at those levels, in competition with selection at individual level	Random mutation	Selection acting on higher-order entities	Major role	Small increments
Punctuated equilibrium	Not a source of new order or information, but a new way of looking at the tempo of evolution	Random mutation	No unique mechanism	–	–
Neo-Darwinism	Incremental changes only from small genetic mutations	Random mutation	Random mutation	Major role	Small increments

other ways? . . . As a common thread, these challenges deny exclusivity to natural selection as the agent of creativity.[92]

If what Gould says is true, then the questions become: What role does natural selection have in explaining evolutionary novelty? Did Darwin identify only a peripheral force? Does novelty result only by mechanisms found *outside* a traditional Neo-Darwinian framework? Answers to these questions vary, depending on whom you ask. Margulis calls into question the very efficacy of natural selection to build novelty; Gould merely questions its importance. The issue, as seen here, is far from being resolved. In the meantime, the question of how life evolved remains open to novel ideas. Which theories ultimately survive remains to be seen.

Although the Meta-Darwinian school is a collection of theories rather than a single unified one, we may still evaluate it with respect to the ten criteria of a good scientific theory. This evaluation is presented in table 8.3.

Table 8.3. Meta-Darwinian school and the ten criteria of a genuine scientific theory

Criterion	Meta-Darwinian school
1. Compactness	Not as compact as Neo-Darwinism but multiple hypotheses to explain evolution is satisfactory if they can be verified.
2. Simplicity	Some theories are quite simple; some verge on ad hoc.
3. Falsifiability	Most can be falsified: experimental tests can be found for most of the hypotheses advanced.
4. Verifiability	Most can but difficult in some cases.
5. Retrodiction	Most of the school's theories were devised to explain (retrodict) already known facts that did not seem to be well explained by Neo-Darwinism.
6. Prediction	Mostly limited to vague predictions.
7. Exploration	Definitely suggests many avenues of exploration.
8. Repeatability	Not a problem because usual scientific method applies.
9. Clarity	Satisfactory although some are vague and subject to interpretation.
10. Intuitiveness	Satisfactory although some are subtle and difficult to grasp.

92. Gould, *Structure of Evolutionary Theory*, 159.

PART 3

POLICY AND OUTLOOK

9

PUBLIC POLICY IMPLICATIONS OF THE EVOLUTION CONTROVERSY

The evolution controversy has important public policy implications for two reasons: the magnitude of the cultural issues associated with the debate and the sheer numbers of laypeople who reject the established theory (Neo-Darwinism). In no other branch of science today does a similar situation exist. It causes the evolution debate to spill over into ongoing legal, cultural, and educational battles. With research money and educational access at stake, there is no possibility of ignoring the public policy dimension of the controversy. In fact, success in the public policy arena is absolutely critical, as evidenced by the substantial investment of each school of thought in this arena. All sides realize that if public policy battles being waged at the level of state education boards or within the courts are lost, their position will be marginalized.

The issues that are debated in public policy battles are often distinct from the scientific issues that we have raised so far. In fact, the public policy war centers on a broader set of issues, which we will examine in this chapter. They include:

1. Who is the legitimate spokesperson for science?
2. What types of evolution research should be funded with public money?
3. Should equal time or any time be given in the classroom for theories other than Darwinism?

4. Should the courts be the primary battleground for the evolution controversy, or should they not be involved at all?

Clearly, all of these issues are related, but it is convenient to concentrate on them individually. In addition, two subsidiary questions are worth addressing since they directly impact the public policy issues:

1. What are the moral and ethical dimensions of the controversy?
2. Do purely naturalistic theories of evolution function as surrogate or virtual religions?

It may be a misnomer to call these two questions "subsidiary" because in reality the issues they raise often drive some of the harsher criticism of evolutionary theory. In fact, the Creationist school, and to a lesser extent the Intelligent Design school, believes that evolution is not only *factually* wrong but also *morally* wrong, insofar as it is inextricably linked to a moral code (or lack thereof) at variance with key Judeo-Christian moral principles:

> The fundamental clash we see in our society at present is the clash between the religion of Christianity with its creation basis and therefore absolutes, and the religion of humanism with its evolution basis and its relative morality that says, "Anything goes."[1]

Given that about 50% of the US population is sympathetic to the Creationist position[2] but is still asked to pay for research and evolution instruction that they see as championing "moral relativism," the stage is indeed set for a full-scale culture war. As a result, the controversy concerns far more than the truth or falsity of some particular theory about how species change. In this climate, questions of public funding and teaching have become political, legal, theological, philosophical, and moral issues—issues that have a tendency to overshadow the actual scientific questions, and in some measure actually dictate possible answers. Indeed, one could easily make the case that many on all sides of the debate hold their *scientific* positions on evolution as a consequence of their positions on the *other* issues.

Of course, Neo-Darwinism and the philosophy of materialism (the basis of the "moral relativism") are different, and one can accept Neo-Darwinism and reject the other, as many theistic evolutionists do. In

---

1. Ken Ham, "Creation Evangelism," *Creation Ex Nihilo* 6, no. 2 (November 1983), special lift-out section, available at www.answersingenesis.org/docs/3359.asp; entire text capitalized in original.
2. See the Gallup poll discussion in chap. 1.

practice, though, they are often found together, and indeed some vocal spokesmen for Neo-Darwinism, such as Richard Dawkins, clearly equate the two by saying that evolution implies materialism and atheism. This plays directly into the hands of the Creationists and other detractors of Neo-Darwinian evolution, with the result that possibly more sophisticated solutions such as those advanced by some theistic evolutionists are drowned out by the heated rhetoric. Such solutions are explicitly rejected by most Creationists and many in the Neo-Darwinian camp.[3] While a complete investigation of the philosophical, theological, and moral aspects of the evolution controversy is beyond the scope of the book, we will touch on these aspects to give readers a taste of the bitter dispute and reasons for that bitterness.[4] With this in mind, let us turn to the major public policy issues being debated.

Policy Issue 1: Who Is the Legitimate Spokesperson for Science?

Whoever is recognized as the spokesperson for science in our society automatically wields great influence because of the prestige that science enjoys among the general population. Such a spokesperson is in a position to affect legislation, court cases, health care decisions, school curricula, and textbook content, to name just a few important areas. Whoever assumes the "bully pulpit," therefore, is of great importance for public policy.

Currently, the Neo-Darwinists, and to some extent Meta-Darwinists, have a stranglehold on that pulpit. When an organization such as the National Academy of Sciences (which advocates by and large the Neo-Darwinist position) speaks out on an issue, such as ecology, people tend to listen. Obviously, they also listen when the subject is evolution. The major scientific organizations such as the NAS, the American Association for the Advancement of Science (AAAS), and the National Science Teachers Association (NSTA) all advocate the teaching and promotion of Darwinian theory and tend to reject calls for discussion of the problems of the theory or examination of alternative theories. In their view, the idea that Creationists, Intelligent Design advocates, or even certain Meta-Darwinian critics of evolution should have a voice is pandering to antiscience.

3. The interested reader is advised to peruse the Answers in Genesis web site (www .answersingenesis.org) to gain a better appreciation for this issue and the positions taken on it.
4. For a detailed discussion of the philosophical aspects of the evolution controversy, and how they can impact the science, see www.evolutionprimer.net.

Critics of the Darwinists, however, reply that in this case, the scientific community has betrayed a sacred trust by allowing extra-scientific agendas to overshadow the usually objective quest for truth that science represents. For them, the very attitude of NAS, AAAS, NSTA, and other groups that want to dismiss objections to Neo-Darwinism out of hand, demonstrates that they have abdicated their role and are themselves exhibiting antiscientific attitudes. For many of these critics, the Neo-Darwinists' credibility has been compromised to such an extent that in their view the school no longer speaks for science, any more than the Republican National Committee speaks for all politicians.

Policy Issue 2: What Types of Evolution Research Should Be Funded with Public Money?

The question at issue here is whether the research efforts, including graduate study,[5] of one of these schools of thought alone should be funded with public money. The US government spends about $130 billion per year on scientific research and development, primarily through grants administered by organizations such as the National Institutes of Health (NIH), the National Aeronautics and Space Administration (NASA), the National Science Foundation (NSF), and others.[6] The system developed to choose priority research areas and select worthy grant recipients has served our country well over many decades. Based primarily on peer review, it functions to weed out unpromising avenues of research and concentrate available funds on those areas that appear most likely to result in new and important discoveries and advancements. While it would be naive and unrealistic to pretend that politics plays no role in the selection of recipients and projects funded, by and large the system works on the basis of disinterested peer review, which is universally regarded as the fairest and most objective method possible.

This method of funding scientific research has enjoyed broad support over the years because the general public believes that science is important to our standard of living, our economic well-being, and our competitive position in the world economy. The public also accepts the knowledge and authority of grant-making organizations with respect to choosing the projects to be funded and is content to hand over to them the responsibility for this function. Most realize that this selection

5. Graduate study is typically funded (at least in part) by federal and state grants and loans, and indirectly by subsidies for libraries and other campus facilities.
6. Marguerite Reardon, "Research Money Crunch in the U.S." www.news.com, November 8, 2005, http://news.com.com/Research+money+crunch+in+the+U.S./2100_1008 _3_5938451.html.

process requires advanced knowledge of the relevant field and acquaintance with research methods and techniques, which neither the general public nor Congress has. Although Congress often designates money for specific research areas, such as cancer or AIDS, it assigns responsibility for allocation of funds and monitoring of research to one of the major government agencies or organizations. Excessive congressional meddling with the funding process—for example, by forcing money to go to certain designated recipients or to research outside of the accepted areas of science—almost always results in dollars wasted on unfruitful and unproductive research efforts.

Our country also has a long-standing tradition, hammered out in the courts and Congress, of separation of church and state. In practice, this limits funding of religious organizations and enterprises to what may be termed "secular areas." Public money cannot be used to pay for religious instruction, but it could be used to fund physics research at a religious college, for example. It is at this point that potential problems begin. No one disputes that those advocating Neo-Darwinism are free to say whatever they choose, and to promote atheism or any other religion, provided that it is on their own time and money, just as Creationists are free to promote their literal interpretation of the Bible and criticize the Neo-Darwinian theory under the same circumstances. The problem arises if there is a concerted effort—or any effort—to establish a particular religious viewpoint, or criticize one, utilizing public money in any form. Given our legal and cultural traditions, either no one is allowed to do this, or every side must be given an opportunity, on an equal basis.

The argument of the Neo-Darwinian and Meta-Darwinian schools is that any public funding given to disprove their theories or establish Creationist theories (1) is wasted money since the outcome (failure) is already certain; (2) is money spent to directly subsidize a religion and therefore unconstitutional; and (3) would likely come out of their own funding, which is already inadequate. They maintain that although their *theories* may be used to support antireligious agendas, they do not support these agendas directly with grant (public) *money*. They steadfastly maintain that religion and science are separate realms and that those who like to cross that line and flaunt their antireligious bias do so on their own time, and hence cannot be regarded as expressing taxpayer-subsidized positions.

Naturally the Intelligent Design school and Creationists denounce this position as hypocritical posturing, intended to deflect public attention from the fact that adherents of the Neo-Darwinian and Meta-Darwinian schools are pursuing a materialistic and antireligious agenda under the guise of objective scientific research. They claim that by hiding behind the mantle of science, the Neo-Darwinists avoid any scrutiny of their position, which might endanger their ongoing funding. They also argue that the experiments Creationists and Intelligent Design adherents are pursuing (described in chaps. 6 and 7), although they may be driven by religious and philosophical positions, fit within the conventional definition of science and therefore are worthy of being funded on the public dime. However, given that the peer-review funding system is in the hands of the Darwinists, they have little hope of getting any federal research funds for their projects. They see this as patently unfair, particularly given the number of taxpayers who support their ideas.

Policy Issue 3: Should There Be Equal Time (or Any Time) in the Classroom for Other Theories?

Equal time in the high school classroom is one of the flash points of the evolution controversy because the position of the majority of educational institutions in this country is that, by and large, only Neo-Darwinian theory (and to some extent Meta-Darwinian theories) should be taught. Creationists and Intelligent Design advocates have responded by means of legal challenges at all levels, and by use of media of all types in an attempt to refute evolution. They also provide instructions to students on how to deal with what they see as an indoctrination effort; specifically, how to pass examinations on evolution and write about evolution without compromising their own beliefs.[7]

Assuming that the goal of education is to produce students who can think for themselves and who believe scientific theories on account of the evidence for them, one has to ask whether it is wise to ignore competing theories and the alleged evidence against the Neo-Darwinian theory. In fact, because of the popularity among the public of competing theories such as Intelligent Design, some evolution textbooks are now devoting a few pages to addressing the perceived deficiencies of these alternative theories.[8]

Given the complexity of this question, there are many possible positions, ranging from teaching only Neo-Darwinian theory (no evidence against it, no competing theories) to mandated teaching of competing theories, to no teaching of evolution at all. As with any other complex and controversial issue, it is impossible to please everyone. Regardless of which alternative is chosen, some group or groups will complain that their "rights" have been violated.

7. See the AiG web site, www.answersingenesis.org, for examples.
8. S. Freeman and J. Herron, *Evolutionary Analysis* (Upper Saddle River, NJ: Pearson-Prentice Hall 2003).

In a well-known anti-Creationist book, Philip Kitcher has argued that teaching of Creationism and evolution as equal theories would be tantamount to saying to students that evidence does not really matter, that ideology rules:

There will be . . . much dredging up of misguided objections to evolutionary theory. The objections are spurious—but how is the teacher to reveal their errors to students who are at the beginning of their science studies? . . . What Creationists really propose is a situation in which people without scientific training—fourteen-year-old students, for example—are asked to decide a complex issue on partial evidence.[9]

This is a valid point, to be sure. As we have noted in chapter 1, science cannot proceed if every crackpot theory has to be discussed. But as we have shown, it is simply not the case that all objections to the Neo-Darwinian theory can be dismissed as spurious, nor can all the alternatives proposed be discarded as mere crackpot theories. Neo-Darwinism may be correct, but current objections to it are not trivial—otherwise there would be no Meta-Darwinian school, for example. Moreover, the fundamental issues are not all that difficult to understand, no more so than other science concepts such as laws of motion or chemical bonding. And as critics of Neo-Darwinism have observed, the same people who object to students making their own decisions about evolution have no qualms about asking them to do just that in what is, perhaps, a more difficult and more far-reaching area, at least with respect to their future lives:

In matters of ethics and morality, fourteen-year-old students are invited to challenge the standards of their parents and make their own decisions. When it comes to evolution, however, the same pupils must be protected from spurious notions that may seem valid to their untutored judgment. . . . [A] great many parents think it would be much wiser to do the reverse: to tell adolescents firmly what limits on behavior they must observe, and to encourage them to practice their critical thinking on more theoretical subjects like evolution, where mistakes are much less likely to cause permanent damage.[10]

Students should be encouraged to critically evaluate all sides of the evolution debate. This practice may require some classroom time but need not imply equal time for all theories. In our modern Internet-enabled society, there is no real possibility that the views and theories of those opposed to Neo-Darwinian evolution can be suppressed or that

9. Philip Kitcher, *Abusing Science* (Cambridge, MA: MIT Press, 1982), 175–76.
10. Phillip Johnson, *Reason in the Balance* (Downers Grove, IL: InterVarsity Press, 1995), 165–66.

they can be kept from students. Attempts to suppress or ignore those views will likely convince students that the Neo-Darwinian school has something to hide. Some members of the Neo-Darwinian school agree. Richard Alexander, professor of zoology at the University of Michigan, comments:

> No teacher should be dismayed at efforts to present creation as an alternative to evolution in biology courses; indeed, at this moment creation is the only alternative to evolution. Not only is this worth mentioning, but a comparison of the two alternatives can be an excellent exercise in logic and reason. Our primary goal as educators should be to teach students to think, and such a comparison, particularly because it concerns an issue in which many have special interests or are even emotionally involved, may accomplish that purpose better than most others.[11]

There is a final reason that makes suppression of problems and critiques of Neo-Darwinism ill-advised, namely the effect that such tactics will have on the next generation of prospective scientists. Given the scientific concerns with the dominant evolution paradigm, it is important for students to understand these early, because it is these young students who will soon become scientists entering their most creative years. They are the main hope to resolve the problems discussed in this book.

Policy Issue 4: Are Courts the Proper Battleground for Evolution?

In the early 1800s, Alexis de Tocqueville observed that in the United States, all disputed questions tend to end up before the courts. The evolution controversy is proving itself to be no exception. A detailed review of the history of litigation related to the evolution controversy may be found elsewhere.[12] Here we give only a few highlights and then discuss whether the courtroom is a sensible venue for resolving evolution-related questions.

Court involvement began with the famous Scopes trial of 1925, when Creationism dominated classroom biology and the issue was whether evolution would be taught as well. During the next fifty years, the tables turned, and evolutionary ideas gradually supplanted and excluded Cre-

---

11. Richard D. Alexander, "Evolution, Creation, and Biology Teaching," in *Evolution vs. Creationism: The Public Education Controversy*, ed. J. P. Zetterberg (Phoenix: Oryx Press, 1983), 91.

12. See www.evolutionprimer.net.

ationist teachings. With the shoe on the other foot, Creationists sought to gain a hearing for their views in the classroom, which led to a series of "equal time" or "balanced treatment" laws. These laws were struck down in the 1980s and 1990s; the most important of them was perhaps *McLean v. Arkansas* in 1982, in which the court tried to spell out what constitutes a legitimate scientific theory.[13] Such a theory must be:

1. Guided by natural law
2. Explained by natural law
3. Testable against the empirical world
4. Tentative
5. Falsifiable

In the judge's opinion, "scientific Creationism" did not qualify as science. On the basis of these criteria though, Creationists (and others) believe that Neo-Darwinian evolution does not qualify as science either because, they argue, the theory is designed to be able to explain any possible piece of evidence that crops up, thereby making the theory unfalsifiable. Regardless of this dispute, the 1982 court case was regarded as a victory for those opposed to Creationism.

In this case, as in all the court battles, the Neo-Darwinian school has consistently (and successfully) argued that science can be separated from philosophy and theology and be taught strictly as science in the classroom. Nonetheless, we find that, in their more candid moments, the Neo-Darwinians admit the essentially religious nature of their outlook:

> Evolution is promoted by its practitioners as more than mere science. Evolution is promulgated as an ideology, a secular religion—a full-fledged alternative to Christianity, with meaning and morality. I am an ardent evolutionist and an ex-Christian, but I must admit that in this one complaint—and [Duane] Gish [a prominent Creationist] is but one of many to make it—the literalists [Creationists] are absolutely right. Evolution is a religion. This was true of evolution in the beginning, and it is true of evolution still today. . . . Evolution therefore came into being as a kind of secular ideology, an explicit substitute for Christianity.[14]

These remarks come from Michael Ruse, a philosopher of science whose testimony in the *McLean v. Arkansas* case was considered to be

13. Taken from "An Attempted Very Brief and Abridged History of Intelligent Design and the Creation-Evolution Controversy," www.ideacenter.org/stuff/contentmgr/files/205b6d0e8df203ccd60e9f3b1687004f/miscdocs/idhistory.pdf.

14. Ruse, Michael, "How Evolution Became a Religion: Creationists Correct?" *National Post*, pp. B1, B3, B7, May 13, 2000.

extremely influential. His remarks suggest that separating religion and science in the classroom may not be so straightforward.

A later case, *Edwards v. Aguillard* (1987), was a further blow to Creationists because it struck down a balanced treatment law on the basis of three requirements:

1. It must not promote any particular religion or religious view.
2. It must not have the primary effect of either advancing or inhibiting religion.
3. It must not result in excessive entanglement of government and religion.

How exactly this should be implemented is far from clear, as the numerous court cases that have emerged since 1987 have demonstrated. Many of these newer cases have involved the introduction of Intelligent Design theory into the classroom, but the courts have remained unsympathetic to this approach as well. A list of legal issues that still need to be resolved has been drawn up by the University of Missouri–Kansas City law school.[15]

While court battles continue to affect the trajectory of the dispute, the ultimate resolution will not come from the courts, because the courts can decide only *legal*, not *scientific* questions. Judges are simply not qualified to determine whether a theory is scientifically valid. They can proceed only by their usual manner of resolving technical disputes, namely by calling in various "expert witnesses" and weighing the opinions expressed in testimony. In this case, the "witnesses" present superficial versions of the opposing views, so for all intents and purposes this approach to resolving the controversy is doomed from the start. As a result, no one who comes out on the losing end will accept the verdict as factually binding, and so court rulings and legislative actions on the matter are merely another football to be kicked around rather than a means to resolve the issue.

On the one hand, rather than relying on the courts, Creationists and Intelligent Design advocates need to prove that they can do legitimate science, with discoveries based on their theories that are clear and undeniable. When they can do that, their ideas will have to be taught. They should not rely on legislative bodies to force their ideas into classrooms. On the other hand, the Neo-Darwinian school should not hide behind the courts—this is unprecedented in the history of science. Science operates

15. This list appears in the university's interesting web site, *Exploring Constitutional Conflicts*, on its page devoted to "The Evolution Controversy," www.law.umkc.edu/faculty/projects/ftrials/conlaw/evolution.htm.

on the basis of theories and evidence and constant criticism. Theories either stand on their own merits and can be defended on the basis of empirical evidence, or they are discarded. They do not need, and should never be given, legal protection.

The Moral and Ethical Dimensions of the Controversy

Public policy inevitably makes moral judgments. Laws and policies in this country increasingly penalize groups and organizations that engage in or promote actions deemed to be against the public good, such as discrimination based on race or sex. The Creationist and Intelligent Design schools have made great efforts to spell out the moral and ethical problems they believe to be a direct result of the philosophical underpinnings of a purely materialistic view of evolution—a view that is shared by many prominent spokespersons for Neo-Darwinism but, as mentioned earlier, is rejected by theistic evolutionists. This is contributing to the acrimony of the controversy. Creationists argue that the ills of modern society are a direct consequence of evolutionary thinking, and that the antitheistic implications of evolution will lead to a complete loss of moral authority:

> The evolutionary philosophy says, "There is no God. All is the result of chance randomness. Death and struggle are the order of the day, not only now but indefinitely into the past and future." If this is true, there is no basis for right and wrong. . . . When there is no absolute authority, you can do whatever is convenient to you. And if people start believing in a world view called evolution, they are going to say, "There is no God. Why should I obey authority? Why should there be rules against aberrant sexual behaviour? Why should there be rules against abortion? After all, evolution tells us we are all animals. So, killing babies by abortion is the same as chopping off the head of a fish or chicken."[16]

This position is echoed by members of the Intelligent Design school:

> The fight over abortion, for example, will not go away until either the Christian worldview or the Epicurean [materialist] worldview goes away. The Christian universe forbids it, while for the Epicurean moral universe it is not even a moral problem. The same is true in regard to other conflictual moral issues.[17]

16. Ken Ham, "The Relevance of Creation," *Creation Ex Nihilo* 6, no. 2 (November 1983), special lift-out section.
17. Benjamin Wiker, *Moral Darwinism* (Downers Grove, IL: InterVarsity Press, 2002), 30.

To some extent, at least, evolutionists agree about the moral code problem. The remarks that follow are from a debate between two advocates of evolution:

> *Jaron Lanier*: There's a large group of people who simply are uncomfortable with accepting evolution because it leads to what they perceive as a moral vacuum, in which their best impulses have no basis in nature.

> *Richard Dawkins*: All I can say is, That's just tough. We have to face up to the truth.[18]

Indeed, there is general agreement that materialistic (or naturalistic) philosophies built upon evolutionary biology leave little room for dignity and values in the traditional sense.[19] The problem is that these philosophies do not flow *by necessity* from Neo-Darwinian theory, despite what critics and advocates of Neo-Darwinian theory might argue. They certainly can be erected upon the edifice of Neo-Darwinian theory, but no one is forced to maintain a materialistic philosophy merely because she or he believes in the scientific validity of Neo-Darwinism. Attaching such extra baggage to Neo-Darwinian theory sets up a false dichotomy between evolution and religion when, in fact, many believe that science and religion are compatible.[20]

This view is echoed by the National Academy of Sciences, which has argued on many occasions that science and religion (and by implication, morality) are completely separate domains:

> Religion and science are separate and mutually exclusive realms of human thought whose presentation in the same context leads to a misunderstanding of both scientific theory and religious belief.[21]

> [It is false] to think that the theory of evolution represents an irreconcilable conflict between religion and science. A great many religious leaders accept evolution on scientific grounds without relinquishing their belief in religious principles.[22]

18. "Evolution: The Dissent of Darwin," *Psychology Today* (January–February 1997): 62.
19. Steven Weinberg, *The First Three Minutes* (New York: Basic Books, 1977), 155.
20. Kenneth Miller, *Finding Darwin's God* (New York: HarperCollins, 1999).
21. Resolution of Council of National Academy of Sciences in 1981, quoted in *Science and Creationism: A View from the National Academy of Sciences* (Washington, DC: National Academy Press, 1984), 6.
22. National Academy of Sciences, *Science and Creationism*, 5–6.

The academy is expressing the view popularized by Gould as "non-overlapping magisteria," or NOMA. While some hold this view in an attempt to avoid the controversy between science and religion, it seems difficult to maintain and overly simplistic:

> Because nature and human nature are necessarily connected, there is no way to escape the interrelationship of science and ethics, and no one should be relieved of the responsibility that this relationship entails.[23]

This interrelatedness implies that in certain instances science can inform religion, while in others religion can inform science. For religious people who believe in evolution, this means incorporating evolution into their faith. These theistic evolutionists maintain that a nonliteral interpretation of Genesis is possible without gutting religious belief. Advocates of this position include many prominent supporters of Neo-Darwinian theory, such as Kenneth Miller, who discusses the issue at great length in his book *Finding Darwin's God*. The belief that evolution and religion can coexist peacefully is accepted even by some members of the Intelligent Design school. In addition, much of the research supported by organizations such as the Metanexus Institute is directed to resolving the perceived conflict of science and religion.[24] The relationship between moral truth and scientific truth remains controversial, but a full examination of the issue is beyond the scope of this book.[25]

A summary of public policy issues and the positions of the four schools are found in table 9.1.

---

23. Wiker, *Moral Darwinism*, 22–23.
24. The Metanexus Institute publishes a journal and sponsors conferences on science and religion; see www.metanexus.org.
25. See www.evolutionprimer.net and the works cited in the bibliography for further discussion.

Table 9.1. Summary of the public policy issues in dispute in the evolution controversy*

Public policy issue		Position of evolution schools			
Issue	Possible positions	Neo-Darwinism	Meta-Darwinism	Intelligent Design	Creationism
Legitimate spokesperson for science	Only officially recognized organizations such as NAS	✓	✓		
	Recognized organizations are biased; others can also speak on evolution			✓	✓
Public funding of evolution research	Status quo: only that research deemed scientific by officially recognized organizations	✓	✓	✓	
	Other schools should be able to receive funding because of quasi-religious teaching of Neo-Darwinism and Meta-Darwinism			✓	✓
	No funding of evolution research				✓
Equal time in the high school classroom	Only Neo-Darwinian theory taught, no objections discussed	✓			
	Only Neo-Darwinian theory taught, some objections discussed	✓			
	Only naturalistic theories taught; no mention of others	✓	✓		
	Only naturalistic theories taught, nonnaturalistic theories acknowledged		✓		
	Both naturalistic and nonnaturalistic theories taught			✓	✓
Courts as appropriate battleground	Courts are important to protect integrity of science and freedom from religious interference	✓	✓		
	Courts should not be involved; matter should be decided at local level			✓	✓
Moral and ethical implications	Moral questions are separable from scientific questions and should not be part of public policy discussion	✓	✓		
	Materialist philosophy is not part of evolutionary science	✓	✓	✓	
	Theistic evolution represents a way to reconcile science and religion	✓	✓	✓	
	Implications of evolution are at variance with established moral truths, thus indicating that theory is wrong			✓	✓

*Check marks in two or more adjacent cells indicate that members of the school have divergent opinions on the issue.

10

SUMMARY AND ASSESSMENT OF THE EVOLUTION CONTROVERSY

The goal of this book is to provide a balanced, in-depth perspective on the evolution controversy. To do this we have reviewed the history of the controversy, the evidence and major points in dispute, the views of the four major schools, and the case for and against each. Undoubtedly, this is a lot of information. To help make sense of it, it is useful to take a step back and assess the key points of the controversy, the difficulties and challenges each school faces, and the road that lies ahead.

Key Points

In the introductory chapter, we introduced the three levels of evolution and stressed that it is important to identify, in any discussion of the issue, which level of evolution is being addressed. If a speaker is proving the validity of the first level, *historical evolution*—the simple belief that there has been change in species over the roughly four billion years that life has existed on earth—this has little bearing on the validity of the third level, *strong Darwinian evolution*, the belief that purely natural mechanisms are the driving force behind historical evolution and can explain all of it. Most of the evidence presented in chapter 2 supported or can be used to support historical evolution and the second level, *common descent*, the belief that all species are linked by common ancestry. Three of the schools assent to historical evolution and the likelihood of common descent in the traditional sense; the fourth school, Creationism, rejects historical evolution as commonly understood, but does accept a limited version of common descent. The Intelligent Design school accepts historical evolution and some common descent but rejects strong Darwinian evolution. The Neo-Darwinian and the Meta-Darwinian schools, however, accept all three levels of evolution, including strong Darwinian evolution.

We have seen that the mechanisms (random mutation and natural selection) posited by the Neo-Darwinists can explain many aspects of life but run into difficulties in other areas, difficulties that may, however, be resolvable. While there is solid evidence that these mechanisms can lead to small-scale change (microevolution), a fact accepted by all schools, there is at present little *direct* evidence that they can explain *all* evolutionary change (macroevolution); what evidence we have is primarily circumstantial. Phrased in other language, the question is whether random mutation and natural selection can work together to create novel complex biological information on a large scale. Neo-Darwinians argue that the available evidence supports their theory, and no other theory has a better explanation of this evidence, so they infer the truth of their theory by *inference to the best explanation*.

In our evaluation of the Creationist school, we spent considerable time investigating the Creationist interpretation of historical evolution and common descent evolution and noted the considerable problems that exist in shoehorning the history of the universe and the earth into roughly 10,000 years. While Creationists hammer Neo-Darwinists and Meta-Darwinists on what they claim are problems with historical evolution and purely naturalistic mechanisms, they have their own difficulties in developing plausible explanations of the fossil record and the age of the earth and the universe in a manner consistent with both observable evidence and the Bible. At the same time, we have found that the Intelligent Design camp has thus far not nailed down their case that natural mechanisms are *insufficient* to explain observed complexity in living organisms, though they are making inroads on the problem of identifying key verifiable criteria.

Below are some of the main points that we have emphasized throughout the book along with the main points that are disputed by the four schools.

- Three meanings of evolution
  - Historical (observed sequence of life-forms over ~4 billion years of earth's history—accepted by all but the Creationist school; Creationists only accept limited change over a much shorter period)

344

- Common descent (all organisms are related to common ancestor—accepted by Neo-Darwinism and Meta-Darwinism, and to a limited extent by Intelligent Design and Creationism)
  - Strong Darwinian (adequacy of natural mechanisms to fully explain the previous two—accepted only by the Neo-Darwinian and Meta-Darwinian schools)
- Key distinctions
  - Between *facts* and *explanations* of facts
  - Between *theory* and *fact*
  - Between *criticism of a theory* and *advocacy of an alternative theory*
- Natural selection also acts as a conservative process, which is a role that is *not* equivalent with evolution in the strong Darwinian (Neo-Darwinian theory) sense
- The major evidence that must be explained, most of which provides circumstantial evidence for some amount of naturalistic evolution (historical and common descent; see chap. 3)
  - Fossil record
  - Geologic features of the earth suggesting a long history
  - Anatomical and physiological similarities
  - Biochemical similarities
  - Genetic similarities
  - Functional and system similarities
  - Geographic distribution of flora and fauna
  - Complexity of biological molecules, systems, and structures
  - Adaptation of flora and fauna to their environment
- The positions of the four schools and how they explain the major evidence (see chaps. 5–8)
  - Neo-Darwinism: random mutation and natural selection acting over eons
  - Meta-Darwinism: other natural mechanisms are required in addition to natural selection and random mutation
  - Creationism: natural selection adapts populations to their environment; random mutation leads only to degenerative speciation; external intervention with common design plan required
  - Intelligent design: observed complexity in some cases is too great for purely naturalistic mechanisms; Intelligent Design or some direct intervention can therefore be inferred.
- The major points in dispute (see chap. 4)
  - Common descent versus common design plan
  - Ability of random mutation to generate new information

- Adequacy of natural selection and random mutation together to account for historical evolution
- Age of the earth and the universe
- Scope of naturalistic explanations in science
- Requirements for a bona fide scientific theory

The Four Schools' Accomplishments, Difficulties, and Areas Needing Improvement

Neo-Darwinian School

The major accomplishments of the Neo-Darwinians are numerous. First and foremost, they have devised a simple yet powerful explanatory framework for all of life that is consistent with evidence as diverse as the presence of pseudogenes to conserved morphologies. They have been successful in integrating data from many different disciplines and have used this evidence to build a powerful case for both historical and common descent evolution.

Despite these successes, the Neo-Darwinian school tends to mix too much extra-scientific material (philosophy) with purely scientific material. This admixture is a principal contributor to the acrimony of the evolution controversy. Advocacy of scientific naturalism in no way demonstrates the efficacy of natural selection and random mutation; only positive evidence will do that. Simply stating that these mechanisms are the best naturalistic option does not make them any more effective. Furthermore, the philosophical positions advocated by prominent members of the school, which are often bundled with the purely scientific positions, not only distort science but actually play directly into the hands of Creationists. A statement such as "evolution tells us that life has no purpose" is rife with enormous metaphysical, religious, and ethical implications. Creationists are quick to point this out and argue that because the philosophy and the science are conjoined, both should be discarded. The Gallup poll statistics quoted in chapter 1 indicate clearly that this line of attack is very effective.

This school also needs to be more candid about all the difficulties it faces, the limits of what it has been able to demonstrate, and in particular what it doesn't know. In terms of how certain complex structures are formed via natural selection, some members such as Robert Pennock are willing to admit that "there remain any number of gaps that have yet to be explained."[1] Unfortunately this admission is not widely disseminated to the public. Instead, we hear statements to the effect that there are

1. Robert Pennock, *The Tower of Babel* (Cambridge, MA: MIT Press, 1999), 171.

no known structures that cannot be "explained" as having emerged via natural selection. This is not the case. Many of these "explanations" are hypothetical just-so stories. Admitting that we don't know something is the first step in scientific inquiry and the first necessary step for the Neo-Darwinian school to fix its credibility problems with the general public (it does not have credibility problems with the majority of scientists).

Finally, the school must refrain from the type of long-range extrapolations that it often advances as evidence. For example, taking the development of antibacterial resistance in bacteria as evidence that complex adaptations such as powered flight can evolve oversteps the bounds of good science. Rather than skirting the details, Neo-Darwinists (and science) would be better served through functional examination of the systems necessary for powered flight. These systems need to be taken apart systematically to better determine if they can be acquired in a gradual fashion and still allow the organism to be viable. This may be difficult, and the results imperfect, but the fact that it is being attempted in certain cases demonstrates its feasibility as a research program.[2]

The accomplishments, issues, and challenges for the Neo-Darwinian school are summarized in table 10.1.

Meta-Darwinian School

The Meta-Darwinian school, although in its infancy, has been successful in pressing Neo-Darwinists to critically evaluate the limits and abilities of natural selection. As a result, it is probably the fastest-growing school of the four, within academia. Members of the school have been successful in staying within the confines of mainstream scientific research while simultaneously expressing doubt with the Neo-Darwinist paradigm.

Unfortunately, many of the problems found in the Neo-Darwinian school regarding the admixing of philosophy and science have found their way into the Meta-Darwinian school. However, because the school is much more open about the problems of natural selection and random mutation as a viable mechanism, it tends to be less reliant on the "it's the only natural mechanism we have, therefore it must be true" stance that can be found in the Neo-Darwinian school. Where such bias exists, it should be recognized and expunged.

Given the diversity of opinions within this school we must separate the members into two camps: those who posit truly novel mechanisms and those who do not. The latter camp comprises the advocates of punctuated equilibrium, the neutral theory, and the concept of exaptations. When

2. See the discussion of this subject in chap. 5.

**Table 10.1. Accomplishments, issues, and challenges
for the Neo-Darwinian school**

Accomplishments	1. Single explanatory paradigm for characteristics, history, and distribution of all life on earth
	2. Integrates material from many disciplines
	3. Completely naturalistic explanation; does not require divine intervention to account for observed phenomena
	4. Utilizes only a very few hypotheses to account for enormous number of observations
	5. Accounts for the presence of genetic similarities/pseudogenes
Issues	1. Confusion of three tiers of evolution and type of proof or demonstration needed to establish each
	2. Use of explain-anything, just-so stories and long-range extrapolation as substitutes for rigorous science
	3. Some fossil record characteristics not accounted for
	4. Criticizes Creationism and Intelligent Design for religious pretensions often without acknowledging its own
	5. Micro-/macroevolution equivalence not adequately demonstrated
	6. Ignores "mainstream" critics; publication and discussion of problems with theory stonewalled
	7. Retreat into unknowability when confronted with difficult problems
	8. Needs to be more upfront regarding predictions that could falsify the theory
Challenges	1. Formulate and run tests that could in principle falsify the theory
	2. Use just-so stories only as starting points to formulate real explanations
	3. Stop papering over problems and weaknesses in the theory, and propose ways of theoretically and experimentally dealing with them
	4. Eschew long-range extrapolations
	5. Scrupulously avoid discussions of the "meaning" of evolution, and the construction of philosophical positions based on interpretations of evolutionary theories
	6. Allow free and open discussion of problems with theory, and present these to students

examined closely, these theories are just variations on the theme that random small-scale genetic change gradually leads to novel structures—a position nearly identical to that of Neo-Darwinism. Given their mechanistic similarities with Neo-Darwinism, all three of these theories have the same problems and must address the same deficiencies described above for Neo-Darwinian theory.

Those Meta-Darwinians who advocate novel naturalistic mechanisms such as large-scale developmental mutations, endosymbiosis (the large-scale exchange of DNA), or self-organization have a different set of tasks. Some of these ideas are relatively speculative and the extent to which they have affected evolution is largely unknown. In fact, little experimental work has been done on them. Certainly there is good circumstantial evidence that endosymbiosis has occurred in rare cases, but since it is posited here as a major driving force in evolution, direct experimental data is needed.

Researchers must examine the extent to which endosymbiosis can occur and the extent to which novel structures can be derived via endosymbiosis. The same holds true for self-organizational principles and large-scale developmental mutations. In the case of developmental mutations, it is necessary to demonstrate that these mutations can be advantageous, build novel structures, and become fixed in a population. Experimentally this is at least testable, since the molecular techniques exist for systematically altering developmental genes. In the case of self-organization, more realistic models are needed, models that take into account such variables as the harsh environment in which an organism must develop and some basic thermodynamic issues that have hitherto been ignored. Basically these approaches need to be put through the experimental ringer.

The accomplishments, issues, and challenges for the Meta-Darwinian school are summarized in table 10.2.

Table 10.2. Accomplishments, issues, and challenges for the Meta-Darwinian school

Accomplishments	1. Recognizes difficulties with Neo-Darwinian paradigm within a scientific context
	2. Theories can be in better agreement with the fossil record
	3. Identifies or proposes natural mechanisms other than natural selection that could play an important role in evolution
Issues	1. Mechanisms proposed may not be adequate to resolve problems of Neo-Darwinism
	2. Theorizing often is too far removed from the real world
	3. Mathematical theories often ignore actual biological problems
	4. Need more predictions that could lead to tests that could falsify the theory
	5. Some theories are hard to distinguish from Neo-Darwinism
Challenges	1. Demonstrate that theories can explain real-world biological phenomena
	2. Develop a rigorous research program demonstrating efficacy of proposed mechanisms
	3. Explain how theories are different from chance (random mutations) with respect to origin of biological information, if indeed they are

Intelligent Design

The Intelligent Design school has had a large effect on the evolution debate by zeroing in on a measurable quantity that has the potential to resolve the controversy, or at least move us further down the road toward that goal. This measurable quantity, of course, is *irreducible complexity*. And the reaction of mainstream science indicates that this has indeed hit a nerve. But the school's work is by no means complete.

An area on which their research should focus is the structure and function of what they consider to be irreducibly complex structures and pathways, for example, flagella, cell transport systems, and blood-clotting pathways. In the case of the flagella, many proteins are involved in allowing this structure to function optimally. Comparing the flagella proteins across various organisms would help separate the components that are absolutely necessary for minimal functioning from those that merely make the flagella function optimally.

This is not enough, though. Intelligent Design researchers must go a step further and take apart these biochemical systems piece by piece, studying their function once certain components or certain combinations of components have been removed. Their task is to determine exactly what each component protein does and which components are absolutely necessary for the flagella to have any minimal functionality. They must also determine if other cellular proteins can fill in for the components they take away. This type of systematic approach will be extremely time consuming, but it is the only way to mount credible evidence that such structures are beyond the reach of natural mechanisms. Merely stating that they are so, based upon isolated and disparate pieces of evidence, is not convincing. If the Intelligent Design researchers progress along these lines, their work will no doubt be recognized—albeit grudgingly—by mainstream science, as they will be doing experiments of the type published in scientific journals.

Another area of research critical to the Intelligent Design school is information theory. The school needs to address the question of what types of information can be derived by purely naturalistic means. If they insist that *no* new information can arise by purely naturalistic processes, they will be marginalized because there is evidence that this can occur. The question then becomes, what type of information can be had by naturalistic processes? While the Intelligent Design school holds that complex specific information is beyond the reach of natural processes, it is unclear how many proteins fall into this category, given the large amount of redundancy in primary and secondary protein structure as well as the modular construction of most proteins. Much more work in applying information theory to real biological proteins is needed before results can be taken very seriously.

The accomplishments, issues, and challenges for the Intelligent Design school are summarized in table 10.3.

Creationists

In recent years, Creationists have made some significant strides in developing sophisticated alternatives to current evolutionary theory to explain observed evidence about the history of life on earth. They have been able to found research institutes and perform a wide range of scientific experiments, albeit on a much smaller scale than the experi-

Table 10.3. Accomplishments, issues, and challenges for the Intelligent Design school

Accomplishments	1. Puts hypothesis of design in nature on an empirical footing
	2. Design filter is an attempt to empirically distinguish theories and subject them to up-or-down test
	3. Directs attention to empirical problems of natural selection and random mutation as the motor for evolution
	4. Recognizes need to ask questions about the limits of natural process and, by implication, of science itself
	5. Focuses on details and probabilities of naturalistic evolution, rather than assuming or explaining away difficulties
Issues	1. Design filter is difficult to apply in practice, because of difficulties in assessing probabilities associated with biological systems and organisms
	2. Unclear how biology would change as a science under Intelligent Design
	3. Lacks a unifying hypothesis of biological causation
	4. Definitively proving that a natural object is designed may be nearly impossible (i.e., ruling out all chance and law hypotheses)
	5. Often criticizes a model for Neo-Darwinian evolution that is not in accord with what that theory actually maintains
	6. Does not address or make falsifiable predictions regarding the fossil record
Challenges	1. Do research and come up with results that mainstream science cannot ignore
	2. Refine design filter to better delineate between designed objects and those that arise through natural processes
	3. Empirically identify complex systems that are impossible for naturalistic mechanisms based on random processes to explain
	4. Get bounds for probability arguments that are better grounded in relevant biology, chemistry, and to some extent physics

ments of mainstream science. But like the Intelligent Design school, Creationists need to do more work that can be described as legitimate science. Since their explanatory paradigm is based on a radically different series of events, they should be able to devise predictions about as-yet undiscovered phenomena that differ significantly from those of orthodox science. One branch of the Creationist school, the Center for Scientific Creation, run by Walt Brown, has made considerable progress toward this goal through its series of thirty-one refutable predictions, several of which have already been verified.

The make-or-break issue for Creationists remains the age of the earth and the universe. If they can succeed in demonstrating a young age, on the order of 10,000 years or so, then the game is over and they have won hands down, because no significant evolution can occur in such a short time frame. But if they fail to do so, their entire theory goes out the window because its raison d'être—to maintain a literal interpretation of the biblical book of Genesis—has been destroyed. As we noted in chapter 6, their three principal approaches to resolving the age-of-the-earth problem all have serious difficulties.

There are additional issues that present stumbling blocks for the school. For example, the presence of nearly identical nonfunctional pseudogenes in different organisms seems unlikely to be the result of creation. Likewise, explanation of the fossil record based on contemporaneous yet isolated ecosystems has serious difficulties that have not been mitigated through any type of experimental demonstration. Another problem that continues to dog the Creationist school is the widespread practice of using only selective evidence to discredit the Neo-Darwinian position.

The accomplishments, issues, and challenges for the Creationist school are summarized in table 10.4. Not all of the items listed apply to all branches of the school.

Table 10.4. Accomplishments, issues, and challenges for the Creationist school

Accomplishments	1. Devised a comprehensive alternative to the Neo-Darwinian explanation for history, distribution, and characteristics of life
	2. Focused public attention on problems of accepted evolution theory
	3. Revealed some of the extra-scientific pretensions of evolution theory
	4. Helped establish a credible catastrophist explanation for the formation of the Grand Canyon
	5. Established research efforts such as the RATE project
	6. In certain cases, have made detailed refutable predictions
Issues	1. Problem of religion dictating answers to science (science not free to find its own answers)
	2. No comprehensive independent research program
	3. Few if any up-or-down tests allowed on scientific portions of theory
	4. No up-or-down tests possible on nonscientific portions of theory
	5. Use of rational explanations when scientific explanations are available
	6. Selective use of facts to support theories
	7. Need for deus ex machina approach of change in physical constants to explain facts such as the apparent age of the universe (equivalent to retreat into unknowability unless other evidence presented)
Challenges	1. Develop research program that will generate results that mainstream science cannot ignore
	2. Explain how science will proceed under a Creationist interpretation
	3. Use all available facts when supporting theories
	4. Present evidence and experimental tests to verify that there was a change in physical constants in past
	5. Address problems such as the presence of pseudogenes and synteny blocks in organisms
	6. Deal with problems of age-of-the-universe theories

The Real Status of the Evolution Controversy

Evolution is widely regarded as a linchpin of modern science, one of its great achievements and organizing principles. It is a theory that can organize and integrate vast amounts of otherwise disparate facts. As we have noted, the sheer volume of circumstantial evidence that can be cited as favoring evolution, particularly historical evolution and common descent, is almost overwhelming. But despite such accolades, evolution has a somewhat checkered history, in terms of the long-range extrapolations that seem endemic to it and the often shaky evidence and arguments adduced to support it, particularly at the strong Darwinian level.

In science, we can reasonably expect that questions regarding the mechanism(s) responsible for historical evolution and common descent should be investigated with the utmost objectivity, and the answers treated with the greatest skepticism. Crucial experiments should be repeated again and again, and even slight deviations in results should be scrutinized. This has not always happened.

These failings seem to occur because extra-scientific considerations often override normal scientific procedures when certain questions are investigated. For example, for someone who believes that the entire universe can be explained by natural causes, then such a cause must be found or proposed for everything, and the notion of Intelligent Design or Creationism becomes ludicrous and is summarily dismissed. Likewise, for someone who believes that a creator was actively involved in the history of life, the out-and-out rejection of such a claim prior to a full evaluation of the evidence seems restrictive and unjustified. Neither belief can be verified scientifically, yet both are held passionately. Civil discourse across this divide is nearly impossible. Adding fuel to the fire is the fact that no natural mechanism has yet been shown to be able to account for all of life. This gives Neo-Darwinians reason to be defensive, while at the same time lending Creationists and Intelligent Design advocates hope that their attacks will one day land a fatal blow—one that even the Neo-Darwinians will have to acknowledge as such.[3]

From the perspective of a typical Neo-Darwinist, the available *evidence* is paramount because it is overwhelming even if circumstantial; thus the theory of evolution is as well established as the heliocentric theory of the solar system—that is, it is an indisputable fact and is defended as such. Any problems for the theory that may exist, such as the ability of

---

3. Actually, many critics of the theory, especially Creationist, feel that they have already done this; obviously, their opinion is not shared by the Neo-Darwinians or most members of the other schools.

natural selection and random mutation to account for specific complex structures, are minor by comparison and can be expected to be resolved through future research.

For Creationists, the evidence adduced by the Neo-Darwinians, including physiological and genetic similarities, can usually be explained in other ways—simpler ways, in their view, but unscientific ways in the eyes of their adversaries—and it is the *problems* that Neo-Darwinian theory faces that are most important.

Is there any hope that these two sides can come together?

Prospects for Resolution

The extra-scientific baggage associated with their respective theories now has members of both Neo-Darwinian and the Creationist schools with their backs to the wall. So when we consider how matters might be resolved, the main obstacle is clear: the ante has been raised so high by the polemical nature of the controversy that resolution in favor of one school will have catastrophic implications for the other. On the one hand, the scientific community by and large, including the National Academy of Sciences, has staked the prestige of science on a particular theory with considerable explanatory power but known problems, in part because it is consistent with a naturalistic philosophy. On the other hand, Creationists have for all intents and purposes staked the truth of their religion on the falsity of that same theory, because of the perceived need for a literal interpretation of the Bible. Clearly, neither the proponents of Creationism nor those of Neo-Darwinism can permit their side to lose or even give ground, regardless of the facts; the extra-scientific stakes for both are just too high.

While it is apparent that Neo-Darwinians and Creationists have much at stake, the future of the Intelligent Design and Meta-Darwinian evolution schools is in the balance as well. Though the Intelligent Design school is often portrayed as an ally of the Creationists (or even assimilated to them for propaganda purposes), and the Meta-Darwinians are viewed as close cousins of Neo-Darwinists, this is an oversimplification. In reality these two schools represent a threat to both Neo-Darwinism and Creationism.

In the case of the Neo-Darwinists, the Intelligent Design school represents the more clear and present danger on account of its direct quantitative challenges. Because of this, the Neo-Darwinians are at great pains to lump this school in with the Creationist school and discredit both as antiscientific and blinded by religious zealotry. In this way they can avoid the need to respond point by point to the issues raised by the

Intelligent Design school. As for the Meta-Darwinists, because they share a commitment to scientific naturalism, they do not pose as much of a threat to the Neo-Darwinian school, as evidenced by the fact that the Meta-Darwinian school contains many former Neo-Darwinists. Nonetheless, a victory by the Meta-Darwinists would certainly diminish the standing of prominent Neo-Darwinists in the eyes of the public and fellow scientists alike. Likewise, Creationists have no friend in the Meta-Darwinians, whose position, for all intents and purposes, is as damaging to the Creationist as the Neo-Darwinian position. But even a victory by the Intelligent Design school would spell doom for the Creationist school and its view of a 10,000-year-old earth coupled with very limited common descent.

This winner-take-all atmosphere leaves little room for compromise, so each side clings tenaciously to its position. At first glance, theistic evolution may seem to offer a way to resolve the problem, but this is not the case. As we noted in chapter 1, theistic evolution is not a new scientific theory to explain observed facts but a theological interpretation of a particular naturalistic scientific theory, either Neo-Darwinism or Meta-Darwinism. Since the four schools differ with regard to what they believe to be adequate scientific explanations of the history of life, theological interpretations are irrelevant to them and do not affect the scientific issues in question. In other words, the dispute is not about which interpretation of a particular scientific theory is the best but about which scientific theory is correct in the first place.

So what is likely to happen? In the short term, very little regardless of the evidence. This is largely due to the fact that the real root of the evolution controversy is not science per se but instead philosophical commitments on the part of the schools that are influencing how science is done and what constitutes a good scientific theory in light of the available evidence. It is hard to imagine any factual discoveries that will move the Creationists—they have already demonstrated that they can (to their own satisfaction at least) explain the evidence as we know it. Likewise, it is difficult to conceive of any facts discoverable in the field that will move the Neo-Darwinians. The other two schools may be more amenable to change in light of new discoveries, but this is not certain.

The only way around this logjam is to *decouple the philosophical (or religious) commitments from the science*—a difficult but not impossible task. The prime requisite for any resolution of the evolution controversy is that the schools all set aside their a priori commitment to extra-scientific philosophies. This will mean that the many "in-your-face" promoters of pro and con arguments must shut up and leave resolution of the controversy to more objective scientists who do not have, or who can completely set aside, extra-scientific commitments. If history is any guide, resolution of the controversy will take several generations, and some of the schools will just fade away, as accumulated evidence and change of the intellectual horizon will make their positions untenable.

The Authors' View of the Controversy

At this point it is appropriate to answer some of the nagging questions that have arisen during the course of the book. The positions the authors have staked out on these questions put them at odds in at least some areas with each of the four schools:

- *Is there evidence in favor of Darwinian and Neo-Darwinian evolution?* Yes.
- *Is there evidence against it?* Yes.
- *Are there problems with it?* Yes.
- *Is it a fraud?* No.
- *Are supporters of it candid about the problems and what needs to be done?* Not always.
- *Is evolution a fact?* This question is ambiguous. Historical evolution at present appears to be a fact; strong Darwinian evolution as of yet is not.
- *Do the theories of the other three schools have problems?* Yes.
- *Is there evidence favoring them?* Yes.
- *Is there evidence against them?* Yes.
- *Can any of the schools be definitively ruled out now?* Not quite, but the Creationists have the most serious problems, followed distantly by the other three schools.
- *Should naturalistic explanations be expected to account for all of natural history?* Not necessarily, although this remains a theoretical possibility.
- *Should nonnaturalistic explanations be admitted into science?* This question is a complete red herring. The important issue is not what we define science to be, but whether purely naturalistic explanations are adequate to explain all observed phenomena. If science is restricted to naturalistic explanations, as traditionally it has been, then the issue is whether there are limits to what science can do. The answer to this question then becomes the jumping-off point for philosophical and religious speculation.

This last point regarding the efficacy of naturalistic explanations remains one of the biggest sticking points in the controversy. Can this question be resolved? Is it possible to determine whether natural causes can explain all of life? While this question is a nonstarter to those ultra-Darwinists like Richard Dawkins, legitimate scientists have nothing to fear from it—that is, nothing to fear if truth is the ultimate goal. The difficulty, however, is not in *posing* the question but rather in *answering* it. While a smoking gun may not be in the offing, there are a variety of tests that would help move the field closer to an experimental resolution of this issue, as opposed to declaring natural causes sufficient by fiat. Possible tests are summarized in table 10.5. Not all points can be resolved with such testing because some differences are due to the use of rational rather than scientific explanations. For example, Creationists advocate that during Creation Week, various events occurred that cannot be duplicated now, such as a change in physical constants.[4] Because no experimental test at present can decide those issues, they do not appear in this table. In many cases the suggested tests are just sketches, which will require additional work to be fleshed out; in some cases there is no obvious test that can be performed. In other cases the tests are long and tedious and would require a concerted effort over many decades. Most of the tests are directed toward discrimination of Creationism from the other schools; however, there are tests that will discriminate among the other three schools as well.

In addition to the scientific concerns regarding the accuracy of any of the four theories, we have seen that the evolution debate includes many public policy issues. Here are the authors' positions on those public policy questions.

- *Should Neo-Darwinism be taught?* Yes; it is the primary scientific theory at this point.
- *Should its problems and the evidence against it also be taught?* Yes. Students have a right to know these problems and contrary evidence, and a need to know if they are going to resolve them.
- *Should the distinctions put forth here between the different tiers of evolution be emphasized?* Yes.
- *Should the logic of scientific explanation be taught?* Yes. This is an essential part of learning what science is, and how to do it correctly.
- *Should the other schools be taught?* Yes. To a limited extent their ideas should be discussed, though not given equal time with the

4. It may, however, be possible to deduce some consequence of this change in physical constants, since it may have left some marks. If so, a test could be devised.

357

Table 10.5. Proposed Tests to Distinguish the Four Schools of Evolution

Branch of science	Fact	Explanation				Crucial test or experiment
		Standard science/ Neo-Darwinian theory	Meta-Darwinian theory	Creationism	Intelligent Design	
Astronomy	Apparent age of the universe, including solar system	Big bang cosmology with no favored center, no edge; apparent age equals actual age; standard model for stellar and planetary formation and evolution		Gravitational time dilation/white hole horizon; earth in favored position; universe has center and edge or speed of light has decreased	Generally accepts standard explanation, but is not wedded to it	Evidence of favored position for earth; possible failure of symmetry or conservation laws in far reaches of space; observation of predicted time dilation
	Relative abundance of elements in universe	Standard model for stellar evolution and catastrophic explosions		Same as above	Generally accepts standard explanation	Same as above
	Quantized galactic redshifts	Speculation about quantum fluctuations		Milky Way galaxy is at or near center of universe	Generally accepts standard explanation	All large-scale structures of universe must exhibit concentricity
Geology	Apparent age of earth as revealed through surface features	Uniformitarianism: present-day forces worked in past but required long time span		Earth in reality is young and features can be accounted for by a global flood	Generally accepts standard explanation	Time required for erosion under severe conditions; discovery of fossilized contemporary organisms
	Geologic column coordinating with biostratigraphy	Uniformity of columns and agreement among dating methods		Separation of biostratigraphic, lithostratigraphic, and chronostratigraphic columns[a]	Generally accepts standard explanation	Evidence that columns do not coordinate; ability of high-energy flood process to sort and bury animals
	Radiometric dating of rocks	Physical constants unchanged; decay rates constant and can be used to extrapolate back in time		Accelerated (by factor of $\sim 10^8$) decay rates due to change in one or more fundamental physical constants during Creation Week and possibly before the global flood	Generally accepts standard explanation	Analysis to determine leftover identifiable effects of physical constant change, e.g., helium deposits in biotite; detailed recording and analysis of all dates obtained to determine distribution
	Fossilization	Occurs slowly over long periods (can occur rapidly in certain cases)		Occurs rapidly	Generally accepts standard explanation	Simulate conditions and observe if fossilization (burial, lithification) can occur over long period or if bones and tissues always disintegrate
	Biostratigraphic separation of fossils	Organisms lived during different time periods and were fossilized over long periods of time in the time order of those periods		Organisms were physically separate, coupled with hydrodynamic sorting and differential escape; during Phanerozoic era, tectonics also resulted in differentiation	Generally accepts standard explanation	Simulate flood conditions and distributions presumed by Creationists to see if observed separation occurs

Explanation

Branch of science	Fact	Standard science/ Neo-Darwinian theory	Meta-Darwinian theory	Creationism	Intelligent Design	Crucial test or experiment
Geology	Quantity of helium in atmosphere, salt in oceans	No accepted explanation at the present time		Earth is young	No position	Exhaustive investigation of all possible sources of salt, helium
	Rotation of earth's core	Constant speed		Decelerating	Generally accepts standard explanation	Measure the acceleration of the core
	Existence of large granite plate beneath floor of western Pacific Ocean	Does not predict existence of such a plate		Sank there as result of global flood activity	Generally accepts standard explanation	Look for the plate
	Fossil progression [historical evolution]	Successive life-forms evolved over the estimated 3.5 billion years from the Precambrian period		Artifact: flora and fauna commonly regarded as old actually were contemporaneous and inhabited different ecosystems; extinct animals were wiped out in global flood; flood also responsible for observed stratigraphy[b]	Generally accepts standard explanation	Discovery of contemporaneous fossils from different periods
Biology	Characteristics of fossil record: Cambrian explosion, absence or paucity of transitional forms	Extremely rapid morphological evolution possible under certain conditions	Mechanism of Neo-Darwinian theory may not account for this phenomenon	In accordance with recent creation of world	Mechanism of Neo-Darwinian theory cannot account for this phenomenon	Specification and test of conditions under which rapid evolution can occur
	Exceptional complexity of biological molecules, processes, and structures	Any degree of complexity can be explained by natural selection operating on random mutations	Mechanism of Neo-Darwinian theory cannot explain this fully but other naturalistic mechanisms can do so	Not explainable by naturalistic processes; requires external intervention; examples of divine creation to show wisdom and impossibility of evolution	Not explainable by naturalistic processes; requires external intervention (Intelligent Design)	Theoretical and experimental investigation of specified complexity notion, and limits of naturalistic processes to produce complexity; exhibit pathway to show that evolution of such characteristics/systems is possible and probable
	New species	Can contain new genetic information through random mutation	Random mutation cannot account for all new information, but other naturalistic mechanisms can	Descended from baramins; no new genetic information, only degeneration	Random mutation cannot account for all new information, intelligent intervention needed	Theoretical and experimental investigation of whether and how much new genetic information can be added to the biosphere by naturalistic means, or only degeneration

a. David J. Tyler and Paul Garner, "The Unformitarian Column and Flood Geology: A Reply to Froede and Reed (1999, CRSQ 36:51–60)," *Creation Research Society Quarterly* 37 (June 2000): 60–61.

b. R. L. Wysong, *The Creation-Evolution Controversy* (Midland, MI: Inquiry Press, 1976), 365; F. L. Marsh, *Life, Man, and Time* (Anacortes, WA: Outdoor Pictures, 1967), 135.

accepted theory. The evolution controversy is an excellent opportunity for students to learn how science works, why science can be controversial, the consequences of mixing science and philosophy and letting philosophy (or religion) drive science, what materialistic and nonmaterialistic explanations are, and whether there are limits to science.

Conclusion

The subject of evolution touches so many fundamental questions, issues, attitudes, and subjects that it will remain controversial for decades. Because of its scope, many citizens will have to make decisions about it at some point in their lives. For many scientists, personal and professional integrity will compel them to take a stand on it as well. It is our hope that this book will enable all readers to make more informed decisions about the subject, staying as close to observable facts as possible and eschewing bias and prejudice. We also hope that the book will prod readers to investigate the subject further and to see past the propaganda from all sides into the heart of the controversy. There are no easy answers, and there may be none for decades or even centuries. But we hope that a broad perspective, clear thinking, sound reasoning, objective analysis, freedom from obsequiousness, and a willingness to speak the truth will ultimately carry the day.

Glossary

abiogenesis. Life arising spontaneously from inanimate matter, by normal physical and chemical processes, without external (human or divine) action.

adaptation. Change of one or more characteristics in response to environmental pressures. For example, animals may grow thicker fur coats in response to colder conditions. Usually applied to populations, where those organisms with the desired characteristics are selected and their relative number grows. For Neo-Darwinism and other schools, adaptation results from beneficial mutations.

allele. A specific version (DNA sequence) of a gene. Individuals may have two different alleles of a gene (heterozygous) or two copies of the same allele (homozygous).

allopatric speciation. Speciation event that occurs when a parent species is separated into two distinct populations by a physical barrier (mountain range, lake, desert, etc.) such that the populations diverge to the point that they no longer can reproduce with each other.

architecture. Highest-level structural and functional organization of something. Tells what major parts or systems will be used and how they will be put together (interact) to achieve the design goals for the thing in question.

baramin. Creationist term that refers to one of the created kinds from which modern-day animals (species) descended. A single baramin may give rise to many modern species and even genuses. Not synonymous with any commonly used biological classification terms such as genus or family but of a similar level of generality. Baramin are assumed to have had a high degree of genetic diversity to account for the different species to which they gave rise, since Creationists deny that new genetic information can arise through mutations.

biological classification. Systematic method of categorizing life-forms utilizing their physiological and genetic characteristics. Dates back to Aristotle but modern system traces to Linnaeus. This process is often used to infer evolutionary relationships.

Cambrian explosion. The abrupt appearance of nearly all major animal phyla

361

in the fossil record during a relatively short period, 10 million years, during the Cambrian period of the earth's history, about 550 million years ago.

clade. A group of organisms grouped and classified together on the basis of homologous features, assumed to be due to a common ancestor.

class. A taxonomic level that is a major subdivision of a phylum.

common descent. Theory that a group of organisms descended from a single progenitor that lived at some time in the past. Sometimes the group is assumed to be all living things, so that the phrase means that all life originated from a single progenitor that lived billions of years ago.

common descent evolution. Second of the three levels of belief in evolution. This level postulates that all organisms descended from a single progenitor but does not make any assumptions about the mechanisms required to account for the changes in organisms over time. Specifically it does not assume the adequacy or inadequacy of the mechanisms of random mutation coupled with natural selection.

complexity. Refers to a system whose components interact in such a manner as to give rise to many possible behaviors; often difficult to explain or predict with mathematical models.

convergent evolution. The evolutionary origin of a structure or behavior in two or more distinct lineages rather than structures or behaviors originating once and being inherited through common descent.

creation science. Theories devised to explain observable facts about living organisms, the physical geology of the earth, and the history of the universe in a manner consistent with a literal reading of the biblical creation story in Genesis but without reference to it

and in accordance with established physical laws.

Creationism. School of thought about evolution that rejects the adequacy of naturalistic processes to account for the diversity of life, instead postulating one or more distinct creation events by God. Often coupled with belief in a young earth (younger than 10,000 years old).

Darwinism. Term used to describe Darwin's theory of evolution, a theory that attempts to explain the origin of new species as the result of natural selection operating over long expanses of time on beneficial changes (mutations) in existing life forms.

deme. A local population of organisms of a species, generally taken to have a similar genetic makeup.

design filter. Algorithm for determining whether an event should be attributed to chance, design, or necessity, and thus to discern instances of design from instances of naturally occurring phenomena, events, or objects.

endosymbiosis. A symbiotic relationship in which one member acquires either part or all of the other symbiont.

epistasis. Situation in which the effect of one gene is modified by one or more other genes. Also termed genetic interaction. An example is a gene that affects the transcription of another gene.

evolution. In biology, there are three meanings: (1) Simple change over time, for example, a shift in relative numbers of light and dark organisms in a population. If extended to all life-forms, it is referred to in this book as "historical evolution." (2) Emergence of new forms of life related to and derived from older forms, without commitment to any particular mechanism, natural or other. In this book, it is called "common descent evolution." (3) Advocacy of a particular natural mechanism, to *explain* (1) and (2); by implication, the entire process that resulted. In this book, it is called "strong Darwinian evolution."

exaptation. A characteristic of an organism that evolved from other usages or had no use at all, that is later used by the organism for a novel, beneficial function. It often arises as a by-product of other structures or systems.

extinction. Death of all members of a given taxonomic level, such as a species. Examples at the species level: dodo bird, woolly mammoth.

fallacy. Logical error in reasoning, so that truth of conclusion is not guaranteed by truth of premises.

fitness. Measure of the ability of an organism (or group of similar organisms) to survive, prosper, and reproduce in a given environment.

genetic drift. Random changes in allele frequencies of a population from one generation to the next. Occurs because offspring are genetic samples of their parents and this sample, because it is random, may not be representative of the parental generation. Such changes are more common in small populations.

genetic homeostasis. Balance that exists between the genes found within a population such that altering one gene can disrupt the fitness of the organisms in the population. This tends to keep the population stable. Pleiotropic and epistatic genes are thought to contribute greatly to genetic homeostasis.

genotype. The set of genes that an organism has.

genus. Taxonomic level above species and below family.

Hardy-Weinberg Law. Genetic law that deals with genotype frequencies at a single locus. According to the law, after each generation with random mating, genotype frequencies will maintain a stable equilibrium in the absence of any external pressures such as natural selection. The law is a cornerstone of population genetics calculations. It is not in dispute in the evolution controversy.

heterozygote. Organism that carries different alleles for a particular gene, one of which is usually dominant over the other.

historical evolution. Theory that life-forms can be arranged in a chronological sequence spanning 3.8 billion years, usually correlated with geologic periods, but without any implication that one descended from another (common descent), and without any assumption about mechanisms of change. This is the first level of generalization from what one can directly observe today by studying modern life-forms and digging up fossils.

homeotic genes. Genes that determine the fate of developing regions of the body. The Hox genes are one of the most well-known examples.

homologous structures. Two or more structures that are thought to be derived from a similar structure in a common ancestor through descent with modification.

homozygote. Organism that carries two identical alleles for a particular gene.

hydroplate theory. Theory developed by Creationist Walt Brown to explain major geologic features of the earth. Assumes that most water was once in interconnected subterranean chambers that ruptured due to increasing pressure, causing granite continental plates (hydroplates) to move.

Intelligent Design. Theory that some biological structures or systems are so complex and improbable that they could not arise by natural processes alone, with the implication that natural mechanisms are unable to account for the history of life. Also postulates that one can identify, through scientific research, proof of the action of an Intelligent Designer in the natural world.

interventionism. Hypothesis that some outside source (usually God) was needed to bring about certain types of change in organisms, or for their initial creation. Opposed to *philosophical naturalism.*

irreducible complexity. A property of a system that performs a given basic function and includes a set of well-matched, mutually interacting, nonarbitrarily individuated parts such that each part in the set is indispensable to maintaining the system's function.

kingdom. Highest taxonomic level.

levels of selection. The hypothesis that natural selection can operate at higher levels rather than just on the individual organisms. For example, natural selection could theoretically operate on entire species, a process known as species selection.

macroevolution. Evolutionary changes above the species level such as the extinction or emergence of new phyla, the rate at which evolutionary change proceeds, and higher-level trends or patterns in lineages.

materialism. Belief that only material things (usually things that can be investigated by science) exist and that they exhaust reality. More or less synonymous with *philosophical naturalism.*

Meta-Darwinism. Loose association of theories that dispute the efficacy of Neo-Darwinism and advocate other natural mechanisms to supplement natural selection. Advocates of these theories diverge in the particular mechanisms that they accept to supplement the perceived deficiencies in the Neo-Darwinian paradigm.

metaphysical. Refers to issues, problems, or theories that go beyond the directly observable and seek to explain reality or some aspect of it in a more comprehensive and integrative manner. Metaphysical theories are not generally testable in the same manner as scientific theories, but science (or at least its interpretation) ultimately rests on a metaphysical foundation.

methodological naturalism. Method of doing science in which supernatural agency is excluded a priori as a causative force. Under this method, science only looks for and studies natural causes.

microevolution. Evolutionary changes in populations and species in response to environmental or other pressures.

natural selection. Description of what happens to a population of organisms when subjected to environmental pressures of some sort (climate change, shortage of food, etc.). Those best able to deal with the situation continue living and produce offspring. Those that have difficulty with the situation may die off or produce fewer offspring.

necessary condition. Something that must be present in order for something else to occur, or some conclusion to follow. Example: water is a necessary condition for life.

negative mutation. A mutation that leaves an organism with a competitive disadvantage in its current environment.

Neo-Darwinism. Theory that advocates that the mechanisms proposed by Darwin and codified in the "new synthesis"—random mutation coupled with natural selection—are adequate to explain, in a naturalistic fashion, the origin of all the organisms we see in the world.

neutral theory. Belief that most molecular differences between organisms are neutral in nature and that mutation and genetic drift, not natural selection, played the predominant role in evolution.

nominalism. Philosophical theory that abstract entities do not exist, only individuals. Thus "species" does not refer to some form or type, but only to an arbitrary collection of individuals. Opposed to *realism.*

ontogeny. Development of an organism over time from its origin as a single cell to its mature phase.

optimization. Modifying the design or parameters of a device or system so as to improve its performance. Usually the goal is to make the performance the best that it can be (optimal), given certain constraints.

order. Taxonomic level that is a major subdivision of a class or subclass.

overdominance. The selective advantage that heterozygotes have over homozygotes.

peripatric speciation. Speciation event that occurs when a small group of members of a parental species is physically isolated in a peripheral locale such that the isolated population diverges to the point that they no longer can reproduce with the parental population.

phenotype. The set of physical characteristics of an organism, corresponding to the expressed genes in its genotype.

philosophical naturalism. Belief that only normal physical and chemical processes are necessary to explain all of life, its origin, and development over time. Opposed to *interventionism.*

phyletic gradualism. Theory regarding the tempo of evolutionary change that advocates that species evolve at a relatively constant slow rate over long expanses of time. This theory is rooted in the principle of uniformitarianism: the slow processes that occur today are the same processes that have fueled evolutionary change in the past.

phylogeny. Development of life-forms (phyla) over the history of the earth.

phylum. Taxonomic level, subdivision of a kingdom.

plate tectonics. In geology, the theory that the continents are really large plates that can move, albeit slowly. Such movements are postulated to account for geologic features of the earth such as mountain ranges, and also events such as earthquakes.

pleiotropy. Situation in which a single gene influences more than one phenotypic trait.

population. All the individuals of a species that inhabit a given area, and that are in contact with one another.

population genetics. Branch of biology that attempts to build mathematical models of evolutionary change, particularly at the level of microevolution.

positive selection. Increased number of organisms with a specific phenotype (or possibly a higher grouping such as a population) because of superior fitness in a particular environment. The phenotype is generally representative of a specific genotype, which is therefore also selected for in such cases.

prediction. A forecast that, under certain circumstances, something presently unknown will be discovered. Usually considered the acid test of a scientific theory.

pseudogene. A DNA sequence that is similar to a functional gene but is not transcribed. Most pseudogenes appear to serve no purpose.

punctuated equilibrium. Theory regarding the tempo of evolutionary change that advocates the alteration of long-term stasis (equilibrium) with geologically rapid change (punctuation).

radiation. Spread of an organism, trait, or gene over a geographic area or through a population.

random mutation. Change in genetic material (DNA or RNA) due to some unpredictable source such as radiation, chemical action, or replication error.

realism. Philosophical theory that abstract entities as well as individuals are real, though not in the same way. Thus "species" would be more permanent than and refer to something beyond an arbitrary collection of individuals. Opposed to *nominalism*.

retrodiction. Explanation of an already known fact by means of a new theory. Important, since theories that cannot explain already known facts are unlikely to be of value, but not as important as *prediction*.

saltationism. Theory of evolutionary change that advocates large rapid jumps rather than the slow and gradual changes of Neo-Darwinian theory.

selection. A process, usually owing to environmental factors interacting with phenotypes, that produces a differential reproductive rate among organisms. Those with a particular phenotype produce more offspring, so that their representation among the population increases. Generally this also increases the representation of a specific genotype or set of genotypes.

species. Lowest taxonomic level. Although many definitions are used, the most common operational definition is the nominalistic biological definition: a group of successfully interbreeding natural populations that are reproductively isolated from other such groups.

stasis. The ability of a species to remain stable or virtually unchanged for long periods of time, often millions of years.

strong Darwinian evolution. Complete explanation of common descent by the hypothesis that natural forces *alone* are responsible for the emergence of all organisms.

sufficient condition. Something that, if it is present, will ensure that something else will occur, or some conclusion will follow. Example: a sufficient condition for getting to Europe is to board the right airplane. This is not a necessary condition, as one can also get there by ship.

symbiosis. A close ecological relationship between two members of distinct species in which one or both members receive some benefit.

sympatric speciation. Speciation event that occurs when members of a parental species become reproductively isolated into two distinct populations within the same range.

taxon, taxa. A classification level for organisms.

theistic evolution. Not a separate theory to explain observed facts about nature, but an interpretation of some theory of evolution, usually Neo-Darwinism, so as to harmonize it with a particular set of theological beliefs. Rejected by young earth Creationists, but accepted by some in other schools.

vestigial structures. Structures that serve no purpose in an organism yet served a purpose in an ancestral organism.

BIBLIOGRAPHY

The breakout of works in this bibliography is intended to help readers do further research and study. It is not intended as a complete list of the voluminous material available on the evolution issue, but it does include many of the works cited in this book. Since some of the works can be put under several headings, the breakout is perforce somewhat arbitrary. The bibliography includes books, articles, web sites (URLs), organizations, and some material published on the Internet. All Internet-related data was accurate at the time of writing but, due to the nature of the Internet, may have changed before this book reaches the reader's hands.

Works on Evolution by Proponents of the Neo-Darwinian Theory

Bonner, John T. *The Evolution of Complexity by Means of Natural Selection*. Princeton, NJ: Princeton University Press, 1988.

Burnie, David. *Get a Grip on Evolution*. Time-Life Books, 1999. [A popular exposition of the ideas of Neo-Darwinian theory.]

Dawkins, Richard. *The Blind Watchmaker*. New York: W. W. Norton, 1996.

Dennett, Daniel C. *Darwin's Dangerous Idea: Evolution and the Meanings of Life*. New York: Simon & Schuster, 1995.

Dobzhansky, Theodosius. *Genetics and the Origin of Species*. 3rd ed. New York: Columbia University Press, 1951.

———. "Nothing in Biology Makes Sense Except in the Light of Evolution." *American Biology Teacher* 35 (1973): 125–29.

Fisher, Ronald A. *The Genetical Theory of Natural Selection*. 2nd rev. ed. New York: Dover, 1958.

Fox, Ronald F. *Energy and the Evolution of Life*. New York: W. H. Freeman, 1988.

Grant, Verne. *The Origin of Adaptations*. New York: Columbia University Press, 1963.

Harvard University's Department of Molecular and Cellular Biology web site, http://mcb.harvard.edu/BioLinks/Evolution.html. [Devoted to evolution.]

Huxley, Julian. *Evolution: The Modern Synthesis*. New York: Harper & Brothers, 1942.

Loomis, William. *Four Billion Years*. Sunderland, MA: Sinauer Assoc., 1988.

Mayr, Ernst. *Systematics and the Origin of Species*. New York: Columbia University Press, 1982.

———. *What Evolution Is*. New York: Basic Books, 2001.

Morris, Simon Conway. *The Crucible of Creation: The Burgess Shale and the Rise of Animals*. Oxford: Oxford University Press, 1998.

———. *Life's Solution: Inevitable Humans in a Lonely Universe*. Cambridge: Cambridge University Press, 2003.

Online Literature Library web site, www.literature.org/authors/darwin-charles. [Has text of several of Darwin's works.]

Osawa, Syozo. *Evolution of the Genetic Code*. Oxford: Oxford University Press, 1995.

Oster, G. F., I. L. Silver, and C. A. Tobias, eds. *Irreversible Thermodynamics and the Origin of Life*. New York: Gordon & Breach, 1974.

Raff, Rudolf A. *The Shape of Life*. Chicago: University of Chicago Press, 1996.

Ridley, Mark. *Evolution*. 2nd ed. Cambridge, MA: Blackwell Science, 1996.

Roff, Derek A. *Evolutionary Quantitative Genetics*. New York: Chapman & Hall, 1997.

Rose, Steven. *Lifelines, Biology beyond Determinism*. Oxford: Oxford University Press, 1998.

Simpson, George Gaylord. *The Major Features of Evolution*. New York: Columbia University Press, 1953.

———. *Tempo and Mode in Evolution*. New York: Columbia University Press, 1984.

Stebbins, G. Ledyard. *Darwin to DNA, Molecules to Humanity*. New York: W. H. Freeman, 1982.

Strickberger, Monroe W. *Evolution*. 3rd ed. Sudbury, MA: Jones & Bartlett, 2000.

Williams, George C. *Natural Selection*. New York: Oxford University Press, 1992.

Works on Evolution by Proponents of the Punctuated Equilibrium Theory

Eldredge, Niles. *Macroevolutionary Dynamics: Species, Niches, and Adaptive Peaks*. New York: McGraw-Hill, 1989.

———. *Reinventing Darwin: The Great Debate at the High Table of Evolutionary Theory*. New York: Wiley, 1995.

Eldredge, Niles, and Ian Tatersall. *The Myths of Human Evolution*. New York: Columbia University Press, 1982.

Gould, Steven J. *Dinosaur in a Haystack*. New York: Harmony Books, 1995.

———. "Evolution's Erratic Pace." *Natural History* 8 (1977): 12–16.

———. "Is a New and General Theory of Evolution Emerging?" *Paleobiology* 6 (1980): 119–30.

———. *The Structure of Evolutionary Theory*. Cambridge, MA: Belknap, 2002.

———. *Wonderful Life: The Burgess Shale and the Nature of History*. New York: W. W. Norton, 1989.

Gould, Steven J., and Niles Eldredge. "Punctuated Equilibria: The Tempo and Mode of Evolution Reconsidered." *Paleobiology* 3 (1977): 115–51.

Stanley, Steven M. *The New Evolutionary Timetable*. New York: Basic Books, 1981.

Works by Other Members of the Meta-Darwinian Evolution School

Carroll, S. *Endless Forms Most Beautiful*. New York: W. W. Norton, 2005.

Goodwin, Brian. *How the Leopard Changed Its Spots: The Evolution of Complexity*. New York: Charles Scribner's Sons, 1994.

Jablonka, Eva, and Marion Lamb. *Evolution in Four Dimensions*. Cambridge, MA: MIT Press, 2005.

Kauffman, Stuart. *At Home in the Universe: The Search for Laws of Self-Organization and Complexity*. Oxford: Oxford University Press, 1995.

———. *Investigations*. Oxford: Oxford University Press, 2000.

Kimura, Motoo. *The Neutral Theory of Molecular Evolution*. Cambridge: Cambridge University Press, 1983.

Margulis, Lynn. *Symbiosis in Cell Evolution*. New York: W. H. Freeman, 1981.

Margulis, Lynn, and D. Sagan. *Acquiring Genomes: A Theory of the Origins of Species*. New York: Basic Books, 2002.

Schwartz, Jeffrey. *Sudden Origins: Fossils, Genes, and the Emergence of Species*. New York: Wiley, 1999.

Works Criticizing Evolution on Mathematical or Information Theory Grounds

Cohen, I. L. *Darwin Was Wrong—A Study in Probabilities*. Greenvale, NY: New Research Publications, 1985.

Hoyle, Fred. *Mathematics of Evolution*. Memphis: Acorn Enterprises, 1999.

Hoyle, Fred, and N. C. Wickramasinghe. *Why Neo-Darwinism Does Not Work*. Cardiff, UK: University College Cardiff Press, 1982.

Moorehead, P. S., and M. M. Kaplan, eds. *The Mathematical Challenges to the Neo-Darwinian Interpretation of Evolution*. Philadelphia: Wistar Institute Press, 1967. [Not all articles in this collection are critical of Neo-Darwinian evolution.]

Spetner, Lee. *Not by Chance! Shattering the Modern Theory of Evolution*. New York: Judaica Press, 1998.

Wilder-Smith, A. E. *The Scientific Alternative to Neo-Darwinian Evolutionary Theory*. Costa Mesa, CA: Word for Today Publishers, 1987.

Yockey, Hubert P. *Information Theory and Molecular Biology*. Cambridge: Cambridge University Press, 1992.

Works Criticizing Darwinian and Neo-Darwinian Evolution Theory by Non-Creationists[1]

Dewar, Douglas. *Difficulties of the Evolution Theory*. London: Edward Arnold & Company, 1931.

Dewar, Douglas, and Frank Finn. *The Making of Species*. London: John Lane, 1909.

Dewar, Douglas, and H. S. Shelton. *Is Evolution Proved?* London: Hollis and Carter, 1947. [This book is a debate with Dewar *contra* and Shelton *pro* evolution.]

Ho, M-W, and P. T. Saunders. "Beyond Neo-Darwinism: An Epigenetic Approach to Evolution." *Journal of Theoretical Biology* 78 (1979): 573–91.

MacBeth, Norman. *Darwinism: A Time for Funerals*. San Francisco: Robert Briggs Assoc., 1982.

———. *Darwin Retried*. Boston: Harvard Common Press, 1971.

Overman, Dean L. *A Case against Accident and Self-Organization*. Lanham, MD: Rowman & Littlefield, 1997.

1. "Creationist" in this context refers to the modern-day Creationist schools, rather than those that existed prior to 1960.

Other Works Criticizing Neo-Darwinian Evolution or Some Aspect of It

Lester, Lane P., and Raymond Bohlin. *The Natural Limits to Biological Change.* Grand Rapids: Zondervan, 1984.

Taylor, Gordon R. *The Great Evolution Mystery.* New York: Harper & Row, 1983.

Works by Creationists

Answers in Genesis web site, www.answersingenesis.org. [A bête noire of the evolutionists, this is a gloves-off, young earth Creationist site.]

Brand, Leonard R. "The Paradigm of Naturalism, Compared with a Viable Alternative: A Scientific Philosophy for the Study of Origins." *Origins* 23, no. 1 (1996): 6–34.

Brown, Walt. *In the Beginning: Compelling Evidence for Creation and the Flood.* 7th ed. Phoenix: Center for Scientific Creation, 2001.

Creation Research Society web site, www.creationresearch.org. [Publishes a journal and has other material available.]

Geoscience Research Institute. *Evidences, the Record, and the Flood.* VHS. Loma Linda, CA: Geosciences Research Institute, 1990.

Gish, Duane. *Creation Scientists Answer Their Critics.* El Cajon, CA: Creation Research Institute, 1993.

———. *Evolution: The Fossils Still Say No!* El Cajon, CA: Creation Research Institute, 1995.

Humphreys, D. Russell. *Starlight and Time: Solving the Puzzle of Distant Starlight in a Young Universe.* Green Forest, AR: Master Books, 1994.

Huse, Scott M. *The Collapse of Evolution.* Grand Rapids: Baker Books, 1986.

Institute for Creation Research. *Grand Canyon, Monument to the Flood.* VHS. El Cajon, CA: Institute for Creation Research, 1994.

Institute for Creation Research. *Thousands not Billions . . .* DVD. Santee, CA: Institute for Creation Research, 2005.

Institute for Creation Research web site, www.icr.org. [Has additional materials.]

Lisle, Jason. *"Distant Starlight" Not a Problem for a Young Universe.* DVD. Florence, KY: Answers in Genesis, 2005.

Menton, David. *Dinosaurs by Design.* DVD. Florence, KY: Answers in Genesis, 2003.

Morris, Henry M. *Scientific Creationism.* Green Forest, AR: Master Books, 1985.

———. *That Their Words May Be Used against Them: Quotes from Evolutionists Useful for Creationists.* San Diego, CA: Institute for Creation Research, 1997.

O'Reilly, Sean. *Bioethics and the Limits of Science.* Front Royal, VA: Christendom Press, 1980.

Perloff, James. *Tornado in a Junkyard.* Arlington, MA: Refuge Books, 1999.

Safarti, Jonathan. *Refuting Evolution—A Response to the National Academy of Sciences' Teaching about Evolution and the Nature of Science.* Green Forest, AR: Master Books, 1999.

———. *Refuting Evolution 2.* Green Forest, AR: Master Books, 2002.

Thaxton, Charles B., Walter L. Bradley, and Roger L. Olsen. *The Mystery of Life's Origin: Reassessing Current Theories.* New York: Philosophical Library, 1984.

TrueOrigin Archive web site, www.trueorigin.org. [Contains responses to various arguments for evolution and against Creationism. It was set up specifically to counter the talkorigins.org web site.]

Vardiman, Larry, Andrew Snelling, and Eugene Chaffin, eds. *Radioisotopes and the Age of the Earth*. El Cajon, CA: Institute for Creation Research, 2000.

Whitcomb, John, and Henry M. Morris. *The Genesis Flood*. Santee, CA: Institute for Creation Research, 1961. [This is the book often considered to mark the birth of the modern Creationist movement.]

Williams, Emmett L., ed. *Thermodynamics and the Development of Order*, St. Joseph, MO: Creation Research Society, 1981.

Woodmorappe, John. *Studies in Flood Geology*. 2nd ed. El Cajon, CA: Institute for Creation Research, 1999.

Works by Members of the Intelligent Design School

Access Research Network web site, www.arn.org. [Focuses on controversial topics including evolution. Much of its current effort is devoted to the Intelligent Design school. ARN also publishes a peer-reviewed journal, *Origins and Design*.]

Behe, Michael. *Darwin's Black Box: The Biochemical Challenge to Evolution*. New York: Free Press, 1996.

Center for the Renewal of Science and Culture at the Discovery Institute web site, www.discovery.org/crsc. [Headquarters for the Intelligent Design movement (among other things).]

Dembski, William. *The Design Inference: Eliminating Chance through Small Probabilities*. Cambridge: Cambridge University Press, 1998.

———. *Intelligent Design*. Downers Grove, IL: InterVarsity Press, 1999.

———. *No Free Lunch: Why Specified Complexity Cannot Be Purchased without Intelligence*. Lanham, MD: Rowman & Littlefield, 2002.

———, ed. *Mere Creation: Science, Faith and Intelligent Design*. Downers Grove, IL: InterVarsity Press, 1998.

———. *Uncommon Dissent: Intellectuals Who Find Darwinism Unconvincing*. Wilmington, DE: ISI Books, 2004.

Denton, Michael. *Evolution: A Theory in Crisis*. London: Burnett Books, 1985.

———. *Nature's Destiny*. New York: Free Press, 1998.

Gonzalez, Guillermo and Jay Richards. *The Privileged Planet: How Our Place in the Cosmos Is Designed for Discovery*. Washington, DC: Regnery, 2004.

Illustra Media. *Unlocking the Mystery of Life: The Scientific Case for Intelligent Design*. DVD. La Habra, CA: Illustra Media, 2004.

Intelligent Design and Evolution Awareness (IDEA) Club at the University of California/San Diego (UCSD) web site, www-acs.ucsd.edu/%7Eidea/index.html. [The links page has links to many pro- and antievolution web sites.]

Intelligent Design Network web site, www.IntelligentDesignnetwork.org. [Mission is to promote evidence-based science education and awareness of evidence for Intelligent Design.]

Johnson, Phillip. *Darwin on Trial*. 2nd ed. Downers Grove, IL: InterVarsity Press, 1993.

———. *Defeating Darwinism by Opening Minds*. Downers Grove, IL: InterVarsity Press, 1997.

———. *Reason in the Balance*. Downers Grove, IL: InterVarsity Press, 1995.

Moreland, J. P., ed. *The Creation Hypothesis*. Downers Grove, IL: InterVarsity Press, 1994.

Wells, Jonathan. *Icons of Evolution*. Washington, DC: Regnery, 2000.

———. *The Politically Incorrect Guide to Darwinism and Intelligent Design*. Washington, DC: Regnery, 2006.

Works Criticizing Creationism

American Association for the Advancement of Science web site. www.aaas.org/spp/dser/evolution. [Information on evolution.]

Berra, Tim M. *Evolution and the Myth of Creationism*. Stanford, CA: Stanford University Press, 1990.

Eldredge, Niles. *The Monkey Business: A Scientist Looks at Creationism*. New York: Washington Square Press, 1982.

———. *The Triumph of Evolution and the Failure of Creationism*. New York: W. H. Freeman, 2000.

Futuyma, Douglas. *Science on Trial*. New York: Pantheon Books, 1983.

Godfrey, Laurie R. *Scientists Confront Creationism*. New York: W. W. Norton, 1983.

Kitcher, Philip. *Abusing Science: The Case against Creationism*. Cambridge, MA: MIT Press, 1982.

National Academy of Sciences. *Science and Creationism*. 2nd ed. Washington, DC: The National Academy Press, 1999. (Text also available on the National Academy web site, www.nas.org.)

National Academy of Science web site. www4.nas.edu/opus/evolve.nsf. [Information on Creationism and evolution.]

National Center for Science Education (NCSE) web site, www.natcenscied.org.

Ruse, Michael. *Darwinism and its Discontents*. Cambridge: Cambridge University Press, 2006.

———. *Darwinism Defended*. Reading, MA: Addison-Wesley, 1982.

Scott, Eugenie. *Evolution vs. Creationism, An Introduction*. Berkeley: University of California Press, 2005.

TalkOrigins Archive web site. www.talkorigins.org. [Generally oriented toward presentation of anti-Creationist and anti-Intelligent Design material. The Creationist response is the trueorigin.org web site given above.]

Was Darwin Wrong? web site. www.wasdarwinwrong.com. [Devoted to reviews of works written by critics of evolution, including those in schools other than Creationism.]

Works Criticizing Intelligent Design

Forrest, Barbara, and Paul Gross. *Creationism's Trojan Horse: The Wedge of Intelligent Design*. Oxford: Oxford University Press, 2004.

Miller, Kenneth. *Finding Darwin's God: A Scientist's Search for Common Ground between God and Evolution*. New York: HarperCollins, 1999.

Pennock, Robert. *Tower of Babel: The Evidence against the New Creationism*. Cambridge, MA: MIT Press, 1999.

See also the works by Eugenie Scott and the National Academy of Sciences listed in the previous section.

Works from Other Areas of Science

Fogel, David B. *Evolutionary Computatio: The Fossil Record*. New York: IEEE Press, 1995.

Michalewicz, Zbigniew. *Genetic Algorithms + Data Structures = Evolution Programs*. New York: Springer-Verlag, 1992.

Nicolis, Gregoire, and Ilya Prigogine. *Self-Organization in Nonequilibrium Systems*. New York: Wiley, 1977.

Oparin, A. I. *Origin of Life*. New York: Dover, 1953.

Prigogine, Ilya. *From Being to Becoming*. San Francisco: W. H. Freeman, 1980.

———. *Thermodynamics of Irreversible Processes*. 3rd ed. New York: Wiley, 1967.

Weinberg, Steven. *Dreams of a Final Theory*. New York: Pantheon Books, 1992.

Works on the History of Evolution Theories

Eiseley, Loren. *Darwin and the Mysterious Mr. X*. New York: Harvest/HBJ, 1979.

Himmelfarb, Gertrude. *Darwin and the Darwinian Revolution*. Garden City, NY: Doubleday, 1962.

Kuhn, Thomas. *The Copernican Revolution*. New York: MJF Books, 1957.

———. *The Structure of Scientific Revolutions*. Chicago: University of Chicago Press, 1962.

Midgley, Mary. *Evolution as a Religion*. London: Methuen, 1985.

———. *Science as Salvation*. London: Routledge, 1992.

Popper, Karl R. *Conjectures and Refutations: The Growth of Scientific Knowledge*. New York: Harper & Row, 1968.

Russell, Jeffrey Burton. *Inventing the Flat Earth*. New York: Praeger, 1991.

Works Dealing with the Evolution Controversy

Alters, B. *Defending Evolution in the Classroom: A Guide to the Evolution/Creation Controversy*. Boston: Jones and Bartlett, 2001.

Eldredge, Niles. *Reinventing Darwin: The Great Debate at the High Table of Evolutionary Theory*. New York: Wiley, 1995.

Larson, Edward J. *Trial and Error: The American Controversy over Creation and Evolution*. Updated ed. New York: Oxford University Press, 1989.

Moreland, J. P., and J. M. Reynolds. *Three Views on Creation and Evolution*. Grand Rapids: Zondervan, 1999.

Numbers, Ronald L. *The Creationists*. Berkeley: University of California Press, 1992.

Pennock, R., ed. *Intelligent Design Creationism and Its Critics: Philosophical, Theological, and Scientific Perspectives*. Cambridge, MA: MIT Press, 2001.

See also the works by Eugenie Scott and the National Academy of Sciences in the section "Works Criticizing Creationism."

Works Seeking to Reconcile Religion and Evolution

American Scientific Affiliation web site, www.asa3.org. [This evangelical Christian organization does not have an official position on evolution but does stand for scientific integrity and the need for well-justified conclusions in all matters.]

Collins, Francis. *The Language of God: A Scientist Presents Evidence for Belief*. New York: Simon and Schuster, 2006.

Durant, John, ed. *Darwinism and Divinity*. Oxford: Basil Blackwell, 1985.

McMullin, Ernan, ed. *Evolution and Creation*. Notre Dame, IN: University of Notre Dame Press, 1985.

Montenat, Christian, Luc Plateaux, and Pascal Roux. *How to Read the World: Creation in Evolution*. Trans. by John Bowden. New York: Crossroad, 1985.

Peacocke, A. R. *Creation and the World of Science*. Bampton Lectures, 1978. Oxford: Clarendon Press, 1979.

See also the book by Kenneth Miller cited under "Works Criticizing Intelligent Design."

INDEX